PHYSICAL ASPECTS OF PROTEIN INTERACTIONS

DEVELOPMENTS IN BIOCHEMISTRY

Volume 1 Mechanisms of Oxidizing Enzymes, edited by Thomas P. Singer and Raul N. Ondarza, 1978

Volume 2 Electrophoresis '78, edited by Nicholas Catsimpoolas, 1978

Volume 3 Physical Aspects of Protein Interactions, edited by Nicholas Catsimpoolas, 1978

PHYSICAL ASPECTS OF PROTEIN INTERACTIONS

Proceedings of the Symposium on Protein Interactions, American Chemical Society Meeting, Miami Beach, Florida, September 12-13, 1978

NICHOLAS CATSIMPOOLAS, *Editor*
Massachusetts Institute of Technology, Cambridge, Massachusetts, USA

ELSEVIER/NORTH-HOLLAND
NEW YORK • AMSTERDAM • OXFORD

The logo appearing on the cover was designed by Dr. Nicholas Catsimpoolas.

Published by:

Elsevier North Holland, Inc.
52 Vanderbilt Avenue, New York, New York 10017

Sole distributors outside of the USA and Canada:

Elsevier/North-Holland Biomedical Press
335 Jan van Galenstraat, PO Box 211
Amsterdam, The Netherlands

Library of Congress Cataloging in Publication Data

Conference on Physical Aspects of Protein
Interactions, Miami, Fla., 1978.
Physical aspects of protein interactions.
(Developments in biochemistry; v.3 ISSN 0165-1714)
Bibliography: p.
Includes index.
1. Proteins—Congresses. 2. Protein binding—Congresses. 3. Molecular association—Congresses. I. Catsimpoolas, Nicholas. II. Title. III. Series.
[DNLM: 1. Proteins—Metabolism—Congresses. W1 DE997VG v. 3/QU55.3 C748p 1978]
QP551.C715 1978 574.1'9245 78-23407
ISSN: 0165-1714
ISBN: 0-444-00304-5 (volume 3)

Manufactured in the United States of America

Contents

Preface

This volume consists of collected papers presented at the American Chemical Society Symposium on "Protein Interactions" which was held in Miami Beach, Florida, September 12-13, 1978. The emphasis of the meeting and this book was placed on the physical aspects of subunit, protein-ligand, and protein-protein interactions as exemplified either in specific biological problems, or in generally useful models. Sedimentation equilibrium and sedimentation velocity, light scattering, membrane osmometry, calorimetry, turbidimetry, volumetric measurements, and affinity chromatography were some of the physicochemical methods employed to study interacting species and to provide data for rigorous thermodynamic treatment of dissociating, binding, self-associating, and polymerizing systems. Although several levels of complexity exist in studying interacting proteins ranging from single polypeptide chains to supra-molecular assemblies, they all represent interrelated manifestations of evolutionary attempts to provide efficient cellular operation, regulation, and recognition. As such, the basic information obtained from the physical study of protein interactions *in vitro* bears important implications in the eventual understanding and harnessing of human disease. In addition, protein interactions play an important role in the functional utilization of conventional and novel protein sources for food, and therefore, the interest of the members of the Division of Agricultural and Food Chemistry who sponsored this Symposium. Their fine cooperation in the organization of this meeting is acknowledged with gratitude.

NICHOLAS CATSIMPOOLAS

Cambridge, Massachusetts
November, 1978

Catsimpoolas, ed. Physical Aspects of Protein Interactions

SELF-ASSOCIATION IN PROTEIN SOLUTIONS

E. T. ADAMS, JR.,[*] LIH-HENG TANG,[+] JERRY L. SARQUIS,[#] GRANT H. BARLOW[§] AND WESLEY M. NORMAN[*]

[*]Chemistry Department, Texas A&M University, College Station, Texas 77843, U.S.A.; [+]Montefiore Hospital and Medical Center, 111 East 210th St., Bronx, New York 10467, U.S.A.; [#]Miami University-Middletown, 4200 East University Blvd., Middletown, Ohio 45042; [§]Molecular Biology Department, Abbott Laboratories, Inc., North Chicago, Illinois 60064, U.S.A.

ABSTRACT

Self-associations are widely encountered in protein solutions. These associations can be rigorously studied by various thermodynamic methods - membrane osmometry, elastic light scattering and sedimentation equilibrium experiments. For self-associations there is an interrelation between the number (M_{nc}) and weight (M_{wc}) average molecular weights, or their apparent values (M_{na} or M_{wa}) under nonideal conditions. In addition it is possible to evaluate the natural logarithm of the weight fraction of monomer ($\ln f_1$) or its apparent value ($\ln f_a$) under nonideal conditions. These relations can be combined in various ways to test for the type of association present (if that is not known beforehand), evaluate the equilibrium constant or constants (K_i) and the nonideal term (BM_1). We have shown how these equations can be used and have related them to experimental examples for a variety of self-associations. Experimentally it has been found that self-associations are influenced by genetic differences in related proteins and by solution conditions; several examples illustrating these various effects are presented. It is also possible to combine the thermodynamic equilibrium data with data obtained from other techniques to obtain more information about the self-associating solute. Finally, the advantages and disadvantages of the three thermodynamic techniques are discussed.

INTRODUCTION

In solution many proteins, as well as many other solutes, undergo self-association[1-4]. The self-association equilibrium is usually dependent on solution conditions - pH, ionic strength, temperature. It can also be affected by changes in the chemical composition of the protein (α_{S1}-caseins[5] or β-lactoglobulins)[6-14] or by chemical modification of the protein (β-lactoglobulin A)[15,16]. Soon after the ultracentrifuge was developed, Svedberg and his associates encountered some self-associating proteins[17]. In 1926 Tiselius[18] published the first theoretical paper dealing with self-associations and sedimentation equilibrium. This treatment dealt only with ideal solution conditions. Several years passed before the next significant advances in the analysis of self-associating proteins were made by Steiner for the light scattering[19] and the osmotic pressure[20] experiments. Rao and Kegeles[21] used the Archibald experiment in their analysis of the self-association of α-chymotrypsin. The first sedimentation equilibrium experiment on a self-associating protein was done by Squire and Li[22], who studied the self-association of ovine adrenocorticotropic hormone (ACTH) in formate buffer (pH 3.5, I=0.2). They did make an attempt to correct for charge effects. Timasheff and Townend[9] encountered a nonideal, monomer-dimer association in their studies with β-lactoglobulin A and β-lactoglobulin B at pH 2.7 by light scattering. They used statistical thermodynamic theory to correct for nonideality. In general most studies carried out on self-associating proteins before 1963 were restricted to ideal, dilute solutions. Prior to 1963 it appeared to be a very difficult matter to analyze a nonideal self-association, such as the one shown in Fig. 1[23]. This pattern, showing an increase in M_{wa} (the

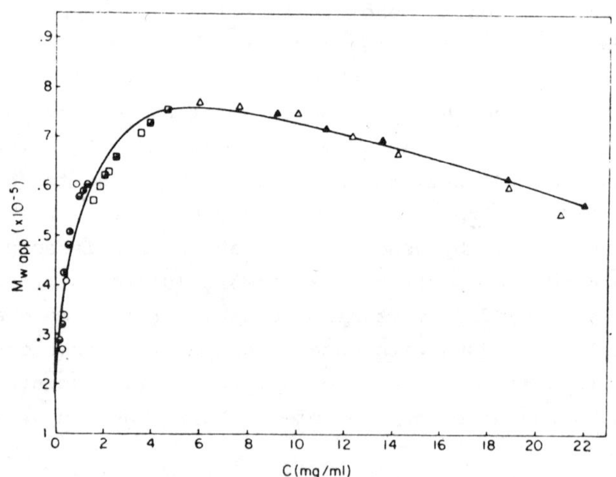

Fig. 1. Self-association of beef heart lactate dehydrogenase at pH 2 and 20 C. The increase in M_{wa} with increasing c up to a maximum near a concentration of 5 mg/ml followed by a decrease in M_{wa} with further increases in c is characteristic of a nonideal self-association. These data were collected from a series of sedimentation equilibrium experiments. The different symbols indicate results from solutions of different initial concentration. Reprinted (by permission of Dr. David B.S. Millar and the American Chemical Society) from ref. 23. Copyright by the American Chemical Society.

apparent weight average molecular weight) with increasing c up to a maximum and then showing a decrease in M_{wa} with higher values of c, is very characteristic of a nonideal self-association.

In 1963 Adams and Fujita[24] published the first paper which showed how to analyze a nonideal monomer-dimer association directly from the experimental data. Subsequently, Adams and Williams[25] showed how to evaluate the apparent weight fraction of monomer (f_a) from sedimentation equilibrium experiments. In 1965 Adams[26] showed that there was an interrelation between the apparent weight (M_{wa}) and number (M_{na}) average molecular weights, so that one could obtain M_{na} from sedimentation equilibrium experiments, if done properly. It was also shown by Adams[27] that one could obtain M_{wa} and $\ln f_a$ from osmotic pressure experiments. These developments opened up the analysis of self-associations to a wide variety of types of self-associations. One could now evaluate equilibrium constants and nonideal terms. The restriction of ideal conditions was now removed. The assumption used by Adams and Fujita[24] in defining the natural logarithm of the activity coefficient of a self-associating solute was critically examined by Ogston and Winzor[28] and found to be a reasonably good one.

In this paper we will show some examples of various types of self-associations that have been encountered experimentally. We will show some ways to analyze various self-associations; these ways will be based primarily on methods developed by Adams and coworkers [1,3,4,11-15,24-27,29,31]. We will also discuss factors influencing self-association, and we shall compare the advantages and disadvantages of various experimental methods for studying self-associations. In this paper we shall not dwell on the derivation of equations, as these are discussed extensively elsewhere; see, for example, Chapter 6 in the Fujita monograph[31]. There are a number of excellent articles and reviews on self-associations which the readers can consult for more details [1,4,30,32], the most recent ones being the review by Kim, Deonier and Williams[3] and the chapter on membrane and vapor pressure osmometry by Adams, Wan and Crawford[29].

EXPERIMENTAL BACKGROUND

There are three basic experiments that can be used to obtain average molecular weights (or their apparent values in nonideal solutions) needed for the analysis of protein self-associations by thermodynamic methods. These are membrane osmometry, elastic light scattering and sedimentation equilibrium. A self-association can be represented by equilibrium of the type

$$nP_1 \rightleftarrows P_n \quad (n = 2,3,\ldots) \tag{1}$$

$$nP_1 \rightleftarrows qP_2 + hP_3 \tag{2}$$

and related equilibria. Here P represents the self-associating solute. At constant temperature the condition for chemical equilibrium for any self-association can be given by

$$n\mu_1 = \mu_n \quad (n = 2,3,\ldots) \tag{3}$$

where μ_i is the molar chemical potential of species i. Before proceeding further one must make a simple assumption about the activity coefficient (y_i) of the self-associating species. Adams and Fujita[24] assumed that $\ln y_i$ can be represented by

$$\ln y_i = iBM_1c \quad (i = 1,2,\ldots) \tag{4}$$

Here B is a constant whose value depends on the temperature and the solute-solvent combination, M_1 is the monomer (or unimer) molecular weight and c is the total solute concentration. By making this assumption one can use the experimental data and data derived from it to evaluate the equilibrium constant or constants (K_i) and the nonideal term (BM_1). The validity of this assumption was tested by Ogston and Winzor[28], who found it to be a good assumption. With this assumption it follows that

$$y_n/y_1^{\,n} = 1 \tag{4a}$$

and hence

$$K_n = a_n/a_1^{\,n} = (c_n/c_1^{\,n})(y_n/y_1^{\,n}) = c_n/c_1^{\,n} \tag{4b}$$

Here $a_n = y_n c_n$ is the activity of species n. It also follows that the total concentration (c) of the self-associating solute is given by

$$c = c_1 + K_2c_1^{\,2} + K_3c_1^{\,3} + \ldots \tag{4c}$$

Braswell[33] has pointed out for ionic self-associating solutes that the limiting form of the Debye-Hückel theory supports eq. (4). Equation (4) has also been supported by light scattering experiments of Edelhoch, Katchalsky et al.[34] on bovine mercaptalbumin monomer and its dimer.

The apparent number average molecular weight (M_{na}) is obtained as the primary data in membrane osmotic pressure experiments from the relation[27,30]

$$\frac{1000\pi}{RT} = \frac{c}{M_{na}} = \frac{c}{M_{nc}} + \frac{B_{*}c^2}{2} \tag{5}$$

Here π is the osmotic pressure in atmospheres, R is the universal gas constant (82.05 ml·atm/deg/mole), T is the absolute temperature, and M_{nc} is the true number average molecular weight. Since $M_i = iM_1$, the number average molecular weight is defined by

$$M_{nc} = c/\Sigma(c_i/M_i) = M_1 \frac{c}{\left[c_1+\frac{k_2c_1^2}{2}+\frac{k_3c_1^3}{3}+\ldots\right]} \qquad (6)$$

and the second virial coefficient B_*M_1 is given by

$$B_* = B + \frac{\bar{v}}{1000M_1} \qquad (7)$$

Here $\bar{v}$ is the partial specific volume of the self-associating solute. Figure 2 shows some plots of π/c vs. c for aqueous solutions of chicken erythrocyte histone F2b[30,35,36].

The apparent weight average molecular weight (M_{wa}) is given as primary data in light scattering experiments from the relation[5,7,9,10,19]

$$\frac{Kc}{\Delta R(o)} = \frac{1}{M_{wa}} = \frac{1}{M_{wc}} + B_*c \qquad (8)$$

The weight average molecular weight (M_{wc}) is defined by

$$M_{wc} = \sum_i c_iM_i/c = \frac{M_1(c_1+2K_2c_1^2+3K_3C_1^3+\ldots)}{c} \qquad (9)$$

In eq. (8) the quantity $\Delta R(O)$ is the excess reduced scattering at zero angle, i.e.,

$$\Delta R(O) = R(O)_{SOLUTION} - R(O)_{SOLVENT} \qquad (10)$$

$$R(\theta) = \frac{i_s(1+\cos^2\theta)}{I_o r^2} \qquad (11)$$

$$R(O) = \lim_{\theta\to o} R(\theta) \qquad (12)$$

and K is an operational instrumental constant defined in the usual manner by

$$K = 2\pi^2 n_o^2(\partial n/\partial c)^2/\lambda^4N \qquad (13)$$

In eq. (11) I_o is the intensity of the incident light, i_s is the intensity of the scattered light, r is the distance from the detector (the photomultiplier tube) to the center of the light scattering cell, and θ is the angle between the incident beam and the detector.

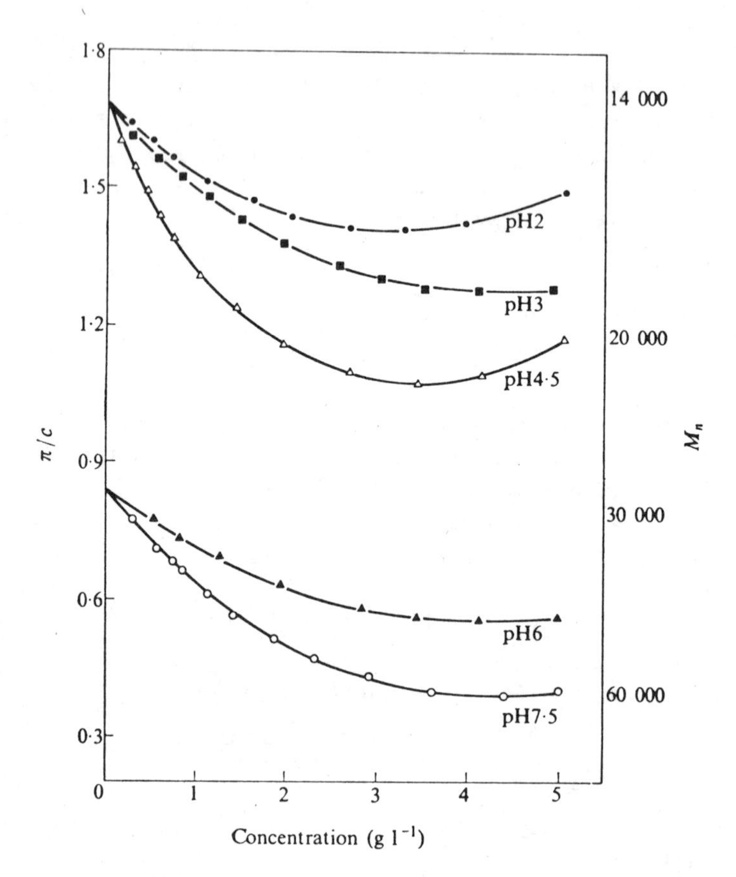

Fig. 2. Self-association of histone F2b from chicken erythrocytes at various pH values. These osmotic pressure experiments were carried out by Diggle and Peacocke[30,35,36] at different pH in various buffers of constant ionic strength (I=0.1). Since $\pi/c = RT/M_{na}$, these are plots of RT/M_{na} vs. c, and it is evident from the minima in some of these plots that nonideal self-associations are encountered. Reproduced from M. P. Tombs and A. R. Peacocke, "The Osmotic Pressure of Biological Macromolecules" (Oxford Univ. Press, London and New York, 1974), by permission of Dr. A. R. Peacocke and the Oxford University Press.

The value of K depends on the temperature, the wavelength of the light used (λ) and on the nature of the solute-solvent combination, since these affect the values of the refractive indices of the solution (n) and the solvent (n_o) as well as the refractive index increment, $(\partial n/\partial c)_{T,P}$. Figure 3 shows some light scattering results on the self-association of apoferritin[37].

The quantity M_{wa} can also be obtained from the sedimentation equilibrium experiment, since [1,3,24,31]

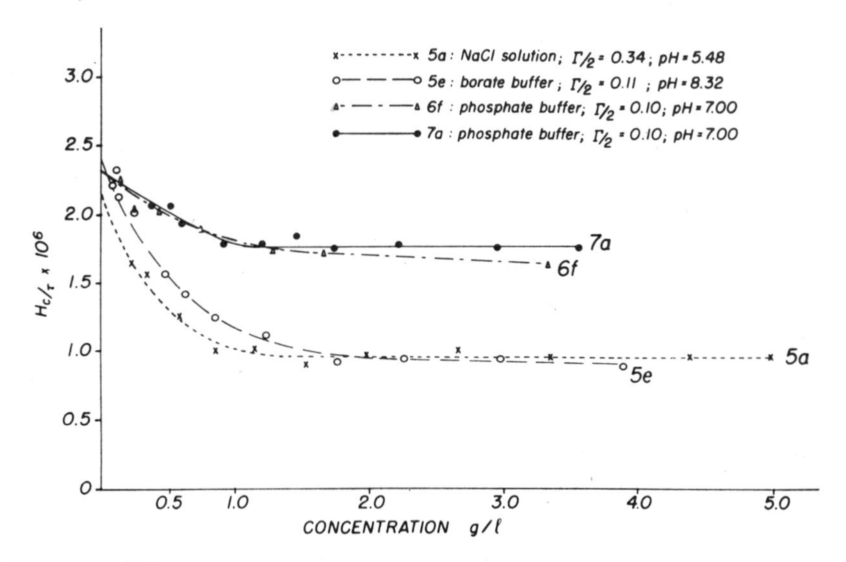

Fig. 3. Self-association of horse spleen apoferritin under various solution conditions. These studies were done by light scattering. The quantity $Hc/\tau = 1/M_{wa}$. Reprinted (by permission of Dr. G. W. Richter and the American Chemical Society) from ref. 37. Copyright by the American Chemical Society.

$$\frac{d\ln c}{d(r^2)} = \frac{d\ln J}{d(r^2)} = AM_{wa} \tag{14}$$

Here

$$A = (1 - \bar{v}\rho)\omega^2/2RT \tag{15}$$

and

$$J = \frac{h(n-n_o)}{\lambda} = \frac{h(\partial n/\partial c)c}{\lambda} \tag{16}$$

is the concentration expressed as interference fringes. Most commonly J refers to interference fringes produced at $\lambda = 546$ nm in a h = 12 mm double-sector centerpiece. Sometimes with multicomponent, ionizable solutes, the quantity A is expressed by

$$A = 1000(\partial\rho/\partial c)_\mu \omega^2/2RT \tag{17}$$

Here $1000(\partial\rho/\partial c)_\mu$ is the density increment, ρ is the density of the solution and ω is the angular velocity of the rotor ($\omega = 2\pi$RPM/60). In order to obtain M_{wa} (or M_{wc} in ideal, dilute solutions) from eqs. (8 or 14) one has to assume that the refractive index increments, $(\partial n/\partial c)_{T,P}$, of each of the self-associating species are equal to each other[1,3,4,31]. Furthermore, in the sedimentation equilibrium experiment it is assumed that the partial specific volumes, $\bar{v}$ (see eq. 15), of the self-associating species are all the same, or that the density increments in eq. (17) are all the same[1,3,4,31]. Figure 1 shows a plot of M_{wa} vs. c for the self-association of beef heart lactate dehydrogenase in 0.1M H_3PO_4-KH_2PO_4, 0.01M dithiothreitol, pH 2.0 at 20°C[23]. These data were obtained from a series of sedimentation equilibrium experiments.

Fortunately, for self-associating solutes there is an interrelation between the apparent average molecular weights. Thus one notes[1,3,4,26,31]

$$\frac{M_1}{M_{na}} = \frac{1}{c}\int_0^c \frac{M_1}{M_{wa}}\,dc \tag{18}$$

and

$$\frac{M_1}{M_{wa}} = \frac{d}{dc}\left(\frac{cM_1}{M_{na}}\right) = \frac{M_1}{M_{na}} + c\,\frac{d}{dc}\left(\frac{M_1}{M_{na}}\right) \tag{19}$$

Furthermore, one can also obtain the apparent weight fraction of monomer, f_a, from the relations[1,3,4,26,31]

$$\ln f_a = \int_0^c \left(\frac{M_1}{M_{wa}} - 1\right)\frac{dc}{c} = \ln f_1 + BM_1c \tag{20}$$

or[26,29,30]

$$\ln f_a = \int_0^c \left(\frac{M_1}{M_{na}} - 1\right)\frac{dc}{c} + \left(\frac{M_1}{M_{na}} - 1\right) \tag{21}$$

These relations can be summarized as follows:

$$\frac{M_1}{M_{wa}} = \frac{M_1}{M_{wc}} + B_*M_1c \tag{22}$$

$$\frac{M_1}{M_{na}} = \frac{M_1}{M_{nc}} + \frac{B_*M_1c}{2} \tag{23}$$

$$\ln f_a = \ln f_1 + B_*M_1c \tag{24}$$

Thus any or all of these quantities can be obtained from a series of experiments with solutions of different protein concentrations by any of the three principal techniques. It is possible to eliminate the second virial coefficient, B_*M_1c, by combining eqs. (22-24) in various ways, thus the quantity ξ is given by[3,4,11,12,29-31,38]

$$\xi = \frac{2M_1}{M_{na}} - \frac{M_1}{M_{wa}} = \frac{2M_1}{M_{nc}} - \frac{M_1}{M_{wc}} \quad (25)$$

and η is defined by[3,4,11,12,29-31,38]

$$\eta = \frac{M_1}{M_{wa}} - \ln f_a = \frac{M_1}{M_{wc}} - \ln f_1 \quad (26)$$

These quantities, ξ and η, can be used to help establish the type self-association present, and for some self-associations they can be used to evaluate the weight fraction of monomer, f_1. Once this can be done, then analysis becomes relatively simple. We will show how these and other relations can be used to analyze various self-associations. For simplicity we will use BM_1c for B_*M_1c.

MONOMER-N-MER ASSOCIATIONS

These associations can be represented by

$$nP_1 \rightleftarrows P_n \quad (n = 2,3,\ldots) \quad (1)$$

where P represents the self-associating solute. For this association the following relations apply:[1,3,4,11-14,29,31,38]

$$c = c_1 + K_n c_1^{\,n} \quad (27)$$

$$f_n = 1 - f_1 \quad (28)$$

$$f_n = K_n c_1^{\,n}/c \quad (29)$$

$$\xi = \frac{2M_1}{M_{nc}} - \frac{M_1}{M_{wc}} = \frac{2+2(n-1)f_1}{n} - \frac{1}{n-(n-1)f_1} \quad (30)$$

and

$$f_1 = \frac{n}{4(n-1)}\left[(\xi+2-\tfrac{2}{n}) - \left((\xi+2-\tfrac{2}{n})^2 - (8/n)(\xi n-1)\right)^{1/2}\right] \quad (31)$$

When a monomer-n-mer association is thought to be present, first assume a value of n (if it is not known beforehand). Examination of a plot of M_1/M_{wa} vs. c will give a good indication of where to start. If the values of M_1/M_{wa} vs. c go beyond 0.5, then it is unlikely that a monomer-dimer association is present. Next solve for f_1 using eq. (31) for each value of ξ. Now use a plot of

$(1-f_1)/f_1^n$ vs. $c^{(n-1)}$ to get K_n, since eq. (29) can be rearranged to give

$$\frac{1-f_1}{f_1^n} = K_n c^{(n-1)} \qquad (32)$$

If the model is correct, then the plot of $(1-f_1)/f_1^n$ vs. $c^{(n-1)}$ will give a straight line (going through or close to the origin) with a slope of K_n. If the model is wrong, then the plot will show curvature. Figure 4 shows plots of M_{wa} vs. c for β-lactoglobulin C in 0.2 M glycine buffer (0.2 M glycine, 0.1 M HCl, I=0.1, pH=2.46)[13]. The maximum in the plots of M_{wa} vs. c indicates nonideal behavior; note that these values of M_{wa} are less than 36,000. This may indicate the presence of a monomer-dimer association; however, since one is dealing with average or apparent average molecular weights, one must test for the presence of other associations.

Figure 5 shows plots based on a modification of eq (2), and it is clear that a monomer-dimer association is the best choice for the three models considered[13]. The monomer-dimer model also describes the self-association of βC in a glycine buffer of twice the ionic strength (0.2 M glycine, 0.1 M HCl, 0.1 M KCl)[13].

For this association once K_2 and f_1 are known, then the nonideal term can be evaluated from eq. (22), since $(M_1/M_{wa}) - (M_1/M_{wc}) = BM_1c$. For a monomer-n-mer association this becomes[12,13]

$$\frac{M_1}{M_{wa}} - \frac{1}{n-(n-1)f_1} = BM_1c \qquad (33)$$

Thus a plot of the left side of eq. (33) vs. c will give a straight line (going through or close to the origin) with a slope of BM_1; Fig. 6 shows such a plot for βC in 0.2 M glycine buffer. It is clear from Fig. 6 that the plot based on n = 2 is the best choice[13]. In some cases the nonideal term may be small, so that the plot based on Fig. 6 will not work as well; in this case one will have to set up an array M_1/M_{wa} values for various values of c and find the best value of BM_1 that fits the array. The best fit is the one for which $\Sigma\delta_i^2$ is a minimum, where $\delta_i = [(M_1/M_{wa}) - \{1/[n-(n-1)f_1]\} - BM_1c]_i$.

Sometimes one can use a plot of η (see eq. 26) and ξ (see eq. 25) as a diagnostic plot to test for the presence of a monomer-n-mer association, as well as some indefinite self-associations. For these cases η and ξ are each functions of f_1, the weight fraction

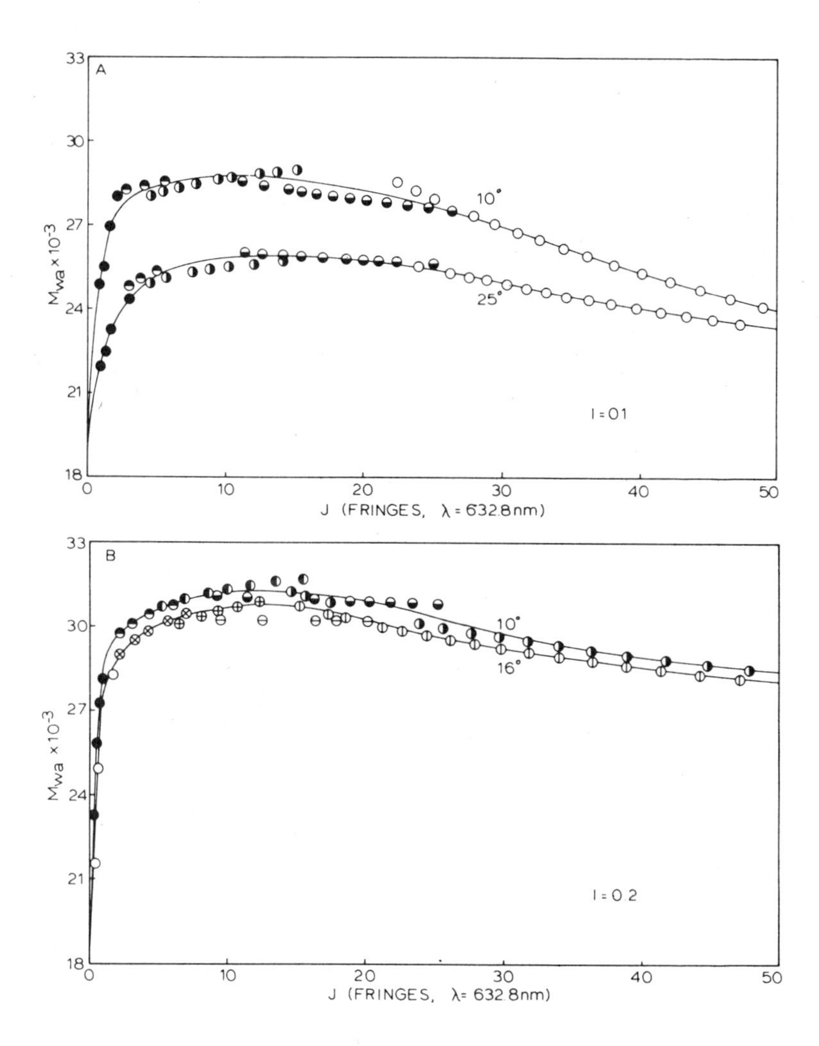

Fig. 4. Plots of M_{wa} vs. C for the temperature-dependent self-association of β-lactoglobulin c in 0.2 M glycine buffer (pH 2.46 at 23°C). For A the ionic strength was 0.1; for B the ionic strength was 0.2, due to the inclusion of 0.1 M KCl in the buffer. The maximum in these plots is a characteristic of nonideal self-associations. Note that the association is greater at lower temperatures, and that the association is greater at the higher ionic strength. From ref. 13 by permission of the authors and the Academic Press, Inc.

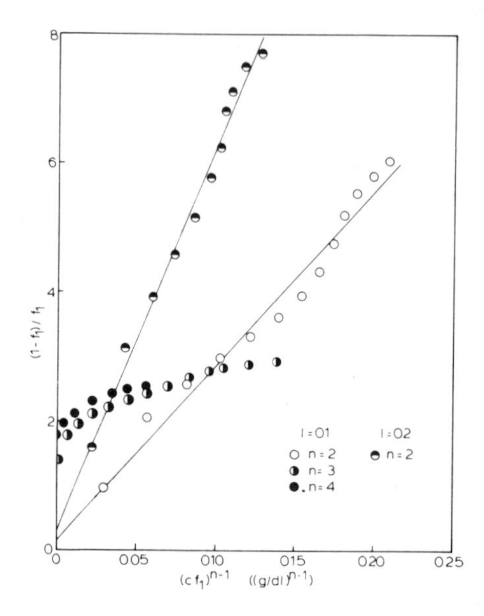

Fig. 5. Test for the presence of a monomer-n-mer association for β-lactoglobulin C in 0.2 M acetate buffer (pH 2.46, I=0.1) for 3 models (n = 2,3 and 4). The only plot based on eq. (32) that seemed to give a straight line going through or close to the origin is the one for n = 2. The curvature in the plots for the other models rules out these self-associations. Note that the plot for n = 2 also describes the self-association of β-lactoglobulin C in 0.2 M glycine buffer of I=0.2. From the slope of these plots one obtains K_2 = 27.2±0.3 dl/g for I=0.1 and K_2 = 60.1±0.2 for I=0.2. From ref. 13 by permission of the authors and the Academic Press, Inc.

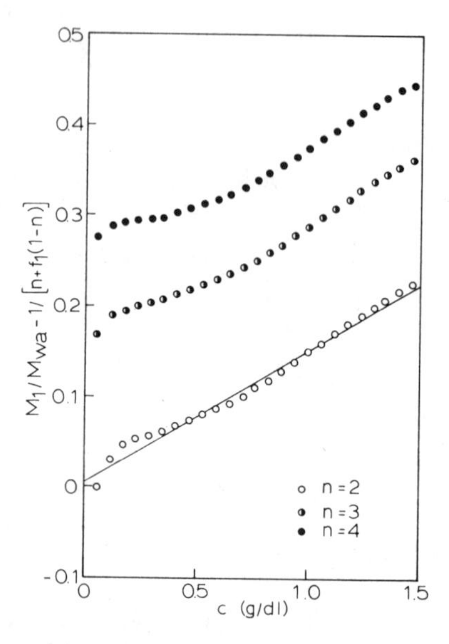

Fig. 6. Evaluation of the nonideal term, BM_1; for the self-association of β-lacto-globulin C in 0.2 M glycine buffer (pH 2.46, I=0.1) at 10°C. Plots based on eq. (33) for the three monomer-n-mer associations tested are shown. The only plot giving a straight line going through or close to the origin is the one for n = 2. From the slope of this line one gets BM_1 = 0.146± 0.009 dl/g. From ref. 13 by permission of the authors and the Academic Press, Inc.

of monomer[38]. Thus one can construct standard plots of η vs. ξ like those shown in Fig. 7. Examples of these standard plots are shown in the paper by Chun, Kim et al.[38], the paper by Adams, Ferguson et al.[4] or in the Fujita monograph[31]. If the experimental values of η and ξ fall on one of these standard plots, this suggests that the self-association is described by that model. For example, the β-lactoglobulin C data of Sarquis and Adams[13] in 0.2 M glycine

Fig. 7. Standard plots of η and ξ for three monomer-n-mer associations (n = 2,3 and 4), and for a Type I SEK Model indefinite (INDEF) self-association. From ref. 4 by permission of the authors and Marcel Dekker, Inc.

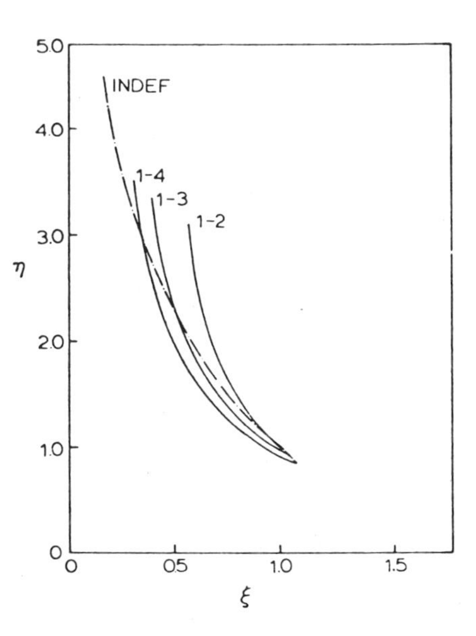

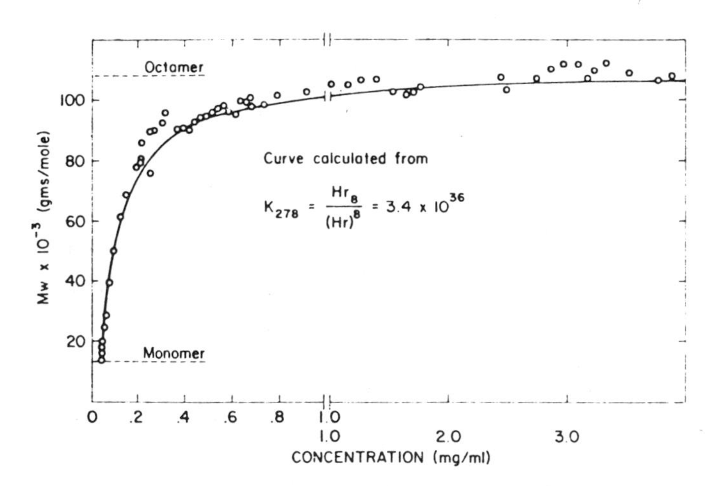

Fig. 8. Plot of M_{wa} vs. c for azidehemerythrin at pH 7 and 5°C in a tris-cacodylate buffer. This self-association is best described as a monomer-octamer self-association. From ref. 40 by permission of the authors and the American Chemical Society. Copyright by the American Chemical Society.

buffer at pH 2.46 seemed to be best described by the η vs. ξ plot for a monomer-dimer association, when compared to other monomer-n-mer models or to a Type I SEK Model indefinite self-association. The presence of the monomer-dimer model was also confirmed by a plot based on eq. (33) for n = 2; this is shown in Fig. 6.

The β-lactoglobulins A, B and C all exhibit a monomer-dimer association at low pH (between pH 2 to 3)[9,11-13,39]. Another protein exhibiting a monomer-n-mer association is hemerythrin, the respiratory protein of a marine worm[40]. Langerman and Klotz[40] have shown that hemerythrin undergoes a monomer-octamer association in a tris-cacodylate buffer (I=0.15, pH 7.0) at 5°C; Fig. 8 shows a plot of M_{wa} vs. c for this association.

MONOMER-N-MER-M-MER (1,N,M) ASSOCIATIONS AND OTHER DISCRETE ASSOCIATIONS[1,3,4,29,30,31]

These associations are represented by

$$nP_1 \rightleftarrows qP_2 + mP_3, \tag{34}$$

$$nP_1 \rightleftarrows qP_2 + hP_4 \tag{35}$$

and related self-associations. It is not as convenient here to use ξ, since it would involve two unknowns f_1 and f_2. For a monomer-dimer-trimer association one notes that

$$\xi = \frac{2+4f_1+f_2}{3} - \frac{1}{3-2f_1-f_2} \tag{36}$$

so other methods need to be explored. One previously used procedure involved combining the equations for c, M_{na}, M_{wa}, and f_a to obtain[1,3,4,29,30,31]

$$\begin{aligned} \frac{6M_1}{M_{na}} &= 5 + 2f_1 + 3BM_1c - \frac{1}{\left[\frac{M_1}{M_{wa}} - BM_1c\right]} \\ &= 5 + 2f_a e^{-BM_1c} + 3BM_1c - \frac{1}{\left[\frac{M_1}{M_{wa}} - BM_1c\right]} \end{aligned} \tag{37}$$

This equation has one unknown, BM_1, and for each value of c one can solve for BM_1 by successive approximations. It may be easier to let $x = BM_1c$, and solve for x at each value of c by successive approximations. Then one could plot x vs c, and the slope of this plot would give BM_1. Alternatively, one can set up an array of values of $6M_1/M_{na}$ corresponding to the different values of c. Now choose various values of BM_1 and calculate $6M_1/M_{na}$, which is the right hand side of eq. (37). Then compare the observed and calculated values of $6M_1/M_{na}$, and choose the value of BM_1 for which $\Sigma\delta_i^2$ is a minimum. Here

$$\delta_i = 6\left\{\left(\frac{M_1}{M_{na}}\right)_{OBSVD} - \left(\frac{M_1}{M_{na}}\right)_{CALCD}\right\}_i \tag{38}$$

If the plot of M_1/M_{wa} vs. c (or a plot of M_1/M_{na} vs. c) shows a minimum, then one knows that $BM_1>0$. One way to note where to choose BM_1 is to start out crudely, say at $BM_1 = 0, 0.05, 0.10$, etc. and observe where δ_i changes sign. One must avoid a value of BM_1 which makes $BM_1c = M_1/M_{wa}$, since for this case $1/\left[\frac{M_1}{M_{wa}} - BM_1c\right]$ is undefined. Once the sign changes have been noted this procedure can be refined to finer intervals about the region of the sign change. One will then have to test these solutions to see if they are real or false solutions; false solutions could lead to negative values of K_2 or K_3 for example. It is possible to use ξ to solve for BM_1c, since

$$f_1 = f_a e^{-BM_1c} \tag{39}$$

and[1]

$$f_2 = 8 - \frac{6M_1}{M_{na}} - \frac{2}{\left[\frac{M_1}{M_{wa}} - BM_1c\right]} + 3BM_1c \tag{40}$$

The insertion of these equations for f_1 and f_2 into eq. (36) leads to an equation in one unknown, BM_1 (or BM_1c), which would have to be solved by successive approximations.

For a monomer-dimer-tetramer the pertinent equations are[1,3,4]

$$\frac{8M_1}{M_{na}} = 6 + 3f_1 + 4BM_1c - \frac{1}{\left[\frac{M_1}{M_{wa}} - BM_1c\right]}$$

$$= 6 + 3f_a e^{-BM_1c} + 4BM_1c - \frac{1}{\left[\frac{M_1}{M_{wa}} - BM_1c\right]} \tag{41}$$

and

$$\xi = \frac{2M_1}{M_{na}} = \frac{M_1}{M_{wa}} = \frac{1+3f_1+f_2}{2} - \frac{1}{f-3f_1-2f_2} \tag{42}$$

In addition to the use of $f_1 = f_a e^{-BM_1 c}$ one can also insert[1]

$$f_2 = 5 - \frac{1}{\left(\frac{M_1}{M_{wa}} - BM_1 c\right)} - \frac{4M_1}{M_{na}} + 2BM_1 c \tag{43}$$

into eq. (42) for ξ in order to get an equation in one unknown, BM_1 (or $BM_1 c$), which is solved by successive approximations.

If more than three associating species are present, then one needs n + 1 relations for n associating species. Such relations are available, for example[1,3,41]

$$M_1^2 \sum \frac{c_i}{M_i^2} = c_1 + \frac{K_2 c_1^2}{4} + \frac{K_3 c_1^3}{9} + \ldots = \int_0^c \frac{M_1^2 dc}{M_{nc} M_{wc}} \tag{44}$$

and[1,3,42]

$$\psi = \frac{c^3 \left[\frac{d}{dc}\left(\frac{M_1}{cM_{wa}}\right)\right]}{\left[\frac{M_1}{M_{wa}} - BM_1 c\right]^3} = -(c_1 + 4K_2 c_1^2 + 9K_3 c_1^3 + \ldots) \tag{45}$$

Thus this analysis could be extended to include four or five associating species. If the system is nonideal then eq. (44) will contain three nonideal terms[41]. Beyond the four or five species one reaches the point of diminishing returns, since the propagation of error due to the combination of the several terms could become quite significant. It is also possible to try using nonlinear least squares to handle this situation; one should see the paper by Lewis and Nutt[43] for more details. Finally, it should be noted that Lewis and Nutt[43], using simulated data, showed that it might be very difficult, if not impossible, to distinguish between a monomer-dimer-tetramer-octamer association and a sequential, indefinite self-association (a Type I or isodesmic association). In addition, Tobolsky and Thach[44] reanalyzed the data of White and Kilpatrick[45] for the self-association of 2-n-butylbenzimidazole and benzotriazole in benzene. White and Kilpatrick[45] used eight equilibrium constants to describe the data, but Tobolsky and Thach,[44] using an ideal, sequential indefinite self-association with two equilibrium constant (a Type III indefinite self-association), found that they could fit the same data within ±1.2% average deviation.

The self-association of α-chymotrypsin in phosphate buffer (pH 6.2, I=0.2) is believed to be a monomer-dimer-trimer association; Fig. 9 shows the plot of M_{wa} vs. c for this association[21]. Millar, Frattali and Willick[23] interpreted their data on the self-association of beef heart lactate dehydrogenase (see Fig. 1) as a monomer-dimer-tetramer-octamer association. Schmidt[5] has interpreted his data on the self-association of α_{S1}-casein B as in terms of a monomer-dimer-trimer-tetramer-pentamer-hexamer association; he did not try to fit his data with an indefinite self-association involving one (Type I SEK) or two (Type III SEK) equilibrium constants. These associations are described in the section that follows this one. Millar, Willick, Steiner and Frattali[46] studied the self-association of a soybean trypsin inhibitor at 20°C and pH 8.35 (0.3 M KCl-0.089 M tris(hydroxymethyl)aminomethane HCl-0.089 M boric acid-0.027 M Na_2 ethylenediaminetetraacetate); they found that their association could be described as a nonideal monomer-dimer-trimer association.

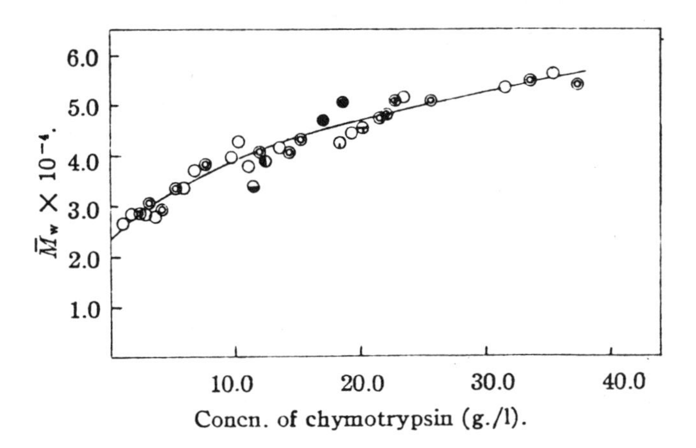

Fig. 9. Plot of M_{wa} vs. c for α-chymotrypsin in phosphate buffer (pH 6.2, I=0.2). These experiments were performed by Rao and Kegeles, using the Archibald method. This association is believed to be a monomer-dimer-trimer association. From ref. 21 by permission of the authors and the American Chemical Society. Copyright by the American Chemical Society.

INDEFINITE SELF-ASSOCIATIONS[1,3,4,29,31,47-49]

Self-associations that appear to proceed without limit, such as

$$nP_1 \rightleftarrows qP_2 + mP_3 + hP_4 + \ldots \tag{2}$$

and related associations, are known as indefinite self-associations. These indefinite self-associations can be grouped into two classes.[48] The first class (the sequential, equal equilibrium constant models or SEK models) requires that the product of the monomer concentration and the equilibrium constant (or constants) be less than one, in order for the various power series that describe C, M_1/M_{na} or M_1/M_{wa} to converge. The second class (the attenuated equilibrium constant models or AK models) do not have this restriction.

For both classes of indefinite self-associations it will be assumed that the convention for the activity coefficient (see eq. 4) still applies; however, eq. (4) is written sometimes as[1,3,4,29,31]

$$\ln y_i = i\hat{B}M_1C, \ i = 1,2,\ldots \tag{46}$$

for C in g/ml. Because M_1/M_{na} and M_1/M_{wa} are independent of the choice of concentration scales (as long as g/volume units are used), it follows that $\hat{B}M_1C = BM_1c$ and $\hat{B}M_1 = 1000\ BM_1$, since $c = 1000C$.

Type I Indefinite Self-Association (SEK Model). This association is made up of a series of associations:[1,3,4,29,31,38,41]

$$\begin{array}{l} P_1 + P_1 \rightleftarrows P_2 \\ P_1 + P_2 \rightleftarrows P_3 \\ \vdots \quad\ \vdots \quad\ \vdots \\ P_1 + P_{i-1} \rightleftarrows P_i \\ \vdots \quad\ \vdots \quad\ \vdots \end{array} \tag{47}$$

For this association it is assumed that $\Delta G°$ for any step (i.e. any $P_1 + P_{i-1} \rightleftarrows P_1$) is the same. This leads to the relations:

$$\begin{array}{l} [P_2] = K[P_1]^2 \\ [P_3] = K^2[P_1]^3 \\ \vdots \qquad\ \vdots \\ [P_i] = K^{(i-1)}[P_1]^i \\ \vdots \qquad\ \vdots \end{array} \tag{48}$$

where $[P_i]$ is the molar concentration of species i (i = 1,2,...). If one converts to practical concentration units C_i, where the C_i are customarily expressed in grams per milliliter (g/ml), then

$$C_i = \left(\frac{1000K}{M_1}\right)^{(i-1)} C_1^{\ i} \quad (49)$$

$$= k^{(i-1)} C_1^{\ i}$$

Here $k = (1000K/M_1)$ is known as the intrinsic equilibrium constant. The total concentration of the associating solute, C, becomes

$$C = C_1 + 2kC_1^{\ 2} + 3k^2C_1^{\ 3} + 4k^3C_1^{\ 4} + \ldots \quad (50)$$

$$= C_1(1 + 2kC_1 + 3k^2C_1^{\ 2} + 4k^3C_1^{\ 3} + \ldots)$$

$$= C_1/(1 - kC_1)^2, \text{ if } kC_1 < 1$$

Similarly one can show that

$$M_1/M_{wa} = (M_1/M_{wc}) + \hat{B}M_1C \quad (51)$$

$$= \frac{1-kC_1}{1+kC_1} + \hat{B}M_1C$$

and

$$M_1/M_{na} = (M_1/M_{nc}) + (\hat{B}M_1C/2) \quad (52)$$

$$= 1-kC_1 + (\hat{B}M_1C/2)$$

For the analysis of a Type I association one notes that the following relations apply:[3,4,12,29,31,49]

$$f_1 = (1 - kC_1)^2 = (M_1/M_{nc})^2 \quad (53)$$

$$\sqrt{f_1} = 1 - kC_1 \quad (54)$$

$$\xi = \frac{2M_1}{M_{na}} - \frac{M_1}{M_{wa}} = \frac{2M_1}{M_{nc}} - \frac{M_1}{M_{wc}} \quad (55)$$

$$= 2(1 - kC_1) - \frac{1-kC_1}{1+kC_1}$$

or

$$\xi = 2\sqrt{f_1} - \frac{\sqrt{f_1}}{2 - \sqrt{f_1}} \quad (56)$$

and

$$\sqrt{f_1} = (1/4)\left[(\xi+3) - \sqrt{(\xi+3)^2 - 16\xi}\right] \quad (57)$$

It also follows from eq. (54) that

$$(1 - \sqrt{f_1})/f_1 = kC = K[P] \qquad (58)$$

where $[P] = 1000\ C/M_1$. Thus a plot of $(1 - \sqrt{f_1})/f_1$ vs. C will give a straight line going through or close to the origin, if a Type I SEK Model indefinite self-association is present. The slope of this plot is k. If one plots $(1 - \sqrt{f_1})/f_1$ vs. [P], then the slope of the straight line will be K. If this model were wrong, then the plot would give curvature, or the plot would give a straight line whose intercept would be quite far from the origin. The self-association of a β-lactoglobulin A sample in 0.2 M acetate buffer (0.1 M HOAc, 0.1 M NaOAc, I=0.1, pH 4.65 at 23°C) at 16°C was reported by Adams and Lewis[41] to be a Type I SEK Model indefinite self-association. Figure 10 shows a plot of M_1/M_{wa} vs. c for the Adams and Lewis data.

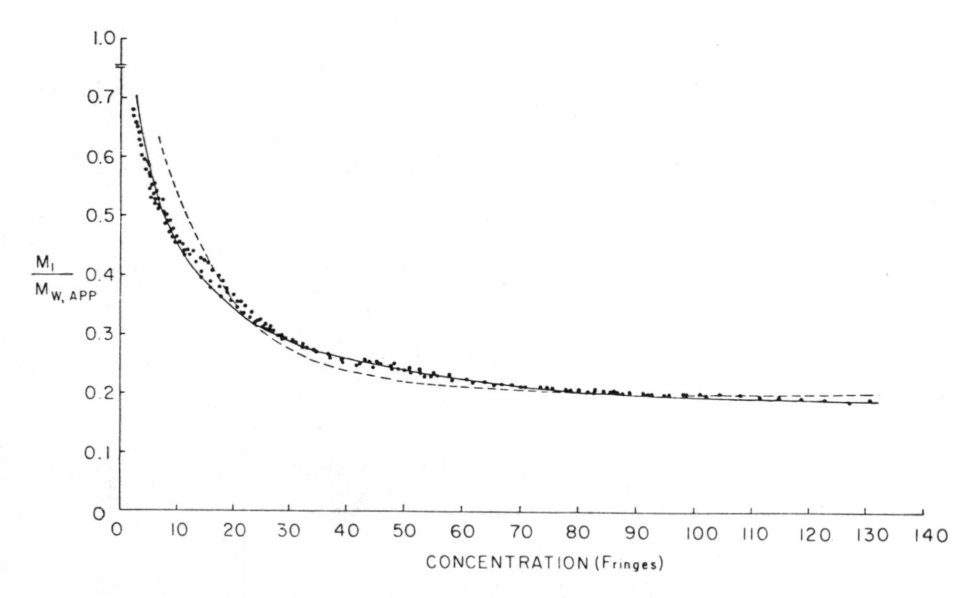

Fig. 10. Plot of M_1/M_{wa} vs. c for β-lactoglobulin A in 0.2 M acetate buffer (pH 4.61, I=0.1) at 16°C. The black dots represent the data obtained from a series of sedimentation equilibrium experiments. Ten solutions of different initial concentrations were used. The dashed line represents an attempt to fit the data as a monomer-dimer-octamer association (K_2 = 24.1 dl/g, K_8 = 4.0x10^6 $(dl/g)^7$, BM_1 = 0.017 dl/g). The solid curve represents an attempt to fit the data as a SEK Model Type I indefinite self-association having k = 4.00x10^2 ml/g and $\hat{B}M_1$ = 1.6 ml/g. It is clear that the indefinite self-association gives a better fit. This is also confirmed in Fig. 11. From ref. 41 by permission of the authors and the American Chemical Society. Copyright by the American Chemical Society.

Figure 11 shows a plot based on eq. (58), which indicates that this is a Type I SEK Model indefinite self-association. Chun, Kim et al.[38] showed that the light scattering studies by Eisenberg and Tompkins[50] on the self-association of bovine liver L-glutamate dehydrogenase could be described by this model.

Once k and f_1 are known, then $\hat{B}M_1$ can be calculated from a modification of eq. (51), if M_{wa} is the primary experimental data. Thus one notes for the Type I association that

$$\frac{M_1}{M_{wa}} - \frac{\sqrt{f_1}}{2 - \sqrt{f_1}} = \hat{B}M_1 C \qquad (59)$$

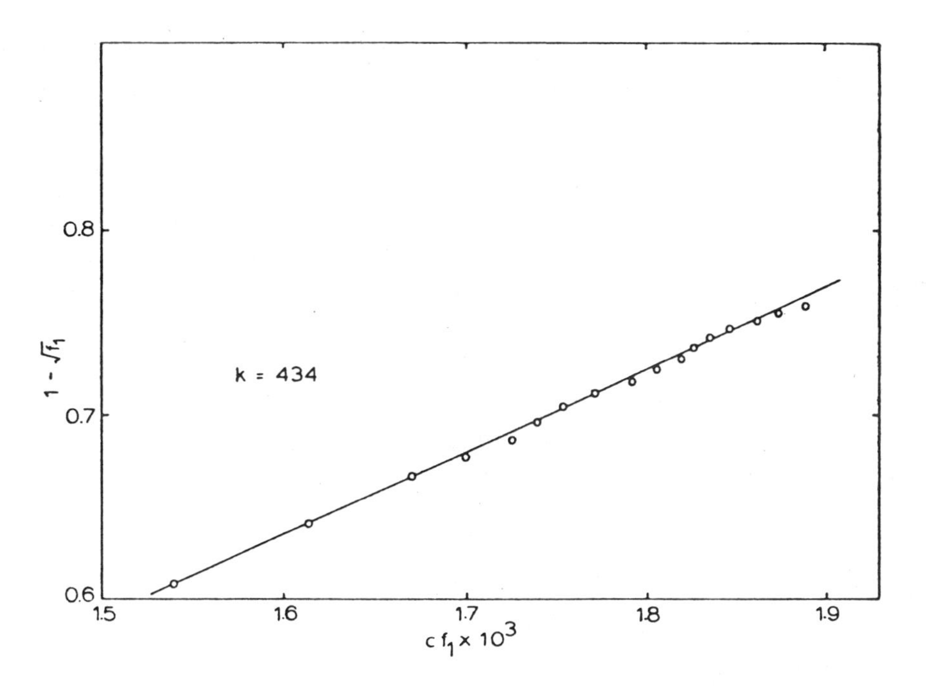

Fig. 11. Plot based on a modification of eq. (58) to test for the presence of an SEK Model Type I indefinite self-association for the Adams and Lewis β-lactoglobulin A data (see Fig. 10). This plot confirmed the presence of an indefinite self-association; from the slope of the plot one obtains $k = 4.34 \times 10^2$ dl/g. From ref. 4 by permission of the authors and Marcel Dekker, Inc.

since $M_1/M_{wc} = \sqrt{f_1}/(2 - \sqrt{f_1})$. An equation analogous to eq. (59) can be obtained from eqs. (52) and (53), when M_{na} is the primary data. Figure 12 shows a plot based on eq. (52) for the Adams and Lewis[41] β-lactoglobulin A data. If $\hat{B}M_1$ is relatively large, then one can use a plot based on eq. (59), such as the one in Fig. 12. When $\hat{B}M_1$ is small, it may be preferable to set up an array of data and find the best $\hat{B}M_1$ by successive approximations. The best $\hat{B}M_1$ will be the one for which $\Sigma\delta_i^2$ is a minimum; here $\delta_i = [(M_1/M_{wa})_{OBSVD} - (M_1/M_{wa})_{CALCD}]_i$. More details are given in ref. 47.

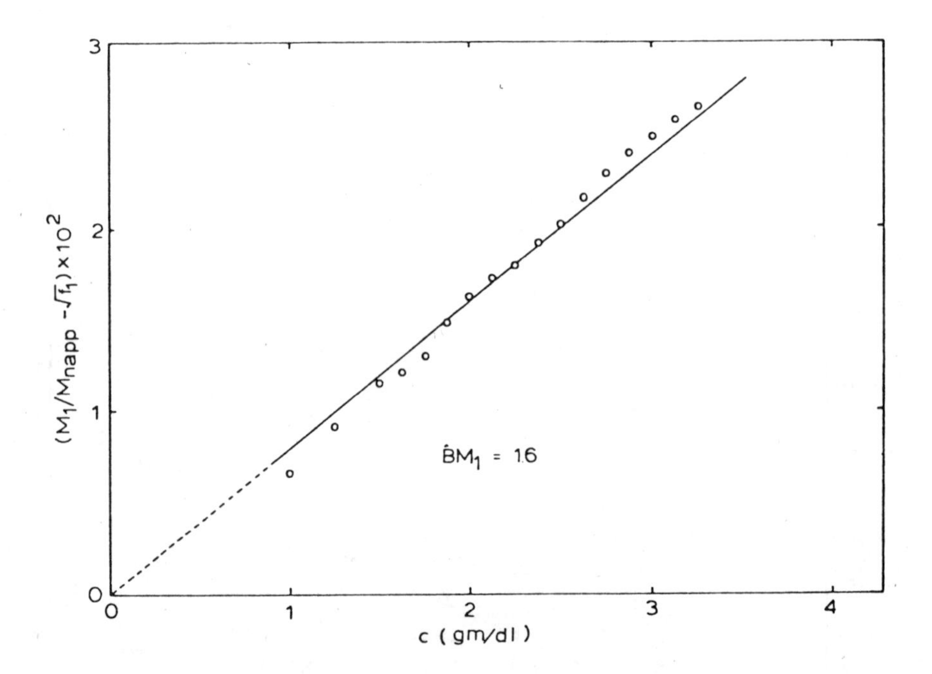

Fig. 12. Evaluation of $\hat{B}M_1$ for the Adams and Lewis β-lactoglobulin A data; this plot is based on eq. (52). From the slope of this plot one obtains $BM_1 = 1.6$ ml/g. Taken from ref. 4 by permission of the authors and Marcel Dekker, Inc.

Type II Indefinite Self-Association (SEK Model)[4,9,28,47] For this association it is assumed that all odd species, except monomer, are absent, so that this association is represented by

$$nP_1 \rightleftarrows qP_2 + hP_4 + mP_6 + \ldots \quad (60)$$

It is also assumed that $\Delta G°$ is the same for $2P_1 \rightleftarrows P_2$, $2P_2 \rightleftarrows P_4$, etc. Thus the total concentration, C, becomes

$$C = C_1\left(1 + \frac{2kC_1}{(1-k^2C_1^2)^2}\right) = C_1\left(1 + \frac{2x}{(1-x^2)^2}\right) \quad (61)$$

if $kC_1 < 1$. Here $x = kC_1$. The equation for ξ (see eq. 25, for a definition of ξ) becomes for this case

$$\xi = \frac{2\left(1 + \frac{x}{(1-x^2)}\right)}{1 + \frac{2x}{(1-x^2)^2}} - \frac{1 + \frac{2x}{(1-x^2)^2}}{1 + \frac{4x(1+x^2)}{(1-x^2)^3}} \quad (62)$$

For this association $\xi = f(x)$ only, so that one can estimate x from a table of ξ vs. x very quickly by successive approximations. Then one notes that since

$$x = kC_1 = kCf_1 \quad (63)$$

it follows that

$$x/f_1 = kC \quad (64)$$

So a plot of x/f_1 vs. C should give a straight line with a slope of k that goes through or close to the origin, if this model is present. The nonideal term can be evaluated from eq. (59) using the appropriate equation for M_1/M_{wc} or from an array of M_1/M_{wa} and C values.[47]

Type III Indefinite Self-Association (SEK Model).[4,29,47-49] This a variant of the Type I association in which it is assumed that $K_{12} \neq K_{23}$, K_{34} etc., but it is assumed that $K_{23} = K_{34} = \ldots = K$. In other words the $\Delta G°$ for the monomer-dimer step is different from the $\Delta G°$ for any other step, and the $\Delta G°$'s for the remaining steps are the same. This association and the Type IV, which is the analogous variant of a Type II association, are sometimes referred to as cooperative associations, whereas the Types I and II associations are noncooperative.

For this association we note that

$$C = C_1\left(1 + \frac{k_{12}C_1(2-kC_1)}{(1-kC_1)^2}\right) \quad (65)$$

$$= C_1\left(1 + \frac{y(2-x)}{(1-x)^2}\right), \text{ if } kC_1 < 1$$

$$y = k_{12}C_1 = (1000K_{12}/M_1)C_1 \quad (66)$$

$$x = kC_1 = (1000K/M_1)C_1 \quad (67)$$

and

$$\xi = \frac{2\left(1 + \frac{y}{(1-x)}\right)}{1 + \frac{y(2-x)}{(1-x)^2}} - \frac{1 + \frac{y(2-x)}{(1-x)^2}}{1 + \frac{y(4-3x+x^2)}{(1-x)^3}} \tag{68}$$

Here the equation for ξ is more complicated, since it seems to contain two unknowns x and y. One can try to solve this equation by an iterative procedure. Alternatively one can use the quantity

$$\ln (f_a/f_{a*}) = \ln (f_1/f_{1*}) + BM_1(C-C_*) \tag{69}$$

The appropriate equation for f_1 is given by eq. (40) of ref 47. The quantity $\ln (f_a/f_{a*})$ is defined by

$$\ln (f_a/f_{a*}) = \int_{C_*}^{C} \left(\frac{M_1}{M_{wa}} - 1\right) dC/C \tag{70}$$

$$= \int_{C_*}^{C} \left(\frac{M_1}{M_{na}} - 1\right) dC/C + \left(\frac{M_1}{M_{na}} - \frac{M_1}{M_{na_*}}\right)$$

More details about these procedures are given in the paper by Tang, Powell, et al.[47] These authors showed that the self-association of β-lactoglobulin A at 16°C in an acetate buffer (0.10 M HOAc, 0.1 M NaOAc, 0.05 M KCl, pH 4.65 at 23°C) of ionic strength 0.15 could be described by this model[47,49]. Figure 13 shows a plot of M_1/M_{wa} vs. C for a βA under these solution conditions.

Type IV Indefinite Self-Association (SEK Model)[4,29,47,49] This is the cooperative variant of a Type II indefinite self-association. Here $K_{12} \neq K_{24}$, K_{46} etc., but $K_{24} = K_{46} = \cdots = K$.

For this association

$$C = C_1\left(1 + \frac{2y}{(1-z^2)^2}\right) \tag{71}$$

$$y = k_{12}C_1 = \left[\frac{1000K_{12}}{M_1}\right] C_1 \tag{72}$$

$$z^2 = k_{12}kC_1^{\,2} = \left(\frac{1000}{M_1}\right)^2 K_{12}KC_1^{\,2} \tag{73}$$

and

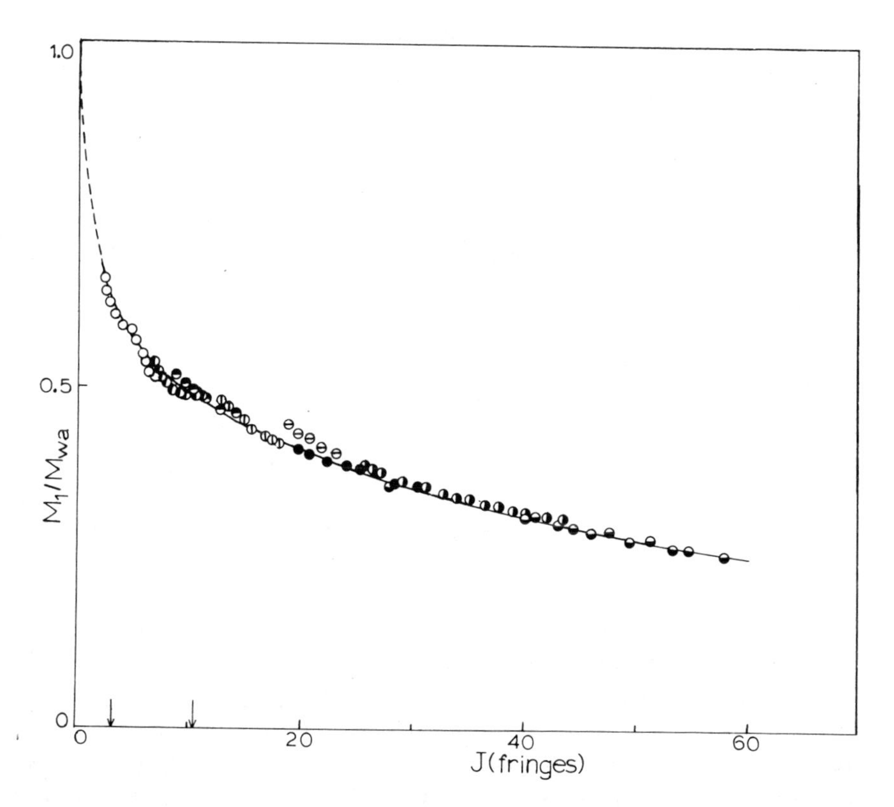

Fig. 13. Plot of M_1/M_{wa} vs. c for the self-association of β-lactoglobulin A in an acetate buffer of ionic strength 0.15 at 16°C (0.1 M HOAc, 0.1 M NaOAc, 0.05 M KCl, pH 4.65 at 23°C). This association is best described as an SEK Model Type III indefinite self-association having $k_{12} = 8.13 \times 10^2$ ml/g, $k = 0.280 \times 10^3$ ml/g and $BM_1 = -3.59$ ml/g. Variable knot spline function were used to put a smooth curve through the data; the arrows indicate the x-positions of the splines or knots. From ref. 47 by permission of the authors and the North-Holland Publishing Co.

$$\xi = \frac{2M_1}{M_{nc}} - \frac{M_1}{M_{wc}} = \frac{2\left(1 + \frac{y}{1-z^2}\right)}{1 + \frac{2y}{(1-z^2)^2}} - \frac{1 + \frac{2y}{(1-z^2)^2}}{1 + \frac{4y(1+z^2)}{(1-z^2)^3}} \tag{74}$$

One can solve for x and z using an iterative procedure as described previously, or by using $\ln(f_a/f_{a*})$, defined by eq. (70); the appropriate expression for f_1 is given by eq. (26) of ref. 47. In addition it can be shown that ξ can be converted to an equation in one unknown, $\hat{B}M_1$ (or $\hat{B}M_1C$), using the relations in appendices A and C of ref. 47. Thus for this association it follows that

$$z^2 = 2R_1 - 1 = \frac{2\left(\frac{M_1}{M_{nc}} - f_1\right)}{(1-f_1)} - 1 \tag{75}$$

$$= \frac{2\left(\frac{M_1}{M_{na}} - \frac{\hat{B}M_1C}{2} - f_a \exp[-\hat{B}M_1C]\right)}{1 - f_a \exp[-\hat{B}M_1C]} - 1$$

and

$$y = (1-z^2)\left(\frac{1}{f_1}\frac{M_1}{M_{nc}} - 1\right) \tag{76}$$

$$= (1-z^2)\left(\frac{1}{f_a \exp[-\hat{B}M_1C]}\left[\frac{M_1}{M_{na}} - \frac{\hat{B}M_1C}{2}\right] - 1\right)$$

A similar procedure can be applied to a Type III SEK Model indefinite self-association.

Indefinite Self-Associations of the Second Class (AK Models)[48]
Studies on the self-association of nucleic acid bases, their derivatives and on nucleosides in aqueous solutions have indicated that these solutes self-associate in stacks that are made up of the planar base moiety, and that the interaction is attributed to hydrophobic interactions rather than hydrogen bonding. In some cases the stacking process has been characterized by a model which assumes that the $\Delta G°$ and also the $\Delta H°$ for the addition of a single base molecule (or mononucleoside) to the stack is independent of the size of the base stacking aggregate. For some materials which associate to a very high degree, this model (essentially the Type I SEK Model indefinite self-association) may over-estimate the degree of association at higher solute concentrations. There have been various attempts to modify this situation, the Type III SEK Model indefinite self-association being one of them.

Another approach has been suggested by Garland and Christian[48], and they have tested it experimentally with thermodynamic and kinetic data. In the SEK model it is assumed for any step ($P_1 + P_{i-1} \rightleftarrows P_i$) that a) $\Delta H_2° = \Delta H_3° \cdots = \Delta H°$ and b) that the $\Delta S°$ of all such reactions is constant. Garland and Christian[48] point out that since the base molecules of various stacks are able to exchange reversibly with one another, that it seems more reasonable that there should be a varying entropy change. Thus for

$$P_1 + P_{i-1} \rightleftarrows P_i \quad (77)$$

$$\Delta S_i^\circ = -R \ln i + \text{constant} \quad (78)$$

and

$$K_i = \exp(-\Delta G_i^\circ/RT) \quad (79)$$

$$= \text{constant} \times \exp(-\ln i)\exp(-\Delta H^\circ/RT)$$

This means that $K_i = K/i$, where K is defined as it was for the SEK model. This procedure still lets $\Delta H_i^\circ = \Delta H_{i+1}^\circ = \ldots \Delta H^\circ$, i.e., the value of ΔH° is independent of the size of the base stack. Also they assume that the volume change for adding monomer to a stack is independent of the size of the stack. This new model has been named by Garland and Christian[48] as the Attenuated Equilibrium Constant (AK) Model. Since this model may apply to some protein self-associations, it seems appropriate to describe it here briefly.

a) Type I Indefinite Self-Association (AK Model). Here it is assumed that the molar equilibrium constants are related as follows:[48]

$$P_1 + P_{i-1} \rightleftarrows P_i, \; i = 2,3,\ldots \quad (77)$$

$$K_i = [P_i]/[P_{i-1}][P_1] = K/i \quad (80)$$

The total concentration of the associating solute (in g/l) becomes

$$c = c_1 + kc_1^{\,2} + (k^2/2)c_1^{\,3} + (k^3/2\cdot 3)c_1^{\,4} + \ldots \quad (81)$$

$$= c_1 \exp(kc_1), \; 0 < kc_1 < \infty$$

$$k = K/M_1 \quad (82)$$

Similarly, one notes that

$$\frac{M_1}{M_{nc}} = \frac{1}{kc_1}[1-\exp(-kc_1)] = \frac{1}{x}[1-e^{-x}] \quad (83)$$

$$\frac{M_1}{M_{wc}} = \frac{1}{1+kc_1} = \frac{1}{1+x} \quad (84)$$

and

$$\xi = \frac{2M_1}{M_{na}} - \frac{M_1}{M_{wa}} = \frac{2}{x}[1-\exp(-x)] - \frac{1}{1+x} \quad (85)$$

Note that $x = kc_1$ can be greater than one, i.e., $0 \leq x < \infty$. For the nonideal case it will be assumed that eq. (4) applies, so that M_1/M_{na} and M_1/M_{wa} are defined in the usual way by eqs. (23) and (22), respectively.

Since ξ is a function of x, one can make up a table or graph of ξ vs. x; the value of x (if this association is present) can be obtained by successive approximations. Once x is known, so is f_1, since

$$f_1 = \exp(-kc_1) = \exp(-x) \tag{86}$$

Furthermore

$$x = kc_1 = kcf_1 \tag{87}$$

so that

$$x/f_1 = x/\exp(-x) = kc \tag{88}$$

A plot of x/f_1 vs. c will give a straight line going through or close to the origin, if this association is present, and the slope of the line would be k. The Garland and Christian[48] treatment was based on ideal solutions, molar concentrations and the ideal osmotic coefficient (M_1/M_{nc} is the osmotic coefficient). We have recast the theory in more practical units, so it can be applied to ideal or nonideal solutions.

b) Other AK Models. One can develop analogous equations for a Type II, Type III or Type IV AK Model indefinite self-association. For the Type II model it is assumed, as we did with the Type II SEK model, that all odd species beyond monomer are absent. The Type III is a variant of the Type I requiring two different equilibrium constants ($K_{12} = K_2$, and $K_i = K/i$ for $i > 2$); the Type IV is the analogous variant of the Type II AK Indefinite Self-Association. We will give more details about these relations and how to analyze them in a future paper. The ideal case for a Type III AK indefinite self-association has also been treated by Garland and Christian[48].

OTHER TYPES OF SELF-ASSOCIATIONS

Pekar and Frank[51], in their studies on the self-association of insulin at pH 7.4 believed that their data could be described by

$$nP_1 \rightleftarrows qP_2 + hP_6 + jP_{12} + lP_{18} + \cdots \tag{89}$$

Thus, the higher aggregates would be multiple aggregates of the hexamer. Teller[52] has published equations for studying discrete self-associations in which all the molar equilibrium constants are equal.

So far our discussion has been based on self-associations involving a homogeneous unimer (or monomer). There is also the possibility of self-association occurring in polymeric solutions. In this case one would have a distribution of unimers which would self-associate to form a distribution of dimers and perhaps higher aggregates. Šolc and Elias[53] have published a very elegant, seminal paper in this area. Quite likely this aspect of self-associations will be explored further in the future.

FACTORS INFLUENCING SELF-ASSOCIATION

There are a number of factors that influence or contribute to self-association: variation in the chemical composition of related proteins, electrostatic factors, temperature, buffer composition, presence of metal ions like Z_n^{+2}, the presence of some amino acids, and perhaps the presence or absence of substrates in the case of enzymes.

With regard to variation in the chemical composition this can be demonstrated by chemical modification[15,16] of the protein or by doing identical experiments with genetic variants. Schmidt studied the self-association of α_{S1}-caseins[5]. There are four genetic variants, Types A, B, C and D, of this protein. Schmidt[5] studied the self-association of three variants (B, C and D) in aqueous solutions using light scattering. Figure 14 shows a comparison of the self-association of these three variants. Under the experimental conditions (21°C, imidazole-HCl-NaCl buffer, pH 6.6, I=0.20) two variants, B and D, associate quite similarly, since the experimental values of M_{wa} vs. c seem to fit the same plot. The other variant, Type C, exhibits a stronger self-association, since higher values of M_{wa} are observed with increasing c. Figure 15 shows the influence of ionic strength on the self-association of α_{S1}-casein B. Note that there is no association at the lowest ionic strength, but there is at the higher values of I[5]. This would suggest that electrostatic (charge) effects are involved, since the increase in the ionic strength screens (reduces) the repulsion of like charged macroions. The self-association of α_{S1}-casein B does exhibit a temperature dependence. At 8°C there is more association than there is at 30°C[5].

Another well known example of the effect of genetic variation on the chemical composition, and hence on the self-association of proteins, is exhibited by the bovine β-lactoglobulins[6-15,39]. There are four genetic variants - Types A, B, C and D, as well as a carbohydrate containing variant of Type A. We will refer to these variants

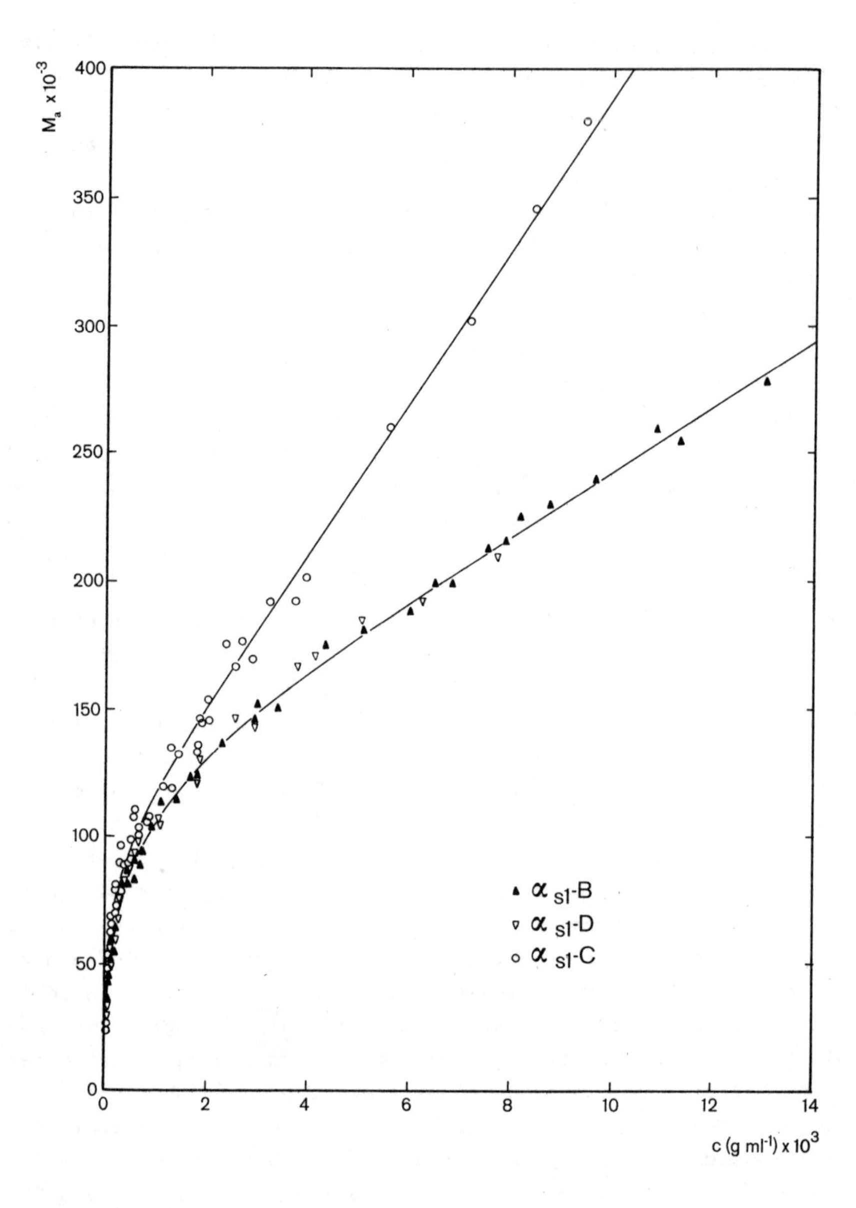

Fig. 14. Self-association of three of the genetic variants of α_{S1}-casein at 21°C in imidazole-HCl-NaCl buffer (pH 6.6, I=0.20). These data were obtained by light scattering. Note that the self-association for two of the variants, α_{S1}-B and D, is essentially the same, since the data points fall on the same M_{wa} vs. c curve. The self-association for the α_{S1}-D variant is much stronger than that of the other two variants. From ref. 5 by permission of Dr. D. G. Schmidt.

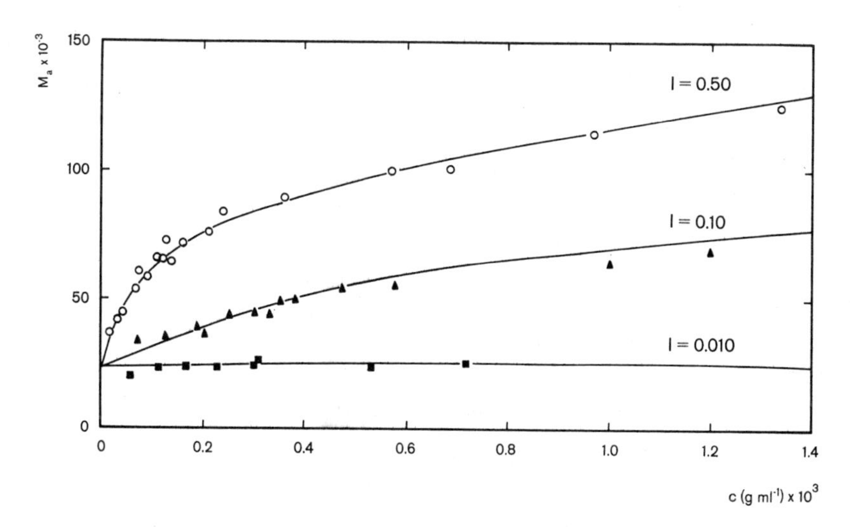

Fig. 15. Effect of ionic strength on the self-association of α_{s1}-casein B at 21°C and pH 6.6. There is essentially no association at I=0.01, whereas there is self-association at I=0.10 and a greater self-association at I=0.50. This would indicate the supporting electrolyte screens charge effects due to the macroion and hence facilitates self-association. From ref. 5 by permission of Dr. D. G. Schmidt.

as βA, βB, βC, and βD. Very elegant studies on the self-association of these proteins have been carried out by Timasheff, Townend and their colleagues[6-10], using light scattering, sedimentation velocity and moving boundary electrophoresis. At pH between 2 - 3.5, βA, βB and βC undergo a reversible monomer-dimer association[9,11-15,39]. If the pH is raised so that it lies between pH 3.5-5.2, the situation is much different. βA undergoes a very strong association that continued beyond dimer[7,10,41,47,49]. McKenzie calls this anomalous, since the other variants do not exhibit this behavior. βB associates slightly beyond dimer[7,10,15], but not as strongly as βA. βC and βD are reported to associate to dimer only[14,15,54]. The carbohydrate containing variant of βA does not exhibit this anomalous association[15]. Armstrong and McKenzie[15,16] have chemically modified the carboxyl groups of βA; the chemical modification had little effect on the optical rotatory dispersion, which would indicate that conformation changes due to chemical modification were minimal. The modified βA did not show the anomaly, as it associated only slightly beyond dimer. These various studies seem to indicate that the aspartic acid residue 64 is involved in the anomalous self-association of βA, since this residue is changed in the other variants and would be affected by chemical modification of the carboxyl groups, if it were present.

The light scattering studies of Timasheff and Townend suggested that βA undergoes a dimer-octamer association, since the minimum molecular weight obtained from the extrapolation of the $1/M_{wa}$ vs. c data was $2M_1$. Additional studies by Kumosinski and Timasheff[10] in 0.2 M acetate buffer (pH 4.65, I=0.10) confirmed this observation; however, they interpreted their association data as a progressive tetramerization (1,2,3,4). Figure 16 shows a plot of $1/M_{wa}$ vs. c for the βA data of Kumosinski and Timasheff.

The sedimentation equilibrium studies of Adams and Lewis[14] with βA at 16°C in acetate buffer (0.1 M HOAc, 0.1 M NaOAc, pH 4.65, I=0.1) indicated that an indefinite SEK Model Type I association described the observed self-association (see Figs. 10, 11 and 12). Furthermore, the trend of the M_{wa} vs. c (or $1/M_{wa}$ vs. c) data indicated that the minimum molecular weight in aqueous solution was M_1 (M_1 = 18,422 g/mole). Additional studies by Tang, Powell et al.[47]

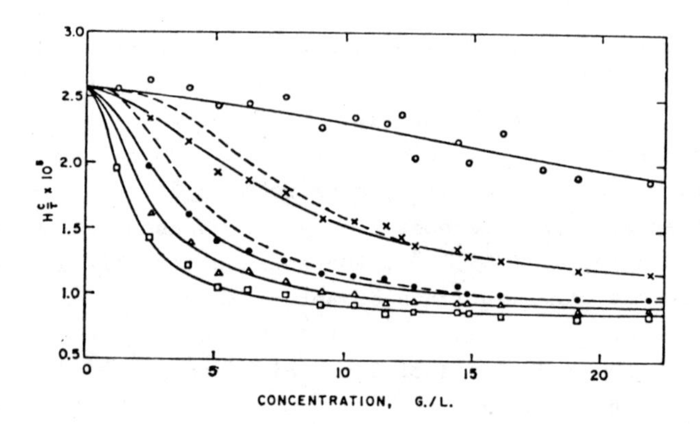

Fig. 16. Light scattering studies on the self-association of β-lactoglobulin A in 0.2 M acetate buffer (0.1 M NaOAc, 0.1 M HOAc, pH 4.65, I=0.1) at various temperatures (O, 25°; x, 15°; ●, 8°; Δ, 4.5°; □, 2°). The common intercept of these plots of $1/M_{wa}$ vs. c is about 2.6 x 10^{-4} which corresponds to M_1 = 38.5 x 10^3. From ref. 10 by permission of the authors and the American Chemical Society. Copyright by the American Chemical Society.

at 16°C in an acetate buffer of higher ionic strength (0.1 M HOAc, 0.1 M NaOAc, 0.05 M KCl, pH 4.65, I=0.15) indicated a slightly stronger association, which could be described as an SEK Model Type III indefinite self-association (see Fig. 13). The range of the M_{wa} values was $27.9_6 \times 10^3$ g/mole at J = 2.49 fringes (12 mm fringes, λ = 546 mm) to $71.9_5 \times 10^3$ g/mole at J = 57.9_8 fringes. The trend of the lower concentration data indicated that M_1 = 18,422 g/mole would be a good choice. For βA J = 4.00c, where c is in g/l.

We can illustrate the effects of genetic differences and also the effects of changes in the chemical environment by comparing the behavior of βA and βC under identical experimental conditions[12,13]. These two proteins differ by three amino acids as Table 1 shows.

TABLE 1
DIFFERENCES IN AMINO ACID COMPOSITION F β-LACTOGLOBULINS A, B, C AND D[a]

Residue No.	A	B	C	D
64	ASP	GLY	GLY	GLY
108	GLU	GLU	GLU	GLN
115	GLN	GLN	HIS	GLN
118	VAL	ALA	ALA	ALA

[a]Based on the work of Braunitzer, G., Chen, R., Schrank, B and Stangl, A. (1972) Hoppe-Seyler's Z. Physiol. Chem., 353, 832. (See also ref. 14).

In 0.2 M glycine buffer (0.2 M glycine, 0.1 M HCl, pH 2.46 at 23°C, I=0.10) both βA and βC undergo a temperature-dependent monomer-dimer self-association. Figure 17 shows the results of sedimentation equilibrium experiments on these proteins in the 0.2 M glycine buffer. With both proteins the association is greater at lower temperatures than it is at higher temperatures. A comparison of the M_1/M_{wa} vs. c curves (see Fig. 17) indicates that the association of βC is somewhat stronger than that of βA; the plots also indicate that both proteins undergo a nonideal self-association, since the plots of M_1/M_{wa} vs. c show minima. If the ionic strength of the glycine buffer is increased to I=0.20, by including 0.1 M KCl in the 0.2 M glycine buffer, then the self-association of βC is stronger, but it still can be described as a monomer-dimer association[13].

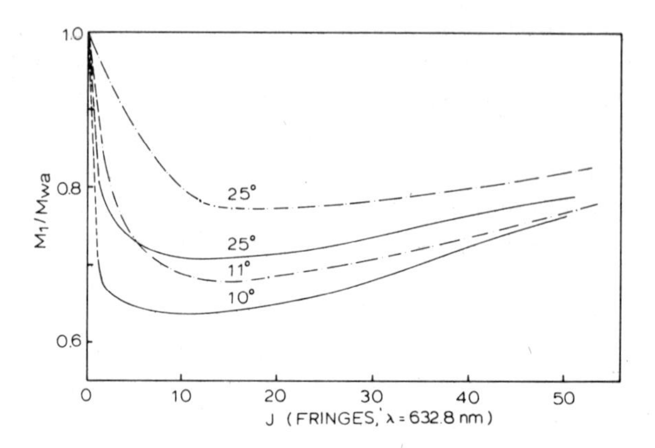

Fig. 17. Comparison of the self-association of β-lactoglobulin A and β-lactoglobulin C in 0.2 M glycine buffer (0.2 M glycine, 0.1 M HCl, pH 2.47 at 23°C, I=0.1). Note how the association for each protein is greater at the lower temperature, and that at comparable temperatures βC undergoes a stronger self-association than does βA. From ref. 14 by permission of the authors and the North-Holland Pub. Co.

When the solution conditions are changed by using 0.2 M acetate buffer (0.1 M HOAc, 0.1 M NaOAc, pH 4.65 at 23°C, I=0.1), the results obtained from sedimentation equilibrium experiments are quite different. The behavior of β-lactoglobulin C in 0.2 M acetate buffer at pH 4.65 is quite remarkable[14]! This behavior is illustrated in Fig. 18. Note that with the buffer having I=0.10 there is no temperature dependence of the self-association of βC, whereas βA does undergo a temperature dependent self-association under these conditions (see Figs. 16 and 20)[14]. In addition if the ionic strength of the buffer is raised to I=0.20 (by the inclusion of 0.1 M KCl in the 0.2 M acetate buffer), the values of M_{wa} vs. c fall on the same curve that the data collected at I=0.10 do[14]. This is true at 10 and 25°C, so there is no temperature dependence of the self-association of βC in the I=0.20 buffer either. The self-association shown in Fig. 18 could be described as a monomer-dimer self-association having $K_2 = 2.10 \times 10^3$ dl/g and $BM_1 = -0.012$ dl/g.

The self-association of βA in 0.2 M acetate buffer (0.1 M HOAc, 0.1 M NaOAc, pH 4.65 at 23°C, I=0.10) is quite different, since it does exhibit a temperature dependent self-association. Furthermore, the association goes beyond a monomer-dimer association. Figure 19 shows a plot of M_{wa} vs. c at 20°C[41,49].

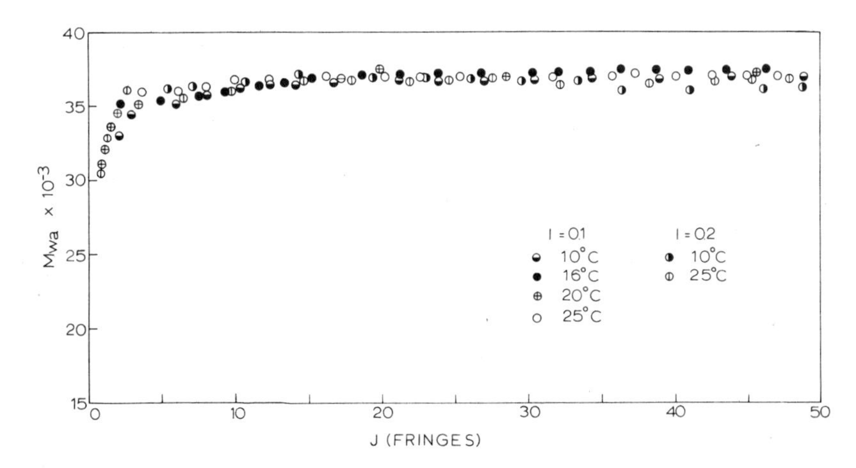

Fig. 18. Self-association of β-lactoglobulin C in acetate buffers (0.1 M HOAc, 0.1 M NaOAc, pH 4.65 at 23°C) having ionic strengths of I=0.10 and I=0.20 (by inclusion of 0.10 M KCl in the buffer). Note that the M_{wa} vs. J data all seem to fall on the same curve, which indicates that the self-association is not affected by temperature changes from 10 to 25°C or by a change in ionic strength of the buffer from I=0.10 to I=0.20. From ref. 14 by permission of the authors and the North-Holland Publishing Co.

These experiments were performed on some βA obtained from Dr. C. N. Pace of the Biochemistry and Biophysics Department of Texas A&M University (TAMU)[49]. Dr. Pace had typed the β-lactoglobulins of the cows in the TAMU dairy herd, and the βA was prepared from cows that were homozygous to βA. The open circles in the low concentration region in Fig. 19 indicate data obtained by Grant Barlow at Abbott Laboratories, Inc. on an ultracentrifuge equipped with a photoelectric scanner. These data were obtained at a wavelength of 280 nm. It is very clear that the trend of the M_{wa} vs. c is to the 18,422 molecular weight unit (M_1) as c goes to zero. The shape of this curve - the sudden rise in M_{wa} vs. c up to about 3 fringes, and then the subsequent slower rise in M_{wa} with increasing c - suggests that there is a strong association to dimer, and then there is a weaker association to larger aggregates. Figure 20 shows plots of M_1/M_{wa} vs. c (here M_1 = 18,422 g/mole) for the same batch of βA at four temperatures - 11,16, 20 and 30 C[49]. The dotted lines represent extrapolation to zero concentration. These data were obtained from an ultracentrifuge equipped with Rayleigh/ schlieren optics; the speeds used ranged from 8,000 to 11,000 RPM.

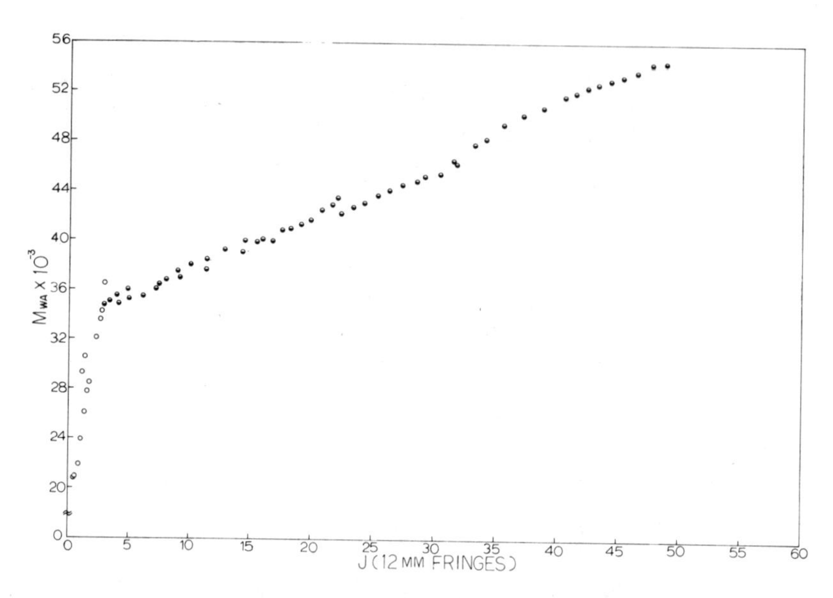

Fig. 19. Plot of M_{wa} vs. J (the concentration in fringes) for β-lactoglobulin A in 0.2 M acetate buffer (0.1 M HOAc, 0.1 M NaOAc, pH 4.65, I=0.10) at 20 C. These data were obtained from sedimentation equilibrium experiments. The open circles represent data obtained from an ultracentrifuge equipped with a photoelectric scanner. The filled circles represent data obtained from an ultracentrifuge equipped with Rayleigh/schlieren optics.

For these data the lowest values of M_{wa} obtained were $M_{wa} = 36.8_7 \times 10^3$ g/mole at J = 1.76 fringes and at 11°C, and $M_{wa} = 33.1_9 \times 10^3$ at J = 1.85 fringes and at 30°C. At 20°C J = 4c for c in g/l. While these data suggest that the trend of M_{wa} vs. c values would go to $M_1 = 18{,}422$ instead of $M_2 = 2M_1$ as c goes to zero, the combination of the 20°C Rayleigh/schlieren data with the scanner data (see Fig. 19) is much more convincing.

The data described here were analyzed for various models to describe the self-association. Using our methods we were able to rule out various discrete self-associations, such as the dimer-octamer or the monomer-dimer-octamer associations. These data could not be described by an SEK Model Type I indefinite self-association[49]. This is evident from the plots of $1-\sqrt{f_1}$ vs. Cf_1 (see eq. 58) at the various temperatures; these plots are shown in Fig. 21. The corresponding plot for βA in the acetate buffer of higher ionic strength (I=0.15, T=16°C) is also shown in this figure. The βA data (at I=0.10) could not be described as a Type II SEK Model

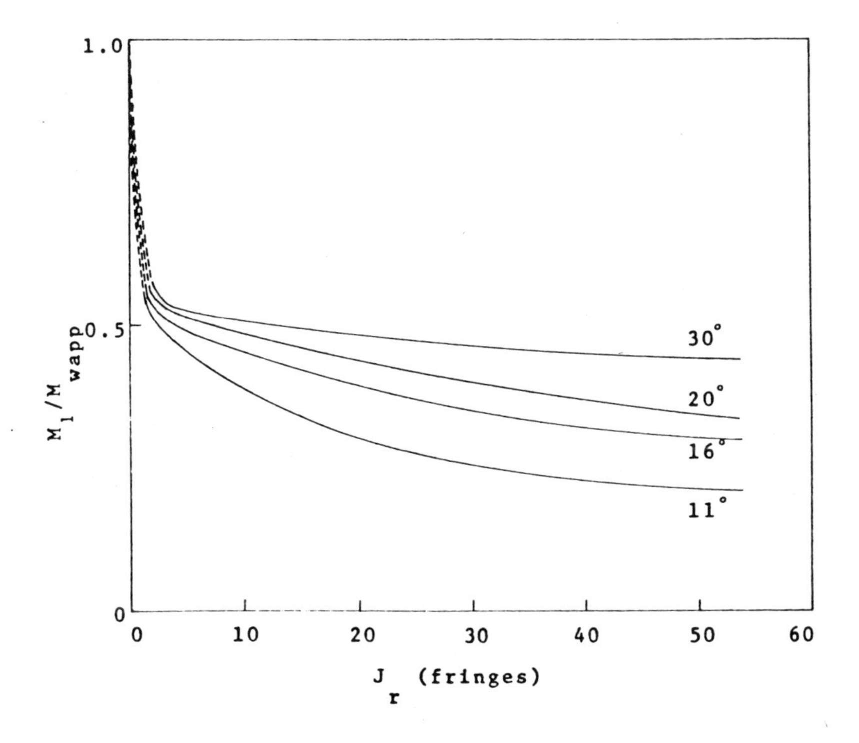

Fig. 20. Plots of M_1/M_{wa} vs. J (the concentration in fringes) for the temperature-dependent self-association of β-lactoglobulin A in 0.2 M acetate buffer 10.1 M HOAc, 0.1 M NaOAc, pH 4.65 at 23°C, I=0.1). Note that the association increases as the temperature decreases. These data were obtained from a series of sedimentation equilibrium experiments. From ref. 49.

indefinite self-association. The data in these experiments would be described by a Type III or a Type IV SEK Model indefinite self-association[49]. Table 2 shows the results of these analyses. Both the Type III and Type IV SEK Model indefinite self-associations seemed to give a good description of the M_{wa} vs. J data for βA at 20°C. Although the scanner (and also the nonscanner) data were collected with great care, it is evident from Fig. 19 that there is a lot of scatter in the M_{wa} vs. J data. The scanner data values were calculated from strip chart recordings. Much better precision could be obtained using a scanner system based on an optical multichannel analyzer, such as the system devised by Rockholt, Royce and Richards[55]. We attempted to do some additional experiments at other temperatures on another ultracentrifuge equipped with a strip chart

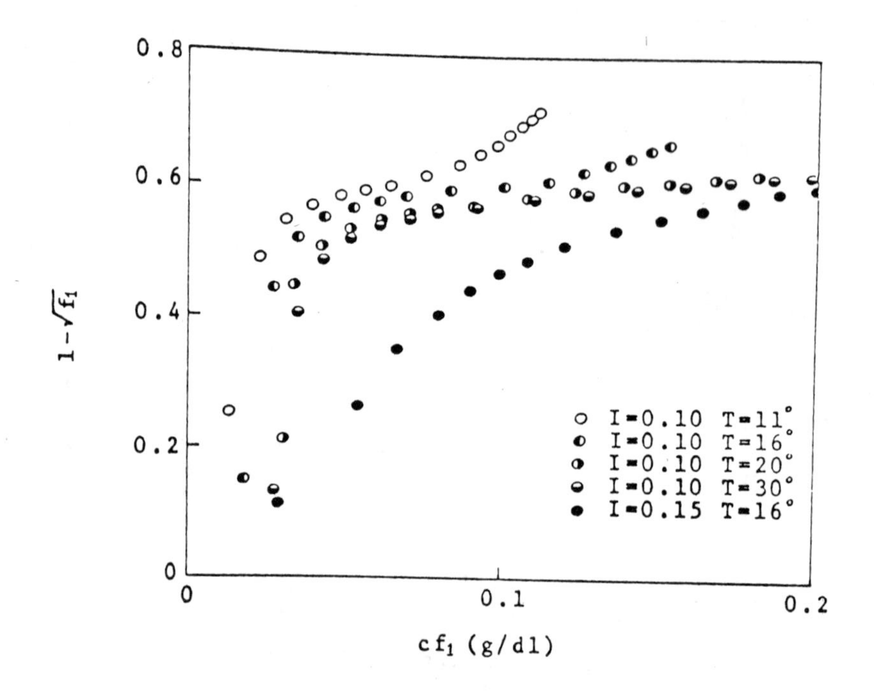

Fig. 21. Test for the presence of an SEK Model Type I indefinite self-association for β-lactoglobulin A samples in 0.2 M acetate buffers (pH 4.65 at 23°C). This plot is based on a modification of eq. (58). The • represent data collected from experiments in a buffer of higher ionic strength (I=0.15); here the USDA sample (see Fig. 13) was used. The other data were collected from experiments in a buffer of I=0.10 using the TAMU sample. From ref. 49.

type of photoelectric scanner. These runs were done over a similar concentration range as that used at 20°C. Although these data were quite scattered, they did indicate that values of M_{wa} much lower than 36,000 were present. The scatter of the data, particularly at lower concentrations, prevented us from really nailing down the model. Nonetheless, the analysis seems to suggest that a cooperative self-association is occurring. If the 1-2-4-6... model is the favored model, then this could suggest that the odd species (3,5,7, etc.) would disproportionate as soon as they were formed. Such a situation has been proposed to explain the behavior of some surfactants in nonaqueous solvents[47]. If these experiments had been carried out to higher concentration (say J = 80 or greater), then the M_1/M_{wa} vs. c curves would begin to diverge, and one might be able to see which model would describe the experimental data better.

TABLE 2

EQUILIBRIUM CONSTANTS, NONIDEAL TERMS AND VARIANCES FOR THE SELF-ASSOCIATION OF β-LACTOGLOBULIN A IN 0.2 M ACETATE BUFFER, pH 4.65, I=0.1[a]

A. SEK Model Type III Indefinite Self-Association

t °C	k_{12} dl/g	k dl/g	BM_1 dl/g	Variance[b]
11	0.54×10^2	8.9±1.2	-0.079±0.042	1.9×10^{-4}
16	0.50	5.1±0.7	-0.052±0.026	1.9
20	0.42	2.8±0.1	-0.054±0.008	1.1
30	0.40	2.3±0.1	-0.011±0.007	0.80

B. SEK Model Type IV Indefinite Self-Association

t °C	k_{12} dl/g	k_*[c] dl/g	k dl/g	BM_1 dl/g	Variance
11	1.05×10^2	14.1±0.7	1.90	-0.007±0.030	0.96×10^{-4}
16	0.85	9.27±0.63	1.01	-0.025±0.031	1.2
20	0.61	6.08±0.20	0.606	0.011±0.011	1.1
30	0.56	5.49±0.22	0.538	0.078±0.012	0.85

[a]From ref. 49.

[b]Variance = $(1/N) \sum_{i=1}^{N} [(M_1/M_{wa})_{OBSVD} - (M_1/M_{wa})_{CALCD}]_i^2$

N = number of data points used

[c] $k_*^2 = k\, k_{12}$

The surprising fact is that there appears to be a difference in the solution behavior of the TAMU βA sample and the USDA βA sample used by Adams and Lewis[41]. Toluene was used as a preservative in the preparation of the TAMU sample[56], and it is known to promote self-association[57]. On a visit by Dr. Timasheff to TAMU, Dr. Adams asked him if he had used toluene as a preservative. He said that it had been used with some samples, but that he and his colleagues obtained similar results in light scattering experiments using samples prepared with and without toluene. It should be noted that some variance was obtained in the studies on the self-association of βB reported by Albright and Williams[39] and those reported by Visser, Deonier, Adams and Williams[11]. These samples came from different sources. Langerman and Klotz[40] carried out the temperature-dependence studies on the self-association of a marine worm hemerythrin on one preparation in order to avoid problems that might arise from using different preparations of hemerythrin. They noted that temperature changes had only a very slight effect on the self-association of hemerythrin.

Finally it should be noted that Roark and Yphantis[58] encountered some peculiar results in their high speed sedimentation equilibrium experiments with a βA sample (from the USDA) in an acetate buffer containing KCl (0.09 M KCl, 0.01 M KOAc, plus HOAc to adjust the pH to 4.55 at 23°C, I=010) at 25°C and also at 4.6°C. Their results indicated that the association was stronger at 4.6°C than it was at 25°C, but the minimum molecular weight for their βA at pH 4.55 was close to 35,000. Some of their experiments were performed on a recrystallized sample. In other experiments they performed column chromatography on the recrystallized βA before doing ultracentrifugation. Two fractions were taken from the chromatographic column - one from near the leading edge and one in the vicinity of the trailing edge. High speed sedimentation equilibrium experiments on the two fractions indicated differences in the association behavior, with the leading edge sample associating more strongly. Perhaps the method of preparation, the subsequent handling and the type buffer used affected the result. This gives one an appreciation of some of the problems that are encountered in the study of some self-associating proteins.

Milthorpe, Jeffrey and Nichol[59] developed a method for obtaining the activity of monomer for any type of self-association. They performed some sedimentation equilibrium experiments on lysozyme in phosphate buffer (pH 6.7, I=0.17) at 15°C and found that a nonideal monomer-dimer equilibrium was consistent with their data. Milthorpe et al.[59] obtained $K_2 = 0.44 \pm 0.01$ dl/g and $BM_1 = -0.02$ dl/g, whereas in an earlier study under the same conditions Adams and Filmer[60] obtained $K_2 = 0.49 \pm 0.03$ dl/g and $BM_1 = -0.03 \pm 0.01$ dl/g. In this case the two separate experiments agreed remarkably, even though different samples of lysozyme were used.

Langerman and Klotz[40] found that the monomer-octamer self-association of hemerythrin was influenced by the kind of ligand that binds with the iron at the active site; this is shown in Fig. 22. The met-tiron-hemerythrin complex seemed to dissociate more readily than other complexes they studied. Perchlorate ion, which binds at a site other than the iron locus of hemerythrin, did not shift the monomer-octamer equilibrium at pH 7, when compared to the association of azidehemerythrin. The plots of M_{wa} vs. c were almost the same in both cases. There was very little effect on the self-association, in the presence or absence of ClO_4^- ion, of azidehemerythrin when the temperature was changed from 5°C to 25°C. This is illustrated in Fig. 23.

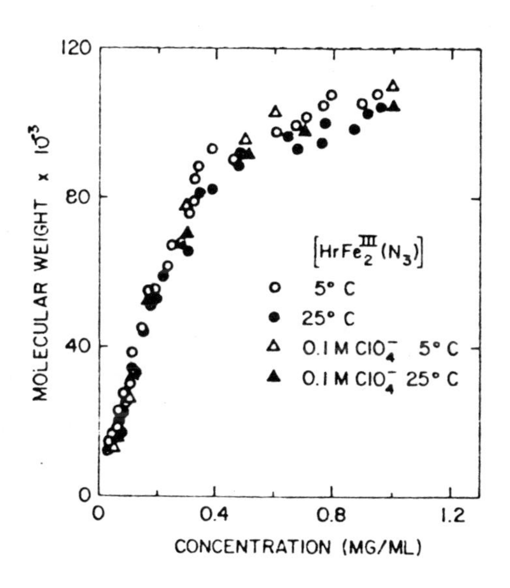

Fig. 22. The effect of temperature on the association of azide hemerythrin in the presence and absence of ClO_4^-. From ref. 40 by permission of the authors and the American Chemical Society. Copyright by the American Chemical Society.

In some cases reducing the ionic strength will increase the extent of self-association. Chymotrypsinogen A exhibits this type of behavior in sodium glycinate buffer at pH 9.3 at 25°C[61]. In order for self-association to occur, the ionic strength must be below I=0.1. As the ionic strength is lowered, the extent of self-association is increased, the self-association at I=0.02 was more pronounced than that at I=0.03; this is shown in Fig. 24. Furthermore, Nichol[61] found that the association seemed to be affected very little when the temperature was changed from 25 to 17.4°C. Hancock and Williams[62] did sedimentation equilibrium experiments on chymotrypsinogen A in veronal buffer (pH 7.9, I=0.03) at 25°C; Fig. 25 shows a plot of their M_{wa} vs. c data. Their data could be interpreted as an SEK

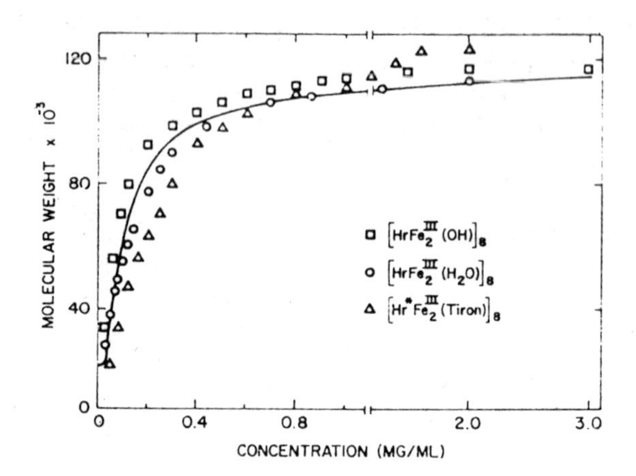

Fig. 23. Effect of different ligands that bind to the iron at the active site on the association of hemerythrin. The symbols represent results obtained with the three different ligands, and the solid curve is the one obtained for azidehemerythrin. From ref. 40 by permission of the authors and the American Chemical Socity. Copyright by the American Chemical Society.

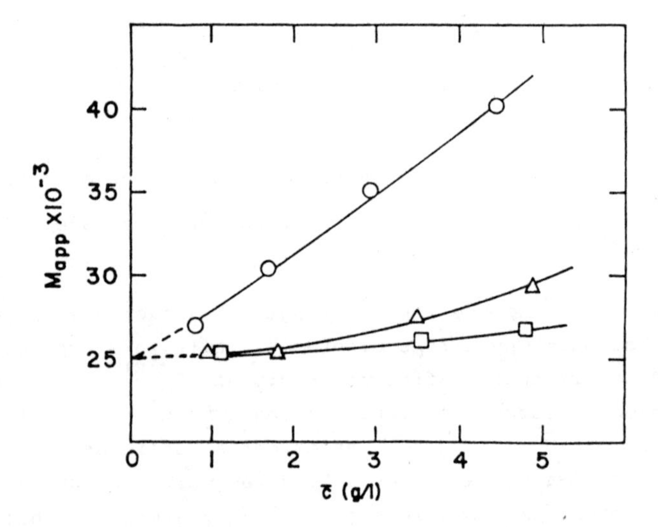

Fig. 24. Concentration dependence of the apparent weight average molecular weight of chymotrypsinogen A in sodium glycinate-glycine buffers (pH 9.30) of various ionic strengths (□ , I = 0.1; △, I = 0.03; o, I = 0.002) at 25°C. Note that there is almost no association at I = 0.1, but the association increases as the ionic strength decreases. From ref. 61 by permission of the author and the American Society of Biological Chemists, Inc.

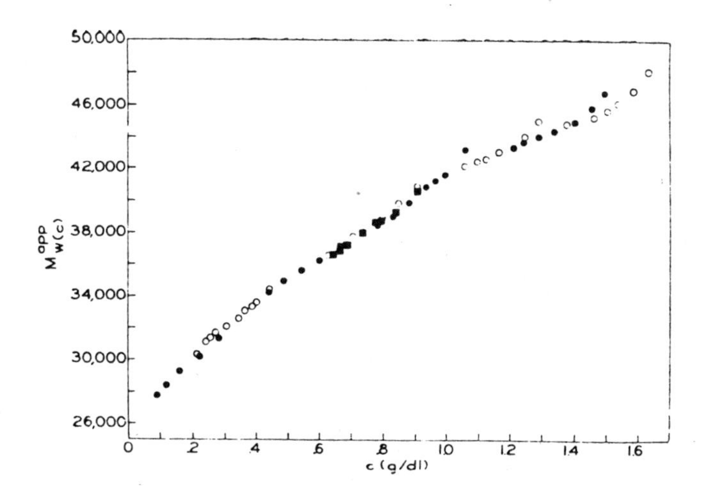

Fig. 25. Plot of M_{wa} vs. c for the self-association of chymotrypsinogen A in veronal buffer (pH 7.9, I = 0.03) at 25°C. From ref. 62 by permission of the authors and the American Chemical Society. Copyright by the American Chemical Society.

Model Type I indefinite self-association. A monomer-dimer-trimer association with a slightly negative BM_1 also seemed to describe their M_1/M_{wa} vs. c data; however, the random error in the value of k for the SEK Model Type I indefinite self-association was lower than that encountered with the equilibrium constants for the monomer-dimer-trimer association. Additional studies on the self-association of chymotrypsinogen A in barbital buffer (pH 8.1, I = 0.021) at 7°C have been carried out by Tung and Steiner[63]. Their results could be interpreted as an SEK Model Type I indefinite self-association.

Chymotrypsin, on the other hand, associates at higher ionic strength[21]. The experiments of Rao and Kegeles[21] were performed at pH 6.2 in a phosphate buffer of ionic strength 0.2 at 25°C. Here a monomer-dimer-trimer association was encountered (see Fig. 9). Diisopropylfluorophosphate (DFP) reacts with the serine at the active site of chymotrypsin and trypsin to form a covalent bond. DFP treated chymotrypsin is enzymatically inactive[64], but it undergoes a weak self-association in phosphate buffer (pH 6.2, I = 0.2) at 25°C[65]. This would indicate that there are other binding sites besides the active site that are involved in the self-association of chymotrypsin.

It has been noted earlier that increasing the ionic strength promotes the self-association of α_{s1}-casein B at pH 6.6 (see Fig. 15) as well as the self-association of β-lactoglobulin C at pH 2.47 (see Fig. 17). On the other hand a change of ionic strength from I=0.1 to I=0.2 has no effect on the self-association of β-lactoglobulin C at pH 4.65 (see Fig. 18). Insulin undergoes self-association at low pH-pH ca. 2. Jeffrey and Coates[66] performed some elegant experiments with insulin at pH 2, using a NaCl-glycine-HCl buffer of I=0.10. There was more association at 15°C than there was at 25°C; Jeffrey and Coates[66] interpreted the self-association of bovine insulin under these conditions as a monomer-dimer-tetramer-hexamer association. Further studies with bovine insulin at 25°C and at pH 2 showed that the association increased with increasing ionic strength over the range I = 0.05-0.20[67]. Figure 26 illustrates this effect of ionic strength on the association of insulin at pH 2.

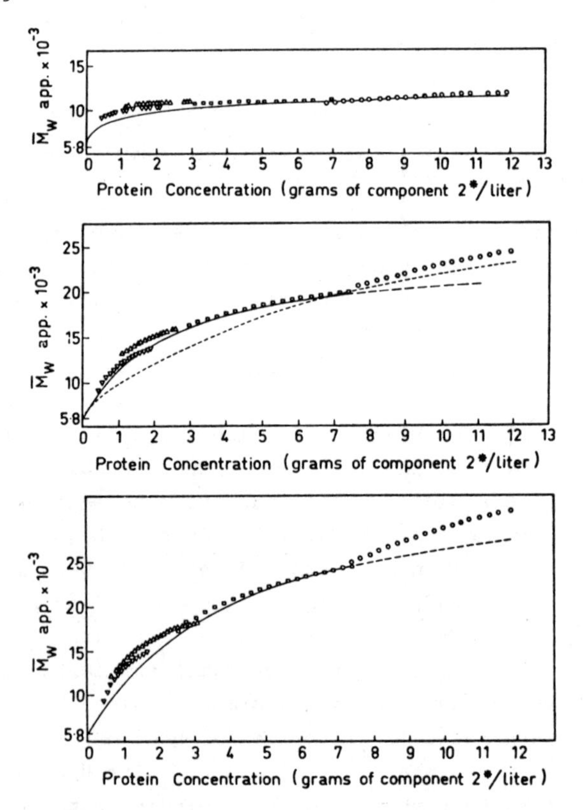

Fig. 26. Effect of changes in ionic strength on the self-association of insulin at pH 2 and 25°C. Top plot, I = 0.05; middle plot I = 0.15; bottom plot I = 0.20. These plots of M_{wa} vs. c were obtained from sedimentation equilibrium experiments. From ref. 67 by permission of the authors and the American Chemical Society. Copyright by the American Chemical Society.

Self-associations of proteins are usually affected by changes in pH of the solution. The behavior of β-lactoglobulin A and C is quite different when the pH is changed from pH 2.47 to pH 4.65; this is evident from Figs. 17, 18 and 20. Figure 2 shows quite clearly that the self-association of chicken erythrocyte histone F2b is affected by pH changes.

The types of ions present or the addition of some small molecules can affect the type of self-association present. Ions could exert their effect presumably through ion binding to the protein. This is quite evident in the light scattering studies on the self-association of apoferritin[37] in various buffer solutions (see Fig. 3). Kakiuchi[68] showed that Zn^{+2} is necessary for the self-association of an amylase obtained from Bacillus subtilis. If the Zn^{+2} is sequestered by a chelating agent, then no association occurs; this is shown in Fig. 27.

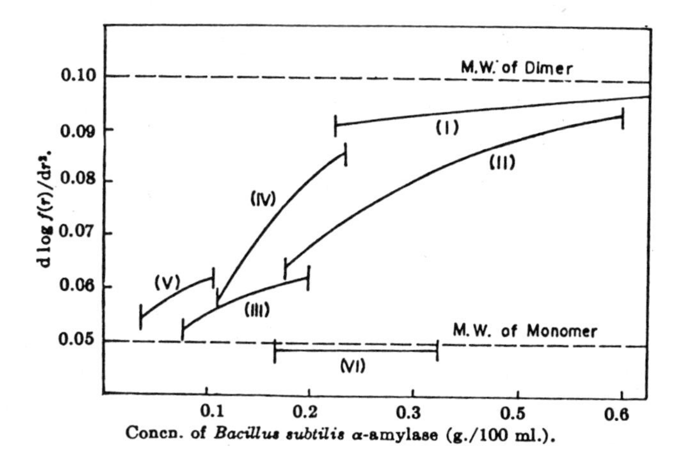

Fig. 27. Plot of M_{wa} vs. c for an α-amylase from B. subtilis in 0.1 M NaCl - 0.005 M $Ca(COOCH_3)_2$ at pH 7. The numbers refer to experiments with different initial protein concentrations (range 0.063 to 0.507 g/dl) and different initial concentrations of Zn^{+2} (range 0.66 to 2.89×10^{-5} M). In curve VI the initial protein concentration was 0.235 g/dl; this solution contained 0.01 M EDTA (ethylenediaminetetraacetate) in 0.1 M NaCl at pH 7.0. It is clear from curve VI that the Zn^{+2} is sequestered by the EDTA and there is no self-association. The varying amounts of Zn^{+2} cause the other curve to nonoverlap. From ref. 68 by permission of the author and the American Chemical Society.

Usually when protein solutions are made up in the presence of denaturants, such as 8 M urea or 4 M guanidinium HCl, there is no self-association; instead the protein is usually dissociated into subunits of minimum molecular weight. However, Kakiuchi and Williams[74] have studied the self-association of a γ-G globulin from multiple myeloma; their solvent was 8 M urea buffered at pH 7. The γ-G globulin underwent a monomer-dimer association. Eisenberg and Tompkins[50], using light scattering, showed that bovine lactate dehydrogenase underwent self-association; the addition of a small amount of diethylstilbesterol to the protein solution prevented the self-association. This would indicate that hydrophobic bonds were involved in the self-association of bovine lactate dehydrogenase[70]. Eisenberg and his associates[70] were able to cross-link the self-associating species and isolate the cross-linked species (dimer, trimer, etc.), which still showed some biological activity. This would indicate that the association sites and the active sites of the bovine lactate dehydrogenase were in different locations on the molecule. Chun, Kim et al.[38] showed that the light scattering data of Eisenberg and Tompkins[50] was an SEK Model Type I indefinite self-association. Using a molecular sieve chromatography (also known as analytical gel chromatography), Chun, Kim et al.[69] were able to show that the indefinite self-association of the bovine lactate dehydrogenase was a linear indefinite self-association. The equilibrium constant obtained from the analysis of the light scattering data[38] agreed remarkably well with the equilibrium constant obtained from the molecular sieve chromatography[69].

Does the presence of substrate affect the extent of self-association of an enzyme? Sarfare, Kegeles and Kwon-Rhee[71] studied the self-association of α-chymotrypsin in sodium phosphate buffer (I=0.2, pH 6.1-6.2 at 25°C) at concentrations 0.17 to 3 g/dl of protein in the presence and absence of a competitive inhibitor, β-phenylpropionic acid, at molar concentrations of inhibitor of 0.001, 0.004 and 0.04 M. Their studies showed that within the experimental error the enzyme polymerized identically in the presence or absence of the competitive inhibitor, β-phenylpropionate. They concluded that the sites for the binding of the inhibitor and the sites for the protein polymerization were mutually exclusive. On the other hand, when diisopropylfluorophosphate (DFP) reacts with the active serine of α-chymotrypsin to form a covalent compound, there is no enzymatic activity[64] and the extent of the self-association is reduced[65]. Here the reduction in the self-association may be due to a steric effect when the covalent bond is formed. Sophianopoulos[72] studied the self-association of lysozyme near pH 8 in phosphate buffer of 0.15 ionic strength in the presence of substrates - N-acetylglucosamine, its dissaccharide (GlcNAc-Glc-NAc) and its trisaccharide (GlcNAc-Glc-NAc-GlcNAc). These substrates greatly reduced the reversible dimerization of lysozyme, which would indicate in this case that the enzymatically active region of lysozyme was one of the sites of contact for the formation of dimer or that it was near this site.

It should be evident from this discussion that there are a number of factors that influence the self-association of proteins, and also that there is a difference in the way the various proteins are affected by these factors. For more information on how to study some of the factors influencing self-association, and on some of the problems that can arise as solution conditions are varied, the reader should consult the excellent paper by Timasheff[73].

COMBINATION OF EQUILIBRIUM DATA WITH OTHER DATA

It is possible to obtain additional information about any self-associating solute by using results from equilibrium data (sedimentation equilibrium, osmotic pressure or light scattering) with results from other techniques, such as sedimentation velocity[74,75] or analytical gel (also known as molecular sieve) chromatography experiments[69], which are used under the same solution conditions. For example, one can obtain the sedimentation coefficient of an n-mer (S_n) in a monomer-n-mer self-association, as well as the ordinary concentration dependence parameter (g) of the sedimentation coefficient. Suppose that the apparent weight average sedimentation coefficient, S_{wa}, is defined by[75]

$$1/S_{wa} = (1/S_{wc}) + gc \tag{90}$$

Here S_{wc} is the weight average sedimentation coefficient; it is defined by[75]

$$S_{wc} = \frac{S_1+K_n c_1^{n-1} S_n}{1+K_n c_1^{n-1}} \tag{91}$$

Now note that[75]

$$\frac{1}{S_{wa}} - \frac{2}{c}\int_o^c \frac{dc}{S_{wa}} = \frac{1}{S_{wc}} - \frac{2}{c}\int_o^c \frac{dc}{S_{wc}} \tag{92}$$

Note also that

$$\lim_{c \to o} S_{wa} = S_1 \tag{93}$$

For a monomer-n-mer association eq. (92) becomes[75]

$$\frac{1}{S_{wa}} - \frac{2}{c}\int_o^c \frac{dc}{S_{wa}} = \frac{1+K_n c_1^{n-1}}{S_1+K_n c_1^{n-1} S_n} - \frac{2}{c}\int_o^c \frac{1+K_n c_1^{n-1}}{S_1+K_n c_1^{n-1} S_n}\, dc \tag{94}$$

If equilibrium experiments have been performed under the same solution conditions and analyzed, then n, c_1 and K_n are known. Furthermore if S_1 is known, then eq. (94) contains only one unknown, S_n, which can be obtained by successive approximations. This is best done on a computer, since the integration on the right hand side will have to be carried at various concentrations for each choice of S_n. Once S_n in known, so is S_{wc}, so that one can obtain the concentration dependence parameter for sedimentation, g, from the relation

$$(1/S_{wa}) - (1/S_{wc}) = gc \tag{95}$$

There are other relations that could be used in this analysis[75]. Kakiuchi and Williams[74] have used a combination of sedimentation equilibrium experiments with sedimentation velocity experiments to study the association of γ-G globulin.

From analytical gel chromatography experiments one can obtain the weight average partition coefficient, σ_{wc}, and the weight average elution volume, V_{ewc}. Clearly, one could tie these data together with other equilibrium data to evaluate σ_n or V_{en}. It should be noted that if a weight average property X_{wc} (where X can be s, σ or V_e) is available, then one can always obtain the z average property, X_{zc}, since[75]

$$X_{zc} = d(cX_{wc})/dc \tag{96}$$

By combining the equilibrium techniques with these other methods, we can learn far more about a self-associating solute than by either technique alone.

ADVANTAGES AND DISADVANTAGES OF VARIOUS METHODS

The theory for membrane osmometry is the simplest theory. With new high speed membrane osmometers, using microcells, it is possible to use 1 ml or less (about 0.03 ml for a measurement, plus additional amounts for rinsing the membrane before the measurement) and obtain an M_{na} in 3-5 minutes[77]. Membrane osmometers having microcells are the Knauer (available from Utopia Instrument Co., P. O. Box 863, Caton Farm Road, Joliet, IL 60434) and the IL 196 Weil Oncometer (available from Instrumentation Laboratory, Inc., 113 Hartwell Ave., Lexington, MA 02173). The big disadvantage to membrane osmometry is to find a suitable membrane. It may be difficult to find a suitable membrane for small proteins, like insulin or ACTH (adrenocorticotropic hormone), although new membranes capable of withholding macromolecules of molecular weight 3-4000 have become commercially available. Perhaps these can be used in osmometry. With osmometry one does not have to determine partial specific volumes ($\bar{v}$) or density increments ($1000\ (\partial\rho/\partial c)_{\mu}$), nor does one need to know the refractive index increment (ψ). There are no pressure effects to be concerned with, since osmotic pressures of macromolecules are usually less than 1 atm. One need not assume that the partial specific volumes of the self-associating species are equal, nor does one have to assume that the refractive index increments of the self-associating species are equal. If enough material is available one can obtain a plot of M_{na} vs. c, so that the values of M_{wa} and f_a can be calculated. The method is nondestructive, and the solute, in principle, can be recovered. Besides finding a membrane, it sometimes may take a while to set up the osmometer. But once this technique has been mastered, it is a simple experiment to carry out.

Diggle and Peacocke[30,35,36] studied the self-association of various histones in aqueous solutions by membrane osmometry. Figure 2 shows plots of π/c vs. c for chicken erythrocyte F2b at I = 0.1 and at various pH values. These associations were analyzed using the procedures developed by Adams[27,29]. In some cases parallel sedimentation equilibrium experiments were performed, which gave values for the equilibrium constants and nonideal terms very close to those obtained from osmotic pressure studies. More details about osmotic pressure experiments and their application to self-associations and mixed associations will be found in the chapter by Adams, Wan and Crawford[29].

Light scattering (more properly elastic light scattering) is the fastest technique, since one introduces the solution into the cell, allows it to come to thermal equilibrium and then measures the intensity of the scattered light at one or more angles. Similar measurements are done on the buffer solution (the solvent). For smaller proteins (proteins whose maximum dimension is less than $\lambda/20$, where λ is the wavelength of light used) the light scattering is usually independent of angle[78]. However, this technique at present requires more material than is needed for osmometry or sedimentation equilibrium experiments, and one must measure the refractive index increment of the associating solute and assume that the refractive index increments of the self-associating species are equal. In addition the solution and the solvent have to be dust free; a small amount of dust can produce a significant amount of light scattering. Thus elaborate steps must be taken to clarify (remove dust) from the solvent and solutions[5]. In addition to obtaining M_{wa} at various concentrations, one can also use light scattering to study kinetics of slow or moderately fast reactions. Edelhoch, Katchalski et al.[34] studied the kinetics of dimerization of mercaptalbunim; this reaction took about two hours to go to completion. Kegeles[79] and his associates have used pressure jump-light scattering to study the association of hemocyanins. With very large molecules, where there is an angular dependence to the light scattering, one can obtain the radius of gyration in addition to molecular weight and second virial coefficient[78]. With a laser light scattering apparatus one can do elastic and also inelastic light scattering. With inelastic light scattering one can get translational and rotational diffusion coefficients[80]. There is also a special cell which can be used to determine electrophoretic mobility from inelastic light scattering experiments[80]. For very strong self-associations, such as have been encountered with β-lactoglobulins A[7,10,41,47] and C[14,54] or with insulin, the elastic light scattering experiment may not be sensitive enough in the low concentration region. It was reported using light scattering[19,81] that the minimum molecular weight of insulin at acid pH (pH ca. 2) was 12,000, whereas the monomer molecular weight is known to be 5,734 for bovine insulin A. Sedimentation equilibrium experiments by Jeffrey and Coates[66,67] and by Adams[24,82] indicated that the minimum molecular weight approached the 5,734 unit at infinite dilution. Similarly there has been a variance between elastic light scattering results with β-lactoglobulins A[7,10,41,47] and C[14,54] at pH 4.7; lower limiting molecular weights have been obtained by sedimentation equilibrium experiments (see Figs. 16, 18, 19 and 20). Light scattering experiments indicated that the β-lactoglobulins C and D existed as a dimer[54] and did not associate in the pH range of 3.5 to 5, whereas sedimentation equilibrium experiments[14] with β-lactoglobulin C at pH 4.7 indicated that a very strong monomer-dimer association was present (see Fig. 18). A photoelectric scanner was used to collect the data at low concentrations (less than 1 g/l) for the βC association[14].

The ultracentrifuge has been used extensively to study self-associations by sedimentation velocity[83] and sedimentation equilibrium experiments[1,3,4,31], and also by Archibald experiments[21]. The theory for sedimentation equilibrium is more rigorous than that for sedimentation velocity experiments. For sedimentation equilibrium experiments, short column (5 mm or less) experiments are usually used; dust is not a problem, since the centrifugal field acts as a dust filter. Since one used $M_{wr\ app}$ (which is the same as M_{wa}) in the analysis, one can cover a wide range of concentrations in a limited series of sedimentation equilibrium experiments - between 4 to 8 different solutions need to be used. Each value of $M_{wr\ app}$ corresponds to one osmotic pressure experiment or to one light scattering experiment, so that 30-50 values of $M_{wr\ app}$ are usually collected from each cell (or compartment in multichannel equilibrium centerpieces). With pulsed or modulated lasers[84] one can run 3 to 5 solutions of different initial concentrations simultaneously, if multihole (4 or 6 hole) rotors and double sector centerpieces are used. With multichannel, equilibrium centerpieces[85] one can run 9 to 15 solutions of different initial concentrations at the same time. Because of the redistribution of solute caused by the centrifugal field, one can study a wide range of concentrations using a limited number of solutions. Many proteins have aromatic amino acids, so that they absorb ultraviolet light, and a photoelectric scanner[86] can be used in many cases to study self-associations at very low concentrations. A more accurate photoelectric scanner that uses an optical multichannel analyzer has been developed by Rochholt, Royce and Richards[55].

The principal disadvantages of the sedimentation equilibrium experiment are that one must know or measure the partial specific volume ($\bar{v}$) of the solute or the density increment ($1000\ (\partial\rho/\partial c)_{\mu}$) as well as the refractive index increment (ψ), in order to obtain $M_{wr\ app}$ (which is the same as M_{wa} for a self-associating solute). One must also assume that the partial specific volumes (or density increments) of the self-associating species are equal, and one must also assume that the refractive index increments of the self-associating species are equal. If these values are not available in the literature, then they must be determined, and this will require dry weight or nitrogen determinations from the protein solutions. If the $\bar{v}$'s of the associating species are unequal, then there will be pressure effects on the self-association, and one will have to correct for pressure effects in order to analyze the self-association. If the ψ's of the associating species are unequal, then the analysis of a self-association becomes more difficult. This would also be true in a light scattering experiment. For self-associating solutes one cannot use the quantities $M_{w\ cell}$ or $M_{z\ cell}$ (or their apparent values under nonideal conditions), which are more precise than the values of $M_{wr\ app}$[87]. If Rayleigh or absorption optics are used in a multicell rotor, then the c vs. r data must be smoothed and differentiated numerically to obtain $M_{wr\ app}$, and this can introduce some error. The supporting electrolytes and the buffer components are redistributed from their original values in a sedimentation equilibrium experiment, so there is always a slight salt and/or pH gradient as distance increases in the cell. If the self-associating solute is very, very sensitive to pH or salt gradients, this could cause problems. In general one should choose an operating speed to minimize these effects. It is also imperative to use clean centerpieces and windows and to handle and prepare solutions with care, as small volumes (usually less than 0.15 ml) are used in short column sedimentation equilibrium experiments.

ACKNOWLEDGMENTS

This work was supported in part by a grant (to ETA, Jr.) from The Robert A. Welch Foundation (Grant A-485). The temperature dependence studies on the self-association of β-lactoglobulin A in 0.2 M acetate buffer (pH 4.65) are taken from the dissertation submitted by Lih-Heng Tang to the graduate school of the Illinois Institute of Technology in partial fulfillment of the requirements for the Ph.D. degree (Dec. 1971). That work was supported by grants (GM 15551 and GM 17611) from the National Institutes of Health. Many thanks are due to Irene Casimiro for typing the manuscript.

REFERENCES

1. Adams, E. T., Jr. (1967) Fractions, No. 3.
2. Reithel, F. J. (1963) Advan. Protein Chem., 18, 123.
3. Kim, H., Deonier, R. C., and Williams, J. W. (1977) Chem. Revs., 77, 659.

4. Adams, E. T., Jr., Ferguson, W. E., Wan, P. J., Sarquis, J. L., and Escott, B. M. (1975) Separation Sci., 10, 175.

5. Schmidt, D. G. (1969) Dissertation, University of Utrecht, Utrecht, The Netherlands.

6. Townend, R., Winterbottom, R. J. and Timasheff, S. N. (1960) J. Am. Chem. Soc., 82, 3161.

7. Townend, R., and Timasheff, S. N. (1960) J. Am. Chem. Soc., 82, 3168.

8. Timasheff, S. N. and Townend, R. (1961) J. Am. Chem. Soc., 83, 464.

9. Timasheff, S. N. and Townend, R. (1961) J. Am. Chem. Soc., 83, 470.

10. Kumosinski, T. F., and Timasheff, S. N. (1966) J. Am. Chem. Soc., 88, 5635.

11. Visser, J., Deonier, R. C., Adams, E. T., Jr. and Williams, J. W. (1972) Biochemistry, 11, 2634.

12. Tang, L.-H. and Adams, E. T., Jr. (1973) Arch. Biochem. Biophys., 157, 520.

13. Sarquis, J. L. and Adams, E. T., Jr. (1974) Arch. Biochem. Biophys., 163, 442.

14. Sarquis, J. L. and Adams, E. T., Jr. (1976) Biophys. Chem., 4, 181.

15. McKenzie, H. A. (1971) in Milk Proteins - Chemistry and Molecular Biology, McKenzie, H. A.,ed., Academic Press, New York, Vol. II, pp. 257-325.

16. Armstrong, J. McD. and McKenzie, H. A. (1967) Biochem. Biophys. Acta, 147, 93.

17. Svedberg, T. and Pedersen, K. O. (1940) The Ultracentrifuge, Clarendon Press, Oxford.

18. Tiselius, A. (1925) Z. Phys. Chem., 124, 449.

19. Steiner, R. F. (1952) Arch. Biochem. Biophys., 39, 333.

20. Steiner, R. F. (1954) Arch. Biochem. Biophys., 49, 400.

21. Rao, M.S.N. and Kegeles, G. (1958) J. Am. Chem. Soc., 80, 5724.

22. Squire, P. G. and Li, C. H. (1961) J. Am. Chem. Soc., 83, 3521.

23. Millar, D.B.S., Willick, G. E. and Frattali, V. (1969) Biochemistry, 8, 2416.

24. Adams, E. T., Jr. and Fujita, H. (1962) in Ultracentrifugal Analysis in Theory and Experiment, Williams, J. W., ed., Academic Press, New York, p. 119.

25. Adams, E. T., Jr. and Williams, J. W. (1964) J. Am. Chem. Soc., 86, 3454.

26. Adams, E. T., Jr. (1965) Biochemistry, 4, 1646.

27. Adams, E. T., Jr. (1965) Biochemistry, 4, 1655.

28. Ogston, A. G. and Winzor, D. J. (1975) J. Phys. Chem., 79, 2496.

29. Adams, E. T., Jr., Wan, P. J. and Crawford, E. J. (1978) Methods in Enzymol., 48, 69.

30. Tombs, M. P. and Peacocke, A. R. (1974) The Osmotic Pressure of Biological Macromolecules, Oxford University Press, London.

31. Fujita, H. (1975) Foundations of Ultracentrifugal Analysis, John Wiley and Sons, New York, Chapter 6.

32. McKenzie, H. A. and Nichol, L. W. (1970) in Milk Proteins - Chemistry and Molecular Biology, McKenzie, H. A., ed., Academic Press, New York, Vol. I, pp. 312-334.

33. Braswell, E. (1968) J. Phys. Chem., 78, 2477.

34. Edelhoch, H., Katchalski, E., Maybury, R. H., Hughes, W. L., Jr. and Edsall, J. T. (1953) J. Am. Chem. Soc., 75, 5058.

35. Diggle, J. H. and Peacocke, A. R. (1968) Fed. Eur. Biochem. Soc. Letts., 1, 329.

36. Diggle, J. H. and Peacocke, A. R. (1971). Fed. Eur. Biochem. Soc. Letts., 18, 138.

37. Richter, G. W. and Walker, G. F. (1967) Biochemistry, 6, 2871.

38. Chun, P. W., Kim, S. J., Williams, J. D., Cope, W. T., Tang, L.-H. and Adams, E. T., Jr. (1972) Biopolymers, 157, 520.

39. Albright, D. A. and Williams, J. W. (1968) Biochemistry, 7, 67.

40. Langerman, N. R. and Klotz, I. M. (1969) Biochemistry, 8, 4746.

41. Adams, E. T., Jr. and Lewis, M. S. (1968) Biochemistry, 7, 1044.

42. Adams, E. T., Jr. (1967) Biochemistry, 6, 1864.

43. Lewis, M. S. and Nutt, G. D. (1976) Biophys. Chem., 5, 171.

44. Tobolsky, A. V. and Thach, R. E. (1962) J. Colloid Sci., 17, 410.

45. White, N. E. and Kilpatrick, M. (1955) J. Phys. Chem., 59, 1044.

46. Millar, D.B.S., Willick, G. E., Steiner, R. F. and Frattali, V. (1969) J. Biol. Chem., 244, 281.

47. Tang, L.-H., Powell, D. R., Escott, B. M. and Adams, E. T., Jr. (1977) Biophys. Chem., 7, 121.

48. Garland, F. and Christian, S. D. (1975) J. Phys. Chem., 79, 1247.

49. Tang, L.-H. (1971) Ph. D. Dissertation, Illinois Inst. of Technology, Chicago.

50. Eisenberg, H. and Tompkins, G. (1968) J. Mol. Biol., 31, 37.

51. Pekar, A. H. and Frank, B. H. (1972) Biochemistry, 11, 4013.

52. Teller, D. C. (1970) Biochemistry, 9, 4201.

53. Šolc, K. and Elias, H.-G. (1973) J. Polym. Sci., Polym. Phys. Ed., 11, 137.

54. Timasheff, S. N. and Townend, R. (1969) Proteides Biol. Fluids, 16, 33.

55. Rockholt, D. L., Royce, C. R. and Richards, E. G. (1976) Biophys. Chem., 5, 55.

56. Pace, C. N., personal communication.

57. Reisler, E. and Eisenberg, H. (1970) Biopolymers, 9, 877.

58. Roark, D. E. and Yphantis, D. A. (1969) Ann. N. Y. Acad. Sci., 164, 245.

59. Milthorpe, B. K., Jeffrey, P. D. and Nichol, L. W. (1975) Biophys. Chem., 3, 169.

60. Adams, E. T., Jr. and Filmer, D. L. (1966) Biochemistry, 5, 2971.

61. Nichol, J. C. (1968) J. Biol. Chem., 243, 4065.

62. Hancock, D. K. and Williams, J. W. (1969) Biochemistry, 8, 2598.

63. Tung, M. S. and Steiner, R. F. (1974) Eur. J. Biochem., 44, 49.

64. Stryer, L. (1975) Biochemistry, W. H. Freeman and Co., San Francisco, pp. 184-185.

65. Adams, E. T., Jr. and Filmer, D. L., unpublished results.

66. Jeffrey, P. D. and Coates, J. H. (1966) Biochemistry, 5, 489.

67. Jeffrey, P. D. and Coates, J. H. (1966) Biochemistry, 5, 3820.

68. Kakiuchi, K. (1965) J. Phys. Chem., 69, 1829.

69. Chun, P. W., Kim, S. J., Stanley, C. A. and Ackers, G. K. (1969) Biochemistry, 8, 1625.

70. Josephs, R., Eisenberg, H. and Reisler, E. (1972) in Protein-Protein Interactions, Jahnicke, R. and Helmreich, E., eds., Springer Verlag, New York, p. 57.

71. Sarfare, P. S., Kegeles, G. and Kwon-Rhee, S. J. (1966) Biochemistry, 5, 1389.

72. Sophianopoulos, A. J. (1969) J. Biol. Chem., 244, 3188.

73. Timasheff, S. N. (1973) Proteides Biol. Fluids, 20, 511.

74. Kakiuchi, K. and Williams, J. W. (1966) J. Biol. Chem., 241, 2781.

75. Weirich, C. A., Adams, E. T., Jr. and Barlow, G. H. (1973) Biophys. Chem., 1, 35.

76. Ackers, G. K. (1970) Advan. Prot. Chem., 24, 343.

77. Armstrong, J. L., personal communication.

78. Kerker, M. (1969) The Scattering of Light and Other Electromagnetic Radiation, Academic Press, New York.

79. Kegeles, G. (1978) Methods in Enzymology, 48, 308.

80. Bloomfield, V. A. and Lim, T. K. (1978) Methods in Enzymology, 48, 415.

81. Doty, P. and Myers, G. E. (1953) Discussions Faraday Soc., No. 13, 51.

82. Adams, E. T., Jr. (1962) Ph. D. Dissertation, University of Wisconsin - Madison.

83. Nichol, L. W., Bethune, J. L., Kegeles, G. and Hess, E. L. (1964) in The Proteins, 2nd ed., Neurath, H., ed., Academic Press, New York, Vol. 2, p. 305.

84. Williams, R. C., Jr. (1978) Methods in Enzymology, 48, 185.

85. Yphantis, D. A. (1964) Biochemistry, 3, 297.

86. Chervenka, C. H. (1971) Fractions, No. 1.

87. Adams, E. T., Jr. (1964) Proc. Natl. Acad. Sci. U.S., 51, 509.

Catsimpoolas, ed. Physical Aspects of Protein Interactions

CHARACTERIZATION OF THE SUBUNIT DISSOCIATION OF YEAST ENOLASE

JOHN M. BREWER, GEORGE J. FAINI*, C. A. WU**, L. P. GOSS,
L. A. CARREIRA and R. WOJCIK
Departments of Biochemistry and Chemistry, University of Georgia,
Athens, Georgia 30602

ABSTRACT

Measurement of relative activities of yeast enolase incubated with magnesium (associated enzyme) or without (partly dissociated enzyme) as a function of protein concentration show that yeast enolase subunits dissociated without magnesium are inactive. This is true at low ionic strength and in 1 M KCl. Sedimentation velocity measurements indicate the enzyme is associated even in 1 M KCl if excess magnesium is present. Consequently, enzymatic activity measurements should be a reliable means of determining subunit dissociation constants. However, dissociation constants obtained from enzymatic activity measurements are higher than from osmotic pressure measurements. This is rationalized, assuming that an "annealing" (activity recovering) reaction subsequent to subunit association does not go to completion. Hence, some associated enzyme is not "annealed" and is counted as inactive monomeric enzyme. However, the pH dependence of the subunit dissociation constant, whether calculated from osmotic pressure or enzymatic activity data, is similar: 0.2-0.5 moles of protons released per mole of enzyme upon dissociation, either at 0.05-0.1 ionic strength or in 1 M KCl. The temperature dependence of the subunit dissociation constant calculated from enzymatic activity measurements also appears identical at 0.05 ionic strength and in 1 M KCl. These data suggest that the enthalpy of subunit association is zero at 20°. This indicates that hydrophobic interactions predominate in producing subunit association in yeast enolase, and that 1 M chloride reduces the favorable entropy of association.

Sedimentation velocity measurements show that the sedimentation constant of partially dissociated enzyme is a function of the rotor speed. This suggests the subunit dissociation is pressure-dependent,

---

*Present address: Merck and Co. Automation and Control, Westpoint, Pennsylvania 19480.

**Present address: Marine Sciences Program, University of Georgia, P. O. Box 517, Brunswick, Georgia 31520.

consistent with the conclusion that hydrophobic interactions between subunits are important.

Laser Raman spectra of the enzyme show no effect of magnesium or 1 M KCl, indicating the effect of the Hofmeister anion chloride may be exerted through an effect on the solvent, water.

Subunits dissociated without magnesium are inactive because the substrate binding is affected. The kinetics of the absorbance change in a chromophoric competitive inhibitor indicate that the development of the absorbance change is limited by the same factor(s), subunit association and annealing, that limit the absorbance change in the enzyme itself.

INTRODUCTION

Measurements of protein association and dissociation reactions provide valuable information, not only about the origins and nature of protein structure but also about the relation between protein structure and function, since proteins generally function through interactions with ligands, large and small. We have studied the characteristics of subunit association in yeast enolase, since the association/dissociation reaction is readily reversible and because it is the simplest possible associating system.

Brewer and Weber[1] showed the enzyme was composed of two subunits of apparently equal molecular weight, which on the basis of the data of Mann *et al.*[2] would be about 45,000. Brewer *et al.*[3] showed by peptide mapping experiments that the subunits are identical.

There is considerable evidence that hydrophobic interactions predominate in producing subunit association in yeast enolase.[4] Brewer and Weber[1] presented evidence that 1 M potassium chloride or removal of magnesium with EDTA facilitated dissociation while 1 M potassium acetate did not. This suggested that electrostatic or hydrogen bond interactions were not involved in maintenance of dimeric structure. This evidence was extended by Brewer[5] who showed that potassium acetate reversed the effects of potassium chloride on fluorescence emission and subunit dissociation and that thiocyanate and sulfate facilitated exposure and burial, respectively, of tryptophans in the protein. That is, the effects of the anions were consistent with their positions in the "Hofmeister series". The anions also appeared to be competitive with one another[5]. These effects were interpreted as a result of differing solubilities of anion-amide complexes[6].

Gawronski and Westhead[7] used enzymatic activity measurements to study the dissociation of the enzyme produced by potassium bromide. They found that relatively few moles of bromide were required for dissociation (or were bound after dissociation) and that the heat capacity of the enzyme changed, indicating exposure of hydrophobic regions, when the subunits separated. This latter observation was consistent with that of Brewer[5] who found that the fluorescence and extent of binding of ANS increased when enolase dissociated.

Brewer[8] found by isopycnic density gradient centrifugation in cesium chloride or sulfate that dissociation was not accompanied by a significant change in water of hydration. This also suggested that subunit dissociation involved largely exposure of hydrophobic surfaces, which would not be heavily hydrated[9]. In addition, titration of isoionic enzyme with chloride and sulfate salts indicated that relatively little difference in anion binding occurred under associating (excess sulfate or magnesium present) or dissociating conditions. Of course, conformational changes subsequent to dissociation could complicate the interpretation of these data.

The relation between subunit structure and activity has been the subject of some controversy. Although Brewer and Weber[1] and Gawronski and Westhead[7] found that subunits dissociated without magnesium are inactive, Keresztes-Nagy and Orman[10], and later Holleman[11] showed that at 40°, in the presence of magnesium and substrate, yeast enolase dissociated into monomers which were fully active. Thus, the data of Brewer and Weber[1] and Gawronski and Westhead[7] were attributed to inhibition by the salts used for dissociation, although Brewer and Weber obtained similar results at low ionic strength. In addition, the reason(s) for the apparent inactivity of subunits dissociated without metal are not known, since Brewer and DeSa[12] and later Brewer[13] showed the subunits could bind magnesium.

More direct evidence about the nature of the subunit association reaction in yeast enolase can be provided by examining the effects of pH and temperature on the subunit association reaction, along with the effect of deuterium oxide[14]. In this paper, we attempt to do this, and provide some new evidence about the relation between subunit structure and enzymatic activity.

MATERIALS AND METHODS

Bakers yeast enolase A was prepared by a modification of the method of Westhead and McLain[15]. Its specific activity was 10-15% higher than that reported by those authors and was homogeneous by disc electrophoresis and other criteria. For routine measurements and calculations, an extinction coefficient, $A_{280,\ 1\ cm}^{1\%}$ of 8.95[16] and a molecular weight of 90,000[2] was used.

Barium 2-phospho-D-glycerate was purchased from Calbiochem Corp., converted to the acid form by treatment with Dowex 50 (H^+), and then crystallized as the tricyclohexylammonium salt from aqueous acetone. Potassium chloride was either reagent grade (J. T. Baker, Inc.) or "Ultrapure" grade (Ventron Corp.). Tris was purchased from the Sigma Chemical Co. and twice recrystallized from aqueous ethanol. Imidazole was obtained from Eastman Kodak and was twice recrystallized from benzene. HEPES, PIPES and tricine were from Calbiochem. Other chemicals were reagent grade. B-19 and B-20 membranes for the osmometer came from Schleicher and Shuell. 3-Aminoenolpyruvate-2-phosphate (AEP) was synthesized according to the procedure of Spring and Wold[17].

Yeast enolase prepared by the method of Westhead and McLain[15] contains strongly bound magnesium[18]. For some experiments, the enzyme was deionized as described by Faller and Johnson[19] then concentrated to 120-170 mg/ml in 1/4" dialysis tubing (from Fisher) by blowing unheated air on the tubing with an Oster Airjet hairdryer.

Residual metal ion content in deionized enzyme preparations was measured by observing the effect of successive addition of excess EDTA and magnesium on the 296 nm absorbance of the enzyme[18]. The absorbance change is produced by nearly all divalent metals studied[20,21] whether these activate or not, and so this assay is useful because of its lack of specificity.

The magnitude of the change, and consequently the sensitivity of the assay, is pH- and buffer-dependent (Figure 1). In tris buffers the magnitude of the 296 nm absorbance change resembles a titration curve with an inflexion point at pH 8.1, close to the pK of tris. In other buffers, the extinction difference is independent of pH but much lower. Examination of the data shows that the 296 nm extinction of the magnesium-enzyme is the same irrespective of the buffer pH; the change in difference extinction is due to an interaction between tris base and apoenzyme which lowers the 296 nm absorbance of the apoenzyme. (The absorbance of the apoprotein is slightly smaller in the absence of EDTA; EDTA itself produces a positive molar difference extinction in the enzyme of 200-400). Such interactions are not confined to tris base, however; the presence of other buffers reduces

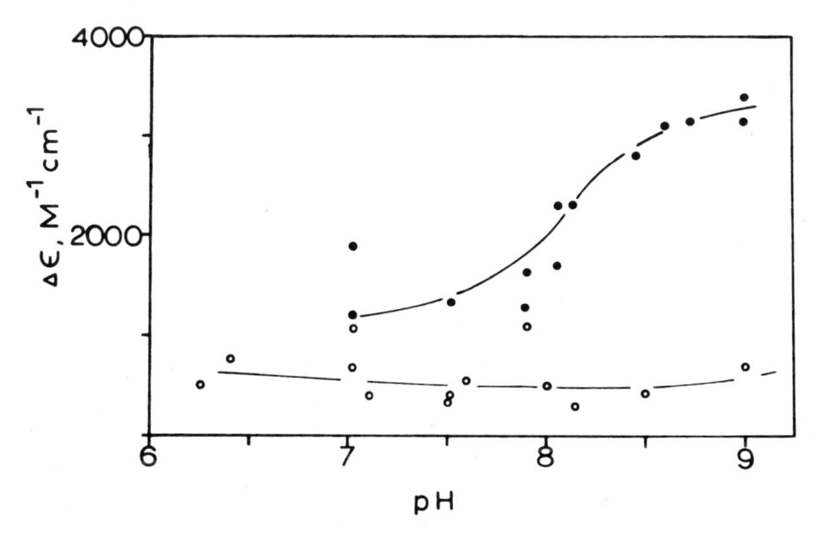

Fig. 1. Effect of buffer on difference extinction coefficient induced by magnesium. The absorbance at 296 nm of 1 or 2 ml of enzyme at either 2.2×10^{-5} or 1.2×10^{-5} M in various buffers was measured after addition of 5μl of 1 M EDTA, then remeasured after addition of 10μl of 1 M $MgCl_2$. In the figure, the filled circles were obtained using 0.05 ionic strength tris-HCl, and the open circles 0.1 M tricine-NaOH, PIPES-HCl, imidazole-acetate, imidazole-HCl, HEPES-NaOH or PIPES-NaOH. Data obtained over a 2.5 year period.

the effect of the tris base, so tricine, PIPES, etc. compete effectively with tris base (not shown). This interaction with apoenzyme may affect measurements of subunit dissociation based on enzymatic activity (see below).

There is also a difference spectral change at 296 nm which is related to subunit dissociation, since one can observe it in 1 M KCl with any buffer on addition of excess EDTA and then excess magnesium (not shown). Its value is approximately 3000 M^{-1} cm^{-1} at pH 7.8 in tris or HEPES-NaOH. The absorbance change due to dissociation would not be a significant factor in the results in Figure 1, since little subunit dissociation would be occurring at the relatively high protein concentrations used (see below).

Stopped-flow measurements were carried out using a modified Gibson-Durrum instrument[12]. Temperature control was provided by a Lauda K2-R refrigerated constant temperature circulating bath. Absorbance measurements were made using a Zeiss PMQ II single-beam spectrophotometer or a Bausch and Lomb Spectronic 200 double-beam instrument. Sedimentation velocity measurements were performed on a Beckman Model

E Analytical Ultracentrifuge fitted with schlieren, interference and absorption optics utilizing a monochromator-scanner. An AN-H rotor was used.

For laser Raman studies, microliter amounts of reagents (buffer, metal, salts) were added to 20 or 40 μl samples of deionized enzyme which were then sealed in glass capillary tubes. These were placed in the beam (488 nm) of a Spex 1401 Laser Raman Spectrophotometer for 20-30 minutes, then the spectra were taken. Assays afterwards showed the enzyme retained full enzymatic activity.

Osmometry was done at 25° with a Hewlett-Packard Model 503 Membrane Osmometer. For these measurements, samples of enzyme (10-20 mg/ml) were made up to the desired final concentration in KCl, EDTA and buffer, then dialyzed overnight against 400 ml of the same solvent. Subunit dissociation constants were calculated from osmotic pressure data using a modification of the equation of Guidotti[22]:

$$\pi = \frac{RTc}{dM} [1 + \frac{K \cdot M}{2c} (-1 + \sqrt{1 + \frac{4c}{KM}})] + RTB'c^2$$

π is the osmotic pressure, c is protein concentration in g/l, d is the density of the solvent, K is the subunit dissociation constant, M is the dimer molecular weight and B' is the average excluded volume term for both monomer and dimer. We used 93,000 for M, in agreement with the values of Mann _et al._[2] which range, depending on the method, from 85,000 to 100,000. The molecular weight of ovalbumin under our conditions is 45,000 in agreement with literature values. The osmotic pressure data were fitted to theoretical curves, calculated assuming different values of K and B'. A Wang-600 computer was used in the curve fitting.

RESULTS

Validity of measurement of subunit dissociation constants using enzymatic activity. Brewer and DeSa[12] presented stopped-flow data on the mechanism of subunit association which they interpreted in terms of the following series of reactions:

$$E + Mg \rightleftarrows EMg \rightleftarrows E^*Mg \rightleftarrows 1/2E^*_2Mg_2 \rightleftarrows 1/2E^{*\prime}_2Mg_2$$

The form of enzyme designated E' is capable of catalysis if sufficient metal and substrate are added. That is, enolase subunits bind magnesium and undergo a conformational change which increases subunit affinity. On associating, the subunits undergo another conformational change which is connected with recovery of enzymatic activity--an "annealing" reaction.

In order for enzymatic activity to accurately reflect the extent of subunit association, three conditions must hold:

1. Monomeric enzyme (E, EMg and E*Mg) must be completely inactive. This question can be answered by measuring relative initial activities of enzyme incubated with excess magnesium (fully associated or at least fully active) or excess EDTA (partly dissociated) as a function of enzyme concentration (Table I).

The dissociation in the absence of magnesium is readily and totally reversible[1], shows no time dependence over at least several hours and so appears to be describable as an equilibrium. If we assume that dissociated subunits are completely inactive, that associated enzyme is fully active and that the enzyme is fully associated in the presence of excess magnesium[13], we can calculate a subunit dissociation constant K_D using the relation:

$$K_D = \frac{4Eo\ (\frac{A+}{A-} - 1)^2}{\frac{A+}{A-}}$$

Eo is the total enzyme concentration (as dimer) in moles per liter, A+ is the initial activity of enzyme incubated with excess magnesium and A- is the initial activity of the enzyme incubated with excess EDTA[13]. If the subunits have some activity, then the ratio of initial activity of enzyme under dissociating conditions to activity when fully associated (incubated with magnesium) should drop to a constant value, the fraction of activity retained by monomeric enzyme, with decreasing protein concentration. In other words, if the subunits were active, K_D should decrease as the protein concentration was reduced.

Table I gives values of K_D calculated from data obtained over a period of several years, at protein concentrations varying over two or three orders of magnitude. Individual values of K_D vary markedly, a total variation of nearly a factor of 4 in 1 M KCl and nearly a factor of 7 in buffer only. Much of the variability comes from the measurements of A-. Note that an error of 10% in either rate will give a 30% error in K_D. However, the important point is that regression analysis indicates there is no trend to the data. These results confirm that the subunits of the enzyme dissociated in the absence of magnesium are inactive, both in the presence and absence of 1 M KCl.

TABLE I

EFFECT OF PROTEIN CONCENTRATION ON RELATIVE ACTIVITIES OF DISSOCIATED YEAST ENOLASE

Enzyme concentration, M (as dimer), (Eo)	Solvent	Relative activity after incubation with: excess EDTA (A-)	excess Mg (A+)	K_D,M
2.5×10^{-8}	buffer only	0.00027	0.0036	1.1×10^{-6}
2.5×10^{-7}	"	0.013	0.043	1.2×10^{-6}
2.5×10^{-7}	"	0.0054	0.018	1.6×10^{-6}
2.5×10^{-7}	"	0.0078	0.0395	3.3×10^{-6}
1.0×10^{-6}	"	0.47	0.89	1.7×10^{-6}
1.0×10^{-6}	"	0.0622	0.17	1.1×10^{-6}
2.5×10^{-6}	"	0.10	0.174	3.1×10^{-6}
2.5×10^{-6}	"	0.25	0.40	2.7×10^{-6}
2.5×10^{-6}	"	0.22	0.42	4.6×10^{-6}
1.5×10^{-5} [a]	"	2.22	2.55	1.2×10^{-6}
3.8×10^{-5} [b]	"	2.35	2.40	0.7×10^{-6}
			average:	$2.0(\pm 1.2) \times 10^{-6}$
2.5×10^{-7}	1M KCl	0.00018	0.0127	0.7×10^{-4}
2.5×10^{-6}	"	0.013	0.141	0.9×10^{-4}
2.5×10^{-6}	"	0.014	0.174	1.0×10^{-4}
2.5×10^{-6}	"	0.0027	0.07	2.5×10^{-4}
1.5×10^{-5}	"	0.10	0.50	1.9×10^{-4}
1.5×10^{-5}	"	0.263	1.05	1.4×10^{-4}
1.5×10^{-5} [a]	"	0.18	0.70	1.2×10^{-4}
3.8×10^{-5} [b]	"	0.72	1.67	1.1×10^{-4}
			average:	$1.3(\pm 0.6) \times 10^{-4}$

Activities were measured as the initial first order rate d(ln A)/dt, with the wavelength at 245 or 248 nm. A- and A+ are activities obtained after incubation with 2-3 mM EDTA or 10 mM $MgCl_2$ respectively. The final magnesium and 2-phosphoglycerate concentrations were 4-8 mM and 0.5-1 mM, respectively. Some of the rates were obtained using different wavelengths and slit widths and are not directly comparable. All activities were obtained at 20° with 0.05 ionic strength tris-HCl, pH 8.3, present.

[a] Reference 13

[b] Calculated from data in reference 12.

2. Subunit association must be relatively slow, since the enzyme can only be assayed in the presence of excess magnesium. Subunit association rate constants are low enough[13] to permit measurement of dimeric enzyme if initial rates are measured.

3. The annealing reaction must go to completion, that is, all dimers must be active. Otherwise, some inactive dimer would be counted as monomer, and K_D calculated from enzyme activities would be too high. At the same time, enzyme incubated in the presence of excess magnesium must be completely associated; if active monomers are present[10,11], then K_D from enzyme activities would be lower than the true value: the enzyme would appear more highly associated than it is.

According to the active enzyme centrifugation data of Holleman[11] the enzyme is completely associated down to assay concentrations (10^{-8} -10^{-9} M) at 0.05 ionic strength and 25°. Figure 2 shows sedimentation constants measured at 20° in 1 M KCl with excess magnesium, obtained using absorption or schlieren optics and from active enzyme centrifugation measurements[11]. The active enzyme centrifugation data strongly suggest dissociation into active monomers at lower protein concentrations in the presence of substrate. However, we can say that the enzyme shows no sign of dissociation in its absence down to 10^{-6} M.

The question of whether all associated enzyme is "annealed" can be answered only by comparison of K_D values calculated from activity measurements with those obtained using a physical technique. This has been done, using osmometry.

<u>Effect of pH on subunit dissociation</u>. Relative initial activities of enzyme incubated at various pH's in various buffers were used to calculate subunit dissociation constants (Figure 3). The scatter of the data is greater than we expected. Replicate determinations of K_D's at one pH gave values close to the initial ones, so it is possible that some of the scatter is actually an effect of the buffer. The PIPES-NaOH, PIPES-HCl and imidazole-HCl buffers seem to give different values from those of the other buffers.

The dissociation constants calculated from osmotic pressure measurements lie somewhat below the activity values (tris buffers). The protein appears more highly associated by a factor of about 3. This can be rationalized if all associated enzyme is not in fact annealed, at least in the tris buffers. We suggest that the equilibrium of the annealing reaction may be shifted to varying extents by different buffer ions, by preferential interaction of one form:

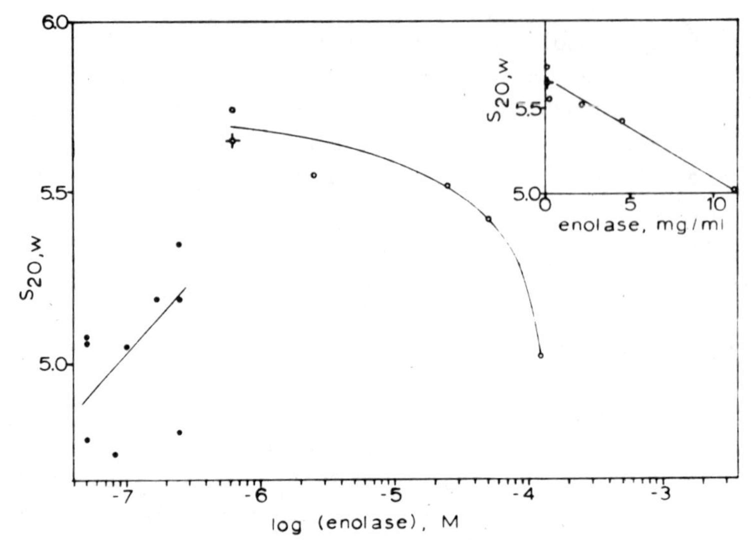

Fig. 2. Concentration dependence of sedimentation constant of enzyme in 1 M KCl with excess magnesium (1-10mM) present. The temperature was 20°. The open circles were obtained using schlieren or absorption optics at 280 nm or 230 nm, and a rotor speed of 68000 rpm. They were calculated from the second moment of the boundary[23]. The open circle marked with a cross is less reliable than the others, since the boundary positions were badly scattered. The closed circles were from active enzyme centrifugation measurements[11] with 2.4mM 2-phosphoglycerate present initially. The rotor speed for these experiments was 60,000 rpm. 0.05 ionic strength tris-HCl, pH 7.8, was used in all solutions.
Inset: Schlieren and absorption optics data replotted.

$$E^*{}_2Mg_2 \underset{}{\overset{K_N}{\rightleftharpoons}} E^{*\prime}{}_2Mg_2$$

with the buffer. In tris buffers the value of K_N need only be about 0.5 to produce the observed effect.

Of more significance, however, is the fact that both types of measurement give the same result: the dissociation reaction is pH dependent, with net release of proton(s) on dissociation. Also, the slopes of all three lines are similar, ranging from 0.23±0.10 (activity measurements in 1 M KCl) to 0.52±0.11 (activity measurements at 0.05-0.1 ionic strength). If only the tris buffer values are counted, in the case of the activity measurements, the slopes are 0.3-0.4.

The slopes are numerically equal to the number of moles of protons released per mole of protein[24], and their values indicate that a non-integral number of protons is released on dissociation. This represents the difference in number of protons bound to associated and

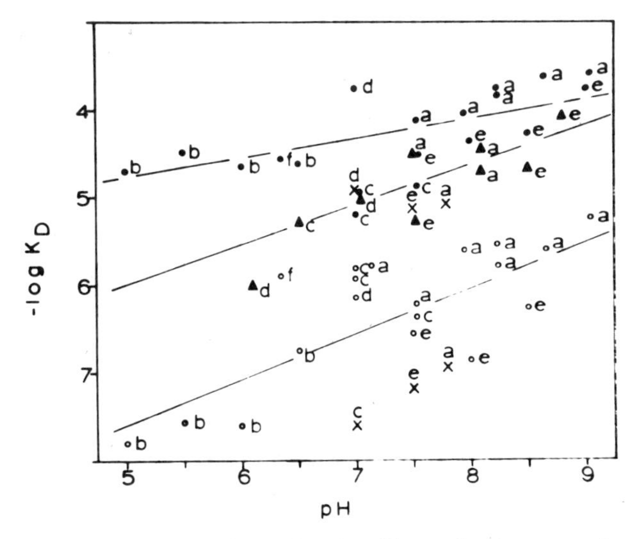

Fig. 3. Effect of pH on the subunit dissociation constant. The values of K_D were calculated from relative enzymatic activity at 0.5-0.1 ionic strength (open circles) or in 1 M KCl (filled circles) or in 1 M KCl from osmotic pressure data (triangles). The letters in the figure refer to the buffer used: a: 0.05 ionic strength tris-HCl; b: imidazole-acetate (0.1 M); c: 0.05 M PIPES-NaOH; d: 0.05 M imidazole-HCl; e: 0.1 M tricine-NaOH; f: 0.1 M PIPES-HCl. The crosses refer to experiments carried out in 90% D_2O. The lines were calculated using a least squares method, using all the data points from either osmometry (slope = 0.46) or relative enzymatic activity in 1 M KCl (slope = 0.23 ± 0.10) or at low ionic strength (slope = 0.52 ± 0.11). Osmotic pressure measurements were carried out using protein concentrations of 1-10 mg/ml at 25°; relative activity measurements were made at 20° using 10^{-6} M enzyme for the low ionic strength measurements or 1.5×10^{-5} M enzyme for the 1 M KCl measurements.

dissociated enzyme, and can in principle be a non-integral value[25].

The crosses in Fig. 3 are from activity measurements made in D_2O solutions. Relative to the values obtained in H_2O, using the same buffers, the values in D_2O are an average of an order of magnitude lower. This suggests that the protein is more associated in D_2O solutions.

Effect of temperature on subunit dissociation. The effect of temperature on subunit dissociation, calculated from activity measurements, is given in Figure 4. In these experiments, reactions were carried out in tris buffers, whose pH is strongly temperature dependent[26]. Consequently, the values of K_D obtained at a given temperature were corrected to a common pH using the tris buffer data in Figure 3. The scatter of the data in Figure 4 is comparable to that in Table I, so there does not appear to be anything unusual occurring.

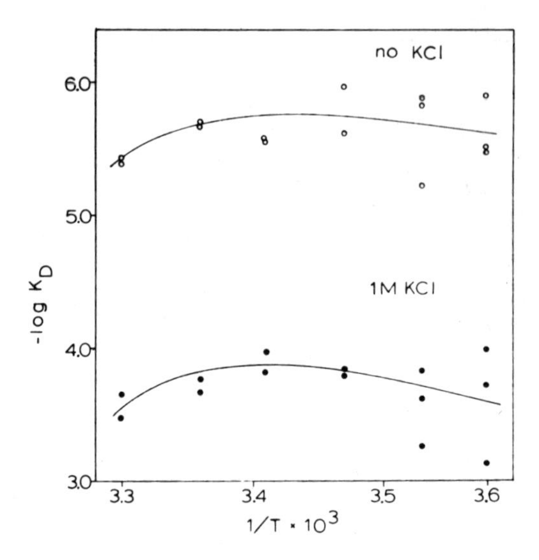

Fig. 4. Effect of temperature on subunit dissociation constant. Values of K_D were obtained from relative activity measurements. The buffers used were 0.05 ionic strength tris-HCl. The published effect of temperature on the pH of tris buffers was used along with the effect of pH on K_D, to correct the K_D values to pH 7.8.

The apparent enthalpy of subunit association at about 20° is zero, indicating a predominance of hydrophobic or ionic interactions[14]. The latter possibility can be ruled out, since high concentrations of sulfate and acetate salts actually promote subunit association[8,13]. The profile also suggests changes occur in heat capacity of the enzyme on subunit association. This was also observed by Gawronski and Westhead[7]. The reciprocal profile was obtained in a sedimentation equilibrium study of the acid-induced dimerization of chymotrypsin[25].

While we have no physical measurements of subunit dissociation constants as a function of temperature, Johnson and Faller[27] obtained exactly the same profile at several salt concentrations measuring difference spectral changes. So the activity measurements appear to be qualitatively reliable (see Discussion).

Note that, like the effect of pH, the profile is essentially the same at 0.05 and 1 ionic strengths. The data in Figures 3 and 4 suggest that the ligands involved in subunit contact do not change in 1 M KCl, which appears to act by reducing the favorable entropy of subunit association.

Effect of rotor speed on the sedimentation constant of enolase. The sedimentation constant of enzyme in 1 M KCl with excess EDTA is a

TABLE II
OBSERVED SEDIMENTATION CONSTANTS OF YEAST ENOLASE IN SEVERAL SOLVENTS AND ROTOR SPEEDS

Protein Concentration mg/ml	Solvent	S_{obs}(S) at: 40,000 rpm	48,000 rpm	60,000 rpm	68,000 rpm
4.5	1 M KCl + EDTA	--	3.93	--	3.68
4.5	1 M KCl + Mg	--	4.80	--	4.67
4.5	1 M KOAc + EDTA	--	3.82	--	3.79
4.5	Buffer only + EDTA	--	5.25	--	5.35
4.5	1 M KCl + EDTA	4.08	--	3.73	--
5.6	1 M KCl + EDTA	--	4.31	--	3.80

The centrifugation was begun at the higher rotor speed, then shifted to the lower after sufficient pictures were taken. Sedimentation constants were calculated from the second moments of the boundary[23].

function of rotor speed (Table II). The higher the rotor speed, the lower the sedimentation constant. This is the case only in solvents where significant subunit dissociation is expected, so it appears to be related to that phenomenon. If this is an effect of pressure, it suggests that either hydrophobic or ionic interactions (or both) are important in intersubunit contact[28].

Laser Raman studies of enzyme conformation. The spectrum of yeast enolase was obtained in several solvents by laser Raman spectroscopy. Figure 5 shows the Raman spectrum of yeast apoenolase. Analysis of the spectrum of the apoenzyme[29-31] suggested an α-helix content of less than 40%, little (less than 5%) β-pleated sheet structure and the rest in a random coil conformation. In addition, the spectrum indicated the tryptophans in the protein are exposed to the solvent (there is no peak at 1361 cm^{-1}) and the tyrosines are weakly hydrogen bonded.

The precision of measurement was ±6% for intense Raman transitions and ±10-15% for the weak transitions. Generally, the precisions obtained using samples of biological materials are approximately ±10%[29-31]. Within these limits, we observed no effect of magnesium or 1 M KCl on the laser Raman spectrum of the enzyme, between 1700 cm^{-1} and 500 cm^{-1}. Table III gives the intensities observed for some representative transitions in the presence or absence of metal or salt.

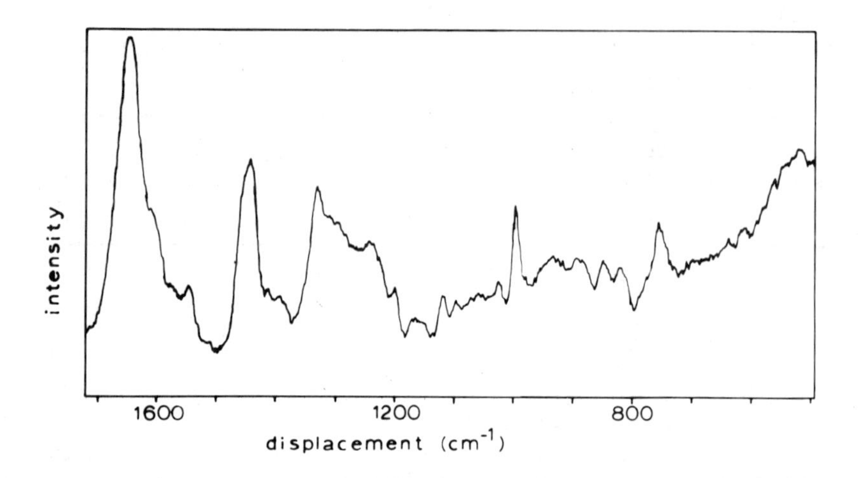

Fig. 5. Laser Raman spectrum of yeast apoenolase. 1.7 mM apoenzyme in 0.025 ionic strength tris-HCl, pH 8.6 (original), (final pH 7.8) was sealed in a capillary tube and after 25 minutes exposure to 488 nm laser light, the Raman spectrum was scanned.

These data show that the metal and salt-induced conformational changes in the enzyme[5] do not involve extensive changes in hydrophobic or amide groups. The suggestion by Robinson and Jencks[6] that the Hofmeister anions function through direct interactions with amide bonds is not supported by these data. Note that these spectra were obtained using protein concentrations above 90 mg/ml; at these concentrations, little dissociation can occur, so the spectra would not show the effect of dissociation, just that of KCl binding.

On the other hand, the salt does change the spectrum of the solvent (water) (Figure 6). A similar effect was observed previously by Walrafen, who studied the effects of a number of electrolytes on water structure using Raman spectroscopy[32]. He suggested that anions such as chloride or bromide disrupted water structure through formation of X. . .HO- complexes. Anions such as fluoride and apparently sulfate tended to be "structure makers."

We suggest that "Hofmeister" anions such as chloride break up water structure and thereby reduce the strength of hydrophobic interactions, which depend on the strength of water interacting with water[33]. Observations that little or no change in ion binding occurs upon subunit dissociation[7,8] are now consistent with the picture of the subunits interacting through hydrophobic residues, not ionic ones.

TABLE III

EFFECT OF MAGNESIUM AND 1 M POTASSIUM CHLORIDE ON RAMAN TRANSITIONS IN YEAST ENOLASE

Transition (position, cm^{-1})	Added: No Additions	2 Moles Mg	1 M CsCl	1 M KCl	1 M KCl + 2 Moles Mg
Amide I, I' (1619)	0.75	0.78	0.74	0.74	0.74
	0.78	0.74	--	0.94	0.90
Amide III (1251)	0.56	0.55	0.61	0.52	0.58
	0.60	0.56	--	0.58	0.57
C-C (948)	0.34	0.34	--	--	0.26
	0.40	0.34	--	0.30	0.37
C-N (1129)	0.24	0.25	0.21	--	--
	0.22	0.22	--	0.19	0.20
N-H (1401)	0.17	0.20	--	--	--
	0.31	0.31	--	0.26	0.27
phenylalanyl-(1005)	0.70	0.69	0.87	0.64	0.68
	0.66	0.66	--	0.65	0.67
tryptophanyl-(1343)	0.76	0.76	0.83	0.81	0.79
	0.85	0.78	--	0.84	0.90
tyrosyl-(858)	0.25	0.25	--	--	0.16
	0.23	0.31	--	0.19	0.23

The numbers are the relative strengths of the transitions, with that at 1459 cm^{-1} taken as 1.00. Values are from two series of spectra (different enzyme preparations).

Effect of subunit dissociation on competitive inhibitor binding. It has been established that the apoenzyme can bind up to two moles of magnesium (one mole per subunit)[19,20,34,35]. This "conformational" metal[18] is strongly bound[19,34,35] and enables up to two moles of substrate to bind[19,36,37]. The binding of the substrate, in turn, enables more magnesium to be bound, and this additional metal enables the enzyme to catalyze its reaction[38].

Spring and Wold[39] showed that two competitive inhibitors of enolases, D-tartronic semialdehyde phosphate (TSP) and 3-aminoenolpyruvate-2-phosphate (AEP), bind more strongly than the substrate or the product, leading to the suggestion that these inhibitors were "transition state analogues". The binding was accompanied by a substantial

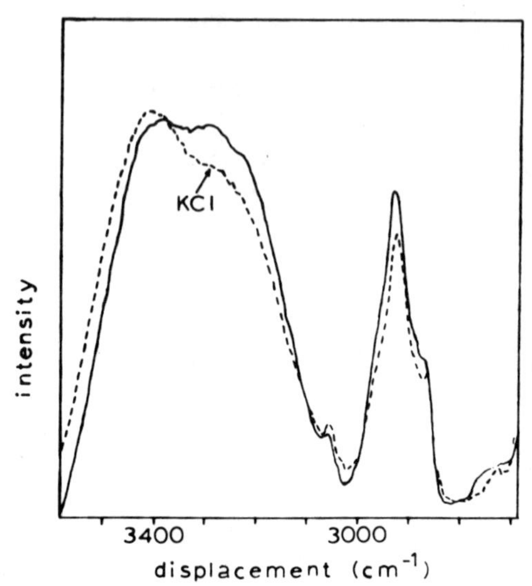

Fig. 6. Effect of 1 M KCl on Laser Raman spectrum of yeast enolase solutions: water region. The conditions of this experiment were the same as those in Figure 5, save for the case of the broken line, which is the spectrum obtained with enzyme in the presence of 1 M KCl. Also, 2 moles of magnesium per mole of enzyme was present in both solutions. The broad peak is largely from the solvent (water).

shift in absorbance of both inhibitors; the difference extinction coefficient of AEP was 21,000 $M^{-1}cm^{-1}$ at 295 nm.

It has been demonstrated that the dissociated subunits can bind the "conformational" magnesium and change in conformation within the dead time of the stopped-flow[12], prior to subunit association[13]. Consequently, the dissociated subunits must be lacking in ability to bind the substrate or the "catalytic" metal, or must require some other conformational adaptation for activity.

If the subunits are inactive because the substrate binding site has been altered, then the appearance of the large 295 nm absorbance change produced by AEP binding to magnesium-enolase would be limited by the rate of subunit association or "annealing". Consequently, we reacted enzyme dissociated in 1 M KCl and EDTA with magnesium and AEP. (It doesn't matter whether the AEP is originally in the metal solution or the enzyme solution.) As a control, we reacted enzyme in 1 M KCl and excess magnesium (associated enzyme) with a solution containing AEP. In addition, enzyme in 1 M KCl was reacted with magnesium alone (the 295 nm absorbance change without inhibitor is much smaller[12]) (Figure 7).

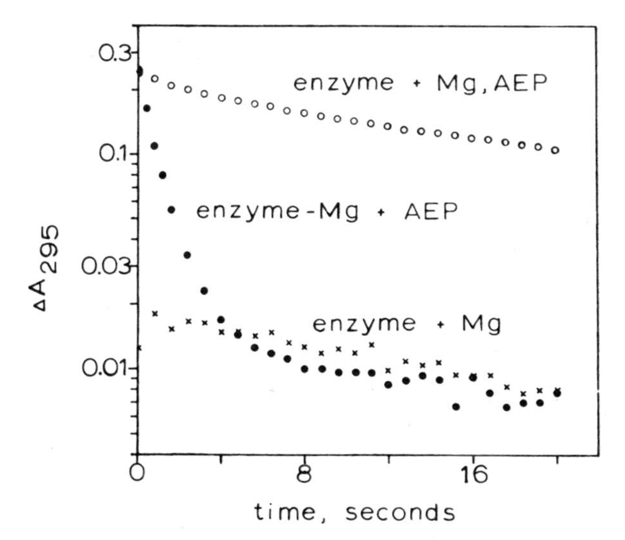

Fig. 7. Kinetics of AEP reaction with associated and dissociated enzyme. 6 x 10^{-6} M yeast enolase with either 5 mM $MgCl_2$ and 1 mM EDTA (solid circles) or 1 mM EDTA (open circles and crosses) was mixed in a stopped-flow spectrophotometer with either 10^{-4} M AEP and 1 mM EDTA (solid circles), 10 mM $MgCl_2$ (crosses) or 10 mM $MgCl_2$ and 10^{-4} M AEP (open circles). 1 M KCl and $\Gamma/2 = 0.10$ tris-HCl, pH 7.9, was present in all solutions. The reaction temperature was 20°. The wavelength was 295 nm and the monochromator half band width corresponded to 1 nm. Data points are averages of 4 replicate experiments.
Reprinted courtesy of Rockefeller University Press

Lane and Hurst[40] showed that the inhibitor binds to the enzyme in a pseudo-first order reaction, provided the inhibitor concentration is at least ten times that of the enzyme, whose rate increases with inhibitor concentration. They also found no evidence for a bimolecular reaction occurring before the absorbance change[40]. We have obtained similar results (Figure 7), though we find a much slower reaction of low amplitude follows inhibitor binding.

However, the rate of appearance of the 295 nm absorbance change seen on adding magnesium and AEP to dissociated enzyme is much slower. A faster initial phase is followed by a slower one constituting the bulk of the reaction. The apparent rate of the slower phase is about 0.034/sec, treating it as a first order reaction. This is lower than the rate of the "annealing" reaction (ca 0.05/sec)[12], but the second order subunit association reaction is probably limiting at the protein concentration used[12,13]. The faster initial part is almost certainly due to AEP binding to the small fraction (see Table I) of dimeric enzyme present at this protein concentration. At higher protein concentrations, the faster initial part becomes more prominent (not shown), which suggests it is due to the presence of relatively more dimeric (and "annealed") enzyme. The kinetics obtained on adding magnesium to dissociated enzyme in the presence of AEP are similar to those obtained on adding magnesium to the enzyme by itself (and identical to those published in reference 12), suggesting that the same process(es) are limiting.

Spectrophotometric titrations of enzyme-AEP solutions with magnesium show that two-thirds of the absorbance change in the inhibitor is obtained on adding two moles of magnesium[21], and that we are seeing all the reaction in the stopped-flow, that is, there is no "dead time" reaction. Since the "conformational" metal binds within the dead time of the stopped-flow[12,13], these data show that the inhibitor binding site in enolase subunits which were dissociated without magnesium has been altered, rather than the catalytic magnesium binding site.

One possible alteration is that the subunits cannot bind the inhibitor until subunit association (and presumably "annealing") has occurred. However, we are not measuring binding directly but a 295 nm absorbance change in the inhibitor occurring upon binding. Spring and Wold[39] showed that the absorbance change produced in the inhibitor TSP on binding to magnesium-enzyme can be duplicated using magnesium and 0.01 M NaOH. The magnesium-enzyme apparently provides a strongly basic group, perhaps deprotonating the inhibitor. It is possible that the inhibitor actually binds to the magnesium-subunits rapidly but does not change in spectral properties; the increase in 295 nm absorbance would not occur until the deprotonating group(s) in the enzyme could perform their function, that is, in a conformational rearrangement occurring after subunit association.

DISCUSSION

The effect of pH on subunit association in yeast enolase has been demonstrated by isopycnic density gradient centrifugation[8] and now using enzyme activity and osmotic pressure. It is not possible to draw any conclusions as to the nature of the group(s) involved in the proton equilibria.

In a study of the effect of ionic strength on the acid dimerization of chymotrypsin, Aune *et al.*[25] interpreted similar findings as an indication that considerable changes in water binding occurred. That cannot be the case here, as the isopycnic density gradient data[8] ruled out the possibility of large changes in water of hydration on dissociation, and because integral changes in proton concentrations of the order 10^{-5} - 10^{-9} M would involve changes of absurdly large amounts of bound water.

The suggestion that the buffer affects the subunit dissociation constant calculated from enzymatic activities is not surprising in view of the results in Figure 1. This suggestion should serve to caution workers in attempts to measure subunit equilibria using enzymatic activity alone.

The apparent qualitative identity between the pH and temperature dependences measured by physical and enzymatic methods suggests that the "annealing" reaction is not strongly affected by either parameter. An ionic interaction, with the pK of the group(s) on the protein being outside the pH range 5-9, is indicated.

The effect of temperature on the apparent subunit association constant, added to the data previously presented - absorption and fluorescence data, ANS fluorescence changes, ion binding measurements - allows only one conclusion: hydrophobic interactions clearly predominate in causing subunit association in yeast enolase. All the evidence is consistent including the possible "pressure effect" (Table II) and the apparent effect of D_2O on subunit association. Because of the stronger association between D_2O molecules, hydrophobic interactions are strengthened[14,33].

The laser Raman data has focused attention on water structure as the possible origin of the effects of chaotropic anions rather than anion-protein interactions. This, too, is consistent with the effects of the anions on subunit interaction thermodynamics, and with the nature of hydrophobic interactions[33].

Whether the inhibitor, AEP, can actually bind to the dissociated subunits or not, it is clear that the inhibitor binding site is affected by dissociation. In view of the fact that AEP is such a strong competitive inhibitor, we feel confident the substrate binding site is affected in the same way. The substrate binding site is presumably lost or changed during a postulated "deannealing" step occurring after the subunits dissociate when magnesium is removed. Associated enzyme appears to be "annealed"[12,13]. Subunits dissociated in the presence of magnesium and substrate by high temperatures[10,11] or with 1 M chloride (Figure 2) may remain active because of the continued presence of the metal and substrate.

Curthoys *et al.*[41], using a partition method, showed that formyltetrahydrofolate synthetase from *Clostridium cylindrosporum* lost the ability to bind folate derivatives when dissociated by removal of the activating cations (K^+ or NH_4^+). Interestingly, the monomers could still bind ATP. We do not measure binding directly as they did, but we do show that the 295 nm absorbance change requires subunit association and not just the binding of magnesium to the subunits or the rapid conformational change that follows binding of the cofactor. Curthoys et al.[41] were forced to make their measurements in the absence of the activating cations.

In the case of formyltetrahydrofolate synthetase, Curthoys et al.[41] concluded that the existence of the folate binding site was either dependent upon a conformational change in the monomers or that each site was formed by the junction of two subunits. In the case of yeast enolase, the data of Keresztes-Nagy and Orman[10] and of Holleman[11] show that the substrate binding region must lie entirely within each subunit, and so the site must be formed in a conformational rearrangement, presumably the "annealing" reaction, that occurs after subunit association.

ACKNOWLEDGEMENTS

We thank Drs. K. C. Aune and R. C. Lord for helpful comments and criticisms. Supported by National Science Foundation grant PCM 76-21378.

REFERENCES

1. Brewer, J.M. and Weber, G. (1968) Proc. Natl. Acad. Sci. U.S.A., 59, 216-223.
2. Mann, K.G., Castellino, F.J. and Hargrave, P.A. (1970) Biochemistry 9, 4002-4007.
3. Brewer, J.M., Fairwell, T., Travis, J. and Lovins, R.E. (1970) Biochemistry 9, 1011-1016.
4. Brewer, J.M. (1976) Tenth Int. Cong. of Biochem. Abstracts (Hamburg, G.F.R.)
5. Brewer, J.M. (1969) Arch. Biochem. Biophys. 134, 59-66.
6. Robinson, D.R. and Jencks, W.P. (1965) J. Am. Chem. Soc. 87, 2462-2470.
7. Gawronski, T.H. and Westhead, E.W. (1969) Biochemistry 8, 4261-4270.
8. Brewer, J.M. (1975) Arch. Biochem. Biophys 171, 466-473.
9. Kuntz, I.D., Jr. and Kauzmann, W. (1974) Adv. Prot. Chem. 28, 239-345.
10. Keresztes-Nagy, S. and Orman, R. (1971) Biochemistry 10, 2506-2508.
11. Holleman, W.H. (1973) Biochim. Biophys. Acta 327, 176-185.

12. Brewer, J.M. and DeSa, R.J. (1972) J. Biol. Chem. 247, 7941-7947.

13. Brewer, J.M. (1976) Eur. J. Biochem. 71, 425-436.

14. Timasheff, S.N. (1973) Protides of the Biological Fluids, (Peeters, H., editor) Twentieth Colloquium, Pergamon Press, Oxford, pp. 511-519.

15. Westhead, E.W. and McLain, G. (1964) J. Biol. Chem. 239, 2464-2468.

16. Warburg, O. and Christian, W. (1941) Biochem. Z. 310, 384-421.

17. Spring, T.G. and Wold, F. (1971) Biochemistry 10, 4649-4654.

18. Brewer, J.M. and Weber, G. (1966) J. Biol. Chem. 241, 2550-2557.

19. Faller, L.D. and Johnson, A.M. (1974) Proc. Natl. Acad. Sci., U.S.A., 71, 1083-1087.

20. Hanlon, D.P. and Westhead, E.W. (1965) Biochim. Biophys. Acta, 96, 537-539.

21. Brewer, J.M. (1978) Biophys. J., in press.

22. Guidotti, G. (1967) J. Biol. Chem. 242, 3685-3693.

23. Gilbert, L.M. and Gilbert, G.A. (1973) Methods in Enzymology (C.H.W. Hirs and S.N. Timasheff, editors), Vol. XXVII, Academic Press, New York, pp. 273-295.

24. Mahler, H.R. and Cordes, E.H. (1966) "Biological Chemistry", Harper and Row, New York, p. 202.

25. Aune, K.C., Goldsmith, L.C. and Timasheff, S.N. (1971) Biochemistry 10, 1617-1622.

26. Sober, H.A. (1968) "Handbook of Biochemistry", Chemical Rubber Co., Cleveland, Ohio.

27. Johnson, A.M. and Faller, L.D. (1977) Fed. Proc. 36, 874.

28. Harrington, W.F. and Kegeles, G. (1973) Methods in Enzymology (C.H.W. Hirs and S.N. Timasheff, editors), Vol. XXVII, Academic Press, New York, pp. 306-345.

29. Chen, M.C. and Lord, R.C. (1974) J. Am. Chem. Soc. 96, 4750-4752.

30. Yu, T.J., Lippert, J.L. and Peticolas, W.L. (1973) Biopolymers 12, 2161-2176.

31. Yu, N.T. and Liu, C.S. (1972) J. Am. Chem. Soc. 94, 5127-5128.

32. Walrafen, G.E. (1971) J. Chem. Phys. 55, 768-792.

33. Tanford, C. (1973) "The Hydrophobic Effect," John Wiley and Sons, New York.

34. Faller, L.D. and Johnson, A.M. (1974) FEBS Letters 44, 298-301.

35. Brewer, J. M. (1974) Arch. Biochem. Biophys. 164, 322-325.

36. Hanlon, D.P. and Westhead, E.W. (1969) Biochemistry 8, 4247-4255.

37. Brewer, J.M. (1971) Biochim. Biophys. Acta 250, 251-257.

38. Faller, L.D., Baroudy, B., Johnson, A.M. and Ewall, R.X. (1977) Biochemistry 16, 3864-3869.

39. Spring, T.G. and Wold, F. (1971) Biochemistry 10, 4655-4660.

40. Lane, R.H. and Hurst, J.K. (1974) Biochemistry 13, 3292-3297.

41. Curthoys, N.P., Straus, L.D. and Rabinowitz, J.C. (1972) Biochemistry 11, 345-349.

Catsimpoolas, ed. Physical Aspects of Protein Interactions

THE VAN'T HOFF AND GIBBS-DUHEM EXPRESSIONS: THERMODYNAMIC COMPENSATION IN INTERACTING PROTEIN SYSTEMS

PAUL W. CHUN

Department of Biochemistry and Molecular Biology, J-245, JHMHC, College of Medicine, University of Florida, Gainesville, Florida, 32610.

ABSTRACT

It is possible to express the gross change in the Gibbs free energy which is the result of all the subprocesses (ΔG°_{motive} and ΔG°_{comp}) contributing to any interaction as:[1,2,3]

$$(\Delta G)^\circ_{gross} = \Delta G^\circ_{motive} + \Delta G^\circ_{comp}$$

The compensation process may be expressed as $\Delta H^\circ_{gross} = \Delta G^\circ_{<T_s>} + <T_c> \Delta S^\circ_{gross}$. Any interacting protein system will have a unique compensatory temperature, <Tc>, at which the contributions of enthalpy and entropy to the association process are balanced, and $\Delta G^\circ_{comp} = 0$. Thus,[1]

$$\Delta G^\circ_{(T)gross} = [\Delta G^\circ_{<T_s>} + <T_c> \Delta S^\circ_{(T)gross}] - T_{exp} \Delta S^\circ_{(T)\,gross}$$

We have found in those associating systems which we examined, specifically bovine liver L-glutamate dehydrogenase (GDH), glucagon, reduced carboxymethylated apoA-II from high density lipoprotein and α-chymotrypsin at low pH, as determined from a linear plot of $\Delta H^\circ_{(T)}$ versus $\Delta S^\circ_{(T)}$, that .

$$<T_c> = T_{exp} + \left[\frac{T_{exp}\ \Delta S^\circ_{(T)gross}}{\Delta C^\circ_{P(T)gross}}\right]$$

as $T_{exp} \rightarrow T_s$, the entropic temperature at which $\Delta S^\circ_{(T)} \rightarrow 0$.

When the Van't Hoff and Gibbs-Duhem expressions are combined by Chun and Saffen[1] to determine the Hofmeister effect, however, it is possible to locate the compensatory temperature in a self-associating protein system perturbed by the presence of a Hofmeister series of anions.

The principal determinants of the linear thermodynamic compensation process operating in any interacting protein system were found to be $\Delta S^\circ_{(T)}/\Delta C^\circ_{P(T)} = <\Delta T'_c> / <T_{exp}>$, where $<\Delta T'_c> = (<T_c> - T_{exp})$.

INTRODUCTION

In the past, researchers have been concerned with obtaining accurate information about the linear Van't Hoff expression, $R \ln K = (-\Delta H^\circ/T) + C$, and the corresponding equilibrium constants as a function of temperature. Earlier investigations of equi - libria of interacting protein systems have generally shown enthalpy changes which are independent of temperature. Most recent studies, however, have revealed that a plot of ln K versus 1/T is not strictly linear, and that neither $\Delta H°$ nor $\Delta S°$ can be considered as independent of temperature[5-12].

A closer examination of the thermodynamics of several self-associating protein systems--bovine liver L-glutamate dehydrogenase, S-carboxymethylated apoA-II protein and glucagon--by analytical ultracentrifugation, molecular sieve chromatography and circular dichroic studies, over a wide range of temperatures, reveals that the association process is affected by several specific thermodynamic temperatures, designated $\langle T_c \rangle$, the temperature of compensation; T_H, the harmonious temperature at which the Gibbs free energy change is at a minimum; $T_{\phi h}$, the temperature at which the enthalpy of the solution becomes zero, and T_S, the temperature at which the entropy approaches zero[4].

Recent studies have reported that bovine liver L-glutamate dehydrogenase undergoes isodesmic association in 0.2 M sodium phosphate buffer, 1×10^{-3} M EDTA at pH 7.0[13-17]. Glucagon, a 29-amino acid residue polypeptide hormone that is bound to target tissues and activates adenyl cyclase, undergoes monomer-trimer association in 0.2 M K_2HPO_4, pH 10.6[11,12,18-20].

The reduced carboxymethylated form of apoA-II protein from human high density lipoprotein has a molecular weight of 8696 g/mole and undergoes monomer-dimer association in 0.01 M phosphate, pH 7.4[10]. In each of these three protein systems, self-association appears to be endothermal at low and exothermal at high temperatures, with a resultant variation in heat capacity.

In considering a non-linear Van't Hoff expression, then, the self-association reaction has a significant effect on the partial molar heat capacity of the interacting protein system, i.e., $\partial(\Delta H°)/\partial T = T\partial(\Delta S°)/\partial T = \Delta C_p°$. The heat capacity change is an essential thermodynamic quantity that expresses the interaction of the various hydrophobic groups with H_2O[2,7,9,21-30].

Perhaps the most striking aspect of the ordering of water molecules about these hydrophobic groups is the thermal lability, which imparts a large, anomalous heat capacity to aqueous solutions of hydrophobic or partially hydrophobic solute, as observed and explained by Edsall[29] and Frank and Evans[21].

The linear relationship between the enthalpy-entropy compensation has been recognized and formulated in some detail by Barclay and Butler[31] and Lumry and Rajender[2].

It is possible to determine the compensating temperature, $\langle T_c \rangle$, at which the contributions of entropy and enthalpy to the self-association process in a biological system are equivalent. When the Gibbs-Duhem and Van't Hoff expressions are combined in the study of an interacting protein system--the enzyme α-chymotrypsin in this case--the effect of a Hofmeister series of anions on the preferential solvation of the protein as influenced by temperature may be assessed.

Klotz[32] and Tanford[33] applied Wyman's treatment of the Gibbs-Duhem expression to a consideration of the preferential binding of a Hofmeister series of anions[34] in self-associating protein systems. Aune and Timasheff[35] and Aune et al.[6], have postulated that the dimerization of α-chymotrypsin is accomplished by the preferential binding of salt, a negative change in heat capacity and a positive entropy at low temperature.

Earlier pH studies[36,37] at 20°C indicated that the pH-dependent sedimentation coefficient of α-chymotrypsin reaches a maximum and a minimum at about pH 4.0. Above pH 6.0, this enzyme undergoes association of a higher order than monomer-dimer[38-42].

In this first section of this paper, we report the thermodynamic analysis of the non-linear Van't Hoff expressions of several self-associating protein systems, specifically in the cases of bovine liver L-glutamate dehydrogenase, glucagon and S-carboxymethylated apoA-II from the human high density lipoprotein complex. We examine the extent to which the second virial coefficient (β) of the interacting solute influences the Gibbs free energy change[3] in such a system,

$$\Delta G^{\circ}_{gross} = \Delta G^{\circ}_{motive} + \Delta G^{\circ}_{comp}$$

and report the effect of thermodynamic temperatures on the equilibrium constants and concomitant changes in the thermodynamic parameters, which suggest that certain changes in the folding of the tertiary structure accompany self-association.

This discussion is followed by an examination of the process of linear thermodynamic compensation, as illustrated by the case of α-chymotrypsin association in the presence of a series of Hofmeister anions.

METHODS

Computational Analysis of Thermodynamic Parameters

1. Evaluation of the weight fraction of interacting monomer from the weight average molecular weight (M_w) as a function of concentration.

The evaluation of the weight fraction of a monomer undergoing isodesmic association has been previously described[15-17]. For isodesmic association, the non-ideality term βM_1 (in ml/g where β is mole ml/g) is evaluated from the slope of a plot of $(M_1/M_{wapp}) - [\sqrt{f_1}/(2-\sqrt{f_1})]$ versus C.[17] In the case of bovine liver L-glutamate dehydrogenase, data for the temperature-dependent molecular weight distribution at different concentrations were taken from Reisler and Eisenberg[8].

2. Evaluation of the weight fraction of monomer from the mean residue ellipticity ($[\theta]_\lambda$) as a function of concentration at a given temperature.

In evaluating equilibrium constants and stoichiometry in cases of two-species association such as that between the glucagon monomer-trimer or apoA-II monomer-dimer, the mean residue ellipticity or weight average ellipticity is related to the weight fraction of monomer of each interacting species as a function of concentration, at a given wavelength λ, i.e.,

$$[\theta]_\lambda = \sum_i f_i(\theta_1 + \beta_i C_i) \qquad (1)$$

TABLE I

TABULATION OF THE WEIGHT FRACTION OF MONOMER OF BOVINE LIVER L-GLUTAMATE DEHYDROGENASE AS A FUNCTION OF TEMPERATURE (ISODESMIC ASSOCIATION MODEL).

The concentration, C, is expressed as X 10^{-3}g/ml. The weight fraction of monomer is computed from the following expression[66]:

$$f_i = \frac{i}{\hat{k}c}\left[1 + \frac{1}{2\hat{k}c}\left(1 - \sqrt{1 + \hat{k}c}\right)\right]^i .$$

The correlation coefficient in all cases was found to be 0.9999. $\hat{k}$ values were taken from Reisler and Eisenberg[8].

Temp, °K	$\hat{k}$ (ml/mg)	C = 0.5	C = 1.0	C = 1.5	C = 2.0	C = 2.5	C = 3.0	C = 3.5	C = 4.0
283	1.29	0.4782	0.3299	0.2556	0.2099	0.1789	0.1563	0.1389	0.1252
288	1.62	0.4273	0.2867	0.2189	0.1783	0.1509	0.1312	0.1162	0.1045
293	1.95	0.3873	0.2542	0.1921	0.1553	0.1309	0.1134	0.1004	0.0898
298	2.10	0.3717	0.2419	0.1820	0.1469	0.1235	0.1069	0.0943	0.0845
303	2.05	0.3767	0.2459	0.1852	0.1496	0.1258	0.1089	0.0962	0.0862
308	1.82	0.4021	0.2660	0.2018	0.1636	0.1381	0.1198	0.1059	0.0950
313	1.50	0.4444	0.3009	0.2309	0.1886	0.1600	0.1393	0.1235	0.1111
BM_1 (ml/mg)		140.0	24.90	3.67	3.99	4.04	5.24	6.46	6.48

where $\sum_i f_i = 1$, the weight fraction of the ith species, and β_i is the second virial coefficient of ith species. The weight fraction of monomer is related to the equilibrium constant, K, by the expression[15,16]:

$$K_i = (Cf_i/C_1^i) - Cf_i/(Cf_i)^i = f_i/(f_1^i c^{i-1}) \qquad (2)$$

The contribution of the third virial coefficient was too small to be considered. Once the second virial coefficient, β in deg-cm^2-ml/decimole/mg, is evaluated from

$$[\theta]_\lambda = f_1\,[\theta]_1 + f_i\,[\theta]_i + \beta C_T \qquad (3)$$

the three interaction parameters, the weight fraction of monomer, equilibrium constant and second virial coefficient, are used in the following expression to regenerate the original data points by the SAS least-square curve-fitting procedures of Barr et al. (Statistical Analysis System, Northeast Regional Data Center, University of Florida).

$$[\theta]_\lambda = f_1\theta_1 + K_i f_1^i \theta_i C_T^{i-1} + \beta C_T \qquad (4)$$

where

$$[\theta]_\lambda C_T = \sum_i K_i \theta_i C_1^i + \beta C_T^2$$

The modes of association for both glucagon and apoA-II protein (Cm apo A-II) have been determined by analytical ultracentrifugation. Data for the temperature-dependent mean residue ellipticity as a function of concentration for these two systems were taken from Osborne et al.[10], and Formisano et al.[11].

TABLE II

ULTRACENTRIFUGAL EQUILIBRIUM SEDIMENTATION STUDIES ON GLUCAGON ASSOCIATION

Except for a variation in BM_1 values, the two model systems fit the data equally well. The sum of the squares of the deviations of K_3 and $\hat{k}$ runs 0.001 to 0.006 from the least square fit of data. Equilibrium constants and BM_1 for the association of glucagon monomer to trimer are determined at 20°C in 0.2 M phosphate buffer, pH 10.0, at a rotor speed of 24,630 rev/min.

	monomer-trimer association				
conc (g/dl)	0.10	0.14	0.18	0.22	
BM_1(dl/g)	-0.68 ± 0.02	-0.89 ± 0.02	-0.92 ± 0.02	-0.90 ± 0.05	
K_3(ml/mg)^2	1.48 ± 0.03	1.51 ± 0.02	1.53 ± 0.03	1.52 ± 0.02	
	isodesmic association				
conc (g/dl)	0.06	0.100	0.140	0.180	0.220
BM_1(dl/g)	-2.02 ± 0.001	-1.41 ± 0.02	-1.09 ± 0.03	-0.88 ± 0.04	-0.73 ± 0.02
$\hat{k}$ (dl/g)	3.86 ± 0.03	4.08 ± 0.07	4.11 ± 0.02	4.25 ± 0.05	4.26 ± 0.01

3. Analytical ultracentrifugation of protein samples

The glucagon samples used for this work were obtained from Eli Lilly Co., Indianapolis, Indiana, Lot NDC0002-1450-series, U.S.P. 1 unit.

The α-chymotrypsin used in these studies was obtained as a 3X crystallized product from Worthington Biochemical Corporation (Lot No. CDS 55J402X) and was used without further purification. It was noted that during the passage of α-chymotrypsin through the gel column scanner[44] in a large zone experiment, that both the leading and trailing edges were biphasic. The corresponding difference profiles exhibited two peaks, with the minor peak moving at a rate corresponding to that of the inclusion volume marker glycylglycine.

All sedimentation equilibrium experiments were performed with a Spinco/Beckman Model E analytical ultracentrifuge equipped with an RTIC unit. The Yphantis six-channel cell[45] with sapphire windows was employed in all experiments. All measurements were made at 20°C. Initial concentrations in terms of fringes were measured separately for each sample, employing a capillary-type synthetic boundary centerpiece.

The speeds were calculated from odometer readings. Type I-N CID-5 spectroscopic plates were used to record the Rayleigh interference patterns, which were analyzed on a Nikon 10X microcomparator.

Determination of the molecular weight distribution as a function of concentration was based on $(M_1/M_{wapp}C) = (\sum_i i K_i C_1^{\,i})^{-1} + BM_1$ and the weight fraction of monomer, f_1, computed based on ξ and η values as previously described[17]. M_{wapp} versus C data were fitted to $^1/M_{wapp} = {}^1/M_{w(c)} + 2\beta_1 C + 3\beta_2 C^2$ by SAS least square analysis[4]. Data for the temperature-dependent sedimentation coefficient as a function of concentration were taken from Aune et al.[6], and extensively compared with our own sedimentation data. The two sets of data were in good agreement.

TABLE III

ULTRACENTRIFUGAL EQUILIBRIUM SEDIMENTATION STUDIES ON α-CHYMOTRYPSIN ASSOCIATION IN 0.2M NaCl, 0.01M SODIUM ACETATE, pH 4.3 at 20°C. RUNS WERE MADE AT 24,030 rev/min FOR 24 HOURS WITH AN INITIAL CONCENTRATION OF 0.8mg/ml and 1.2mg/ml. SAMPLES USED FOR THIS WORK WERE OBTAINED FROM WORTHINGTON BIOCHEMICAL CORPORATION, CDS 55J402X, 3X CRYSTALLIZED.

NOTE: Sum of the square of deviation of K_2 and $\hat{k}$ runs 0.001 to 0.006 from the least square fit of data. Non-ideality (BM_1) was analyzed based on $[M_1/M_{wapp}C] = [\sum_i i K_i C_1^{\,i}]^{-1} + BM_1$ and weight fraction of monomer, f_1, was based on ξ and η[17]. A standard plot of $(1-\sqrt{f_1})$ versus Cf_1 was applied to isodesmic association.

Conc (mg/ml)	Monomer-Dimer K_2(ml/mg)	Monomer-Dimer BM_1(ml/mg)	Isodesmic Association $\hat{k}$	Isodesmic Association BM_1 (ml/mg)
0.5	0.98	-0.038	0.51	0.130
1.0	1.17	-0.036	0.48	0.127
1.5	1.24	-0.023	0.47	0.097
2.0	1.26	-0.020	0.46	0.086
2.5	1.26	-0.015	0.46	0.078
3.0	1.28	-0.010	0.47	0.061
mean	1.20 ± 0.01	0.020	0.48 ± 0.05	0.100
STD deviation	1×10^{-3}	3×10^{-3}	6×10^{-3}	0.0133

4. Thermodynamic Analysis of the Non-linear Van't Hoff Expression.

The free energy of association is expressed as a series expansion as a function of temperature, namely, $\ln Keq = -\frac{\Delta H^\circ}{R}(\frac{1}{T}) + \alpha + \ell T + \gamma T^2$ (non-linear Van't Hoff expression). Multiplying by (-RT) yields the standard free energy change ΔG° as a function of temperature, i.e.,

$$-RT\,\ell n\,Keq = \Delta H^\circ_0 - \alpha RT - \ell RT^2 - \gamma RT^3 \qquad (5)$$

Letting $A = \Delta H^\circ_0$, $B = -R\alpha$, $C = -R\ell$ and $D = -R\gamma$, this equation (5) may be expressed as,

$$\Delta G^\circ = A + BT + CT^2 + DT^3 \qquad (6)$$

A linear, optimized regression analysis of equation (5) is performed in order to obtain the free energy values, a procedure which may be used with some success provided that the equilibrium constant and second virial coefficient, β, are precisely evaluated as a function of temperature at different concentrations.

The first derivative of equation (5), $\Delta G°$, as a function of temperature will yield the entropy,

$$\frac{\partial \Delta G°}{\partial T} = -\Delta S° = \left(\frac{\partial \Delta H_0^o}{\partial T}\right) - \alpha R - 2\ell RT - 3\gamma RT^2 \qquad (7)$$

noting that $(\partial \Delta H_0^o/\partial T) = 0$.

The second derivative of equation (5) as a function of temperature is:

$$\frac{\partial(\Delta G°/T)}{\partial(1/T)} = \Delta H° = \frac{\partial}{\partial(1/T)}\left[\frac{\Delta H_0^o}{T} - R\alpha - R\ell T - R\gamma T^2\right] \qquad (8)$$

Thus, $$\Delta H° = \Delta H_0^o + R\ell T^2 + 2R\gamma T^3 \qquad (9)$$

The second derivative of equation (8) as a function of temperature may be evaluated for the change in heat capacity term,

$$\frac{\partial \Delta H°}{\partial T} = \Delta C_p^o = \left(\frac{\partial \Delta H_0^o}{\partial T}\right) + 2R\ell T + 6R\gamma T^2 \qquad (10)$$

It should be noted that in dealing with the second derivative of a series expansion of free energy changes as a function of temperature, it is essential to perform residual error analysis at each step of the procedure in order to ascertain the validity of the resulting values. It is imperative to cover a temperature range in which there is adequate curvature in the dependence of the free energy of association as a function of temperature. Furthermore, since the heat capacity change as a function of temperature is determined from the second derivative of equation (8), extremely precise values of K are needed to depict the dependence of ΔC_p^o on temperature.

5. Thermodynamic Analysis of the Gibbs-Duhem and Van't Hoff Expressions

The Gibbs free energy of association, as a function of the activity of the salt (A_x), activity of water (A_w), activity of the hydrogen ion concentration (A_{H+}), temperature (T) and pressure (P), is expressed as $G = f(\mu_{a_x}, \mu_w, \mu_{H^+}, T, P)$.

The dependence of the equilibrium constant on the free energy change, $\Delta G°$, the solvent composition, pH, pressure and temperature may be expressed in general form as[46]:

$$d\,\ell n\,K_{eq} = \left.\frac{\partial\,\ell nK}{\partial\,\ell nA_x}\right|_{T,P,A_{H+}A_w} d\,\ell n\,A_x + \left.\frac{\partial\,\ell nK}{\partial\,\ell nA_w}\right|_{T,P,A_x,A_{H+}} d\,\ell n\,A_w + \left.\frac{\partial\,\ell nK}{\partial\,\ell nA_{H+}}\right|_{T,P,A_x,A_w} d\,\ell n\,A_{H+}$$

$$+ \left.\frac{\partial\,\ell nK}{\partial\,T}\right|_{P,A_x,A_w,A_H^+} dT + \left.\frac{\partial\,\ell nK}{\partial\,P}\right|_{A_x,A_w,A_H^+,T} dP$$

A) The Gibbs-Duhem expression at constant A_{H+}, T and P is $d\,\ell n\,Keq$

$$d\,\ell n\,Keq = \left.\frac{\partial\ell nK}{\partial\ell nA_x}\right|_{A_w} d\,\ell n\,A_x + \left.\frac{\partial\,\ell nK}{\partial\,\ell nA_w}\right|_{A_x} d\,\ell n\,A_w$$

where $(\partial\ell nK/\partial\ell nA_x) = \Delta\bar{\nu}_x$, $(\partial\,\ell nK/\partial\ell nA_w) = \Delta\bar{\nu}_w$ and $d\,\ell n\,A_w = -\left(\frac{n_x}{n_w}\right) d\,\ell n\,A_x$;

n_x and n_w are designated as the total number of moles of Hofmeister salt (x) and water (w) in the solution.

B) The Van't Hoff expression at constant A_{H^+}, A_x, A_w and P is $d\,\ell n\ Keq = (\frac{\partial \ell nK}{\partial T})dT$ where $(\partial\,\ell nK/\partial T) = (\Delta H°/RT^2 + \beta + 2\gamma T\ )$

C) At constant A_{H^+}, A_{w_2} and P,[47,48] $d\,\ell n\ Keq = (\frac{\partial \ell nK}{\partial T})_{A_x} dT + (\frac{\partial \ell nK}{\partial \ell nA_x})_T\ d\,\ell n\ A_x$ where $(\partial \ell n\gamma_{\pm}/\partial T) = -\bar{L}_x/\nu RT^2$. $\bar{L}_x = \lambda_{(H)}\sqrt{C}$. Knowing $\bar{L}_x$ equals the relative partial molal heat content of the Hofmeister anion or salt, one can obtain the mean molal activity coefficient ($\gamma_{\pm}$) as a function of temperature, where $\lambda_{(H)}$ = the limiting slopes for the relative partial molal heat content and $\nu = \nu_+ + \nu_-$, ionic constituents.

D) At constant T,P,A_x and A_w,[48,49] $d\,\ell n\ Keq = (\frac{\partial\ \ell nK}{\partial\ \ell\ nA_{H^+}})d\,\ell n\ A_{H^+}$, where $(\partial \ell nK/\partial \ell nA_{H^+}) = \Delta\bar{\nu}_{H^+}$, noting that the pH dependence of temperature is evaluated from $pH(S) = [\frac{A}{T} + B + CT + DT^2]$[49].

$$d\,\ell n\ K_{eq} = (\frac{\partial \ell n\gamma\pm}{\partial T}) - \frac{1}{2.303}\ \frac{\partial}{\partial T}\ [\ \frac{A}{T} + B + CT + DT^2]$$

E) At constant T, A_x,A_w, and A_{H^+}, $d\,\ell n\ Keq = -\ (\frac{\Delta V}{RT})dP$ [50].

F) The combined Gibbs-Duhem and Van't Hoff expression[1] is

$$d\ \ell n\ Keq = (\frac{\partial \ell nK}{\partial \ell nA_x})_{A_w,T}\ d\ell n\,A_x + (\frac{\partial \ell nK}{\partial \ell nA_w})_{A_x,T}\ d\ \ell n\ A_w + (\frac{\partial \ell nK}{\partial T})_{A_x,A_w}\ dT \tag{11}$$

We are concerned only with the three terms shown in this equation as applied to 1:1 electrolytes. Robinson and Harned[47] defined

$$d\ \ell n\ A_x = [\frac{A}{T^2} - \frac{B}{T}]\ dT \tag{12}$$

for computation of the activity coefficient of a NaCl solution from 0°C to 100°C, where $d\,\ell n\ C_x = 0$. Therefore $d\,\ell n\ A_w = -\ (\frac{n_x}{n_w})\ d\,\ell n\ A_x$ as a function of temperature yields,

$$d\ \ell n\ A_w = -\ (\frac{n_x}{n_w})\ [\frac{A}{T^2} - \frac{B}{T}]\ dT \tag{13}$$

Note that A and B are temperature coefficients of 1:1 electrolytes, substitution of equations (12) and (13) into equation (11) gives,

$$d\ \ell n\ \ Keq = \Delta\bar{\nu}_x[\frac{A}{T^2} - \frac{B}{T}]dT - (\frac{n_x}{n_w})\Delta\bar{\nu}_w[\frac{A}{T^2} - \frac{B}{T}]\ dT + [\frac{\Delta H°}{RT^2} + \beta + 2\gamma T]dT \tag{14}$$

Integration of equation (14) yields,

$$\ell n\ Keq = \Delta\bar{\nu}_{pref}\ [-\ \frac{A}{T} - B\ \ell n\ T + \xi] + [-\ \frac{\Delta H°}{RT} + \alpha + \beta T + \gamma T^2] \tag{15}$$

where $\xi = [\Delta\bar{\nu}_{pref}\ (I + \ell n\ C_x\ )]$. Therefore,

$$\Delta G° = (-RT)\Delta\bar{\nu}_{pref}\ \ell n\ A_x\ + \Delta H^\circ_0 + (-\alpha R)T + (-\beta R)T^2 + (-\gamma R)T^3 \tag{16}$$

Hence,
$$\ell\,n\ Keq = [\Delta\bar{\nu}_x - (\frac{n_x}{n_w})\Delta\bar{\nu}_w]\ \ell\,n\ A_x + \int_{T_i}^{T_f} (\frac{\partial \ell nK}{\partial T})\ dT \tag{17}$$

$$\begin{aligned} \Delta H° &= \ell(BT - A) + \Delta H^\circ_0 - \beta' T^2 - 2\gamma' T^3 \\ \Delta S° &= \ell[B(\ \ell n\ T\ + 1) - I - \ \ell n\,(C_x)]\ -2\beta'\ T - 3\gamma'\ T^2 - \alpha' \\ \Delta C^\circ_p &= \ell B - 2\beta' T - 6\gamma'\ T^2 \end{aligned} \tag{18}$$

where $\ell = -R\Delta\nu_{pref}$, $\alpha' = -R\alpha$, $\beta' = -R\beta$ and $\gamma' = -R\gamma$

It should be noted in dealing with equation (16), which is a function of temperature, that it is essential to establish the activity coefficient of the ion pairs ($\gamma_{\pm}$) as a function of temperature. Such a determination has been limited to the examination of only a few 1:1 electrolytes to date[47,48].

6. Effect of a Hofmeister Series of Anions on the Self-association Reactions

Since the value of $\Delta\bar{\nu}_{pref}$ is due to the change in the equilibrium constants between a Hofmeister series of anions and the associating protein, in principle it should be possible to relate and evaluate with a high degree of accuracy the preferential binding of salt and water. Based on equation (17)

$$\ell n\ Keq = \Delta\bar{\nu}_x\ \ell n\ A_x + \Delta\bar{\nu}_w\ \ell n\ A_w + C \qquad (19)$$

where $C = \int_{T_i}^{T_f} (\partial \ell nK/\partial T)\ dT$. C remains constant when $\frac{\partial\ \ell n\ K}{\partial \ell n\ A_{H^+}}$ and dT are zero, fitting $\ell n\ A_x$ and $\ell n\ A_w$ to $\ell n Keq$ (where $\ell n Keq$ is a function of $\ell n\ A_x$ at a given temperature). $\ell n\ A_x$ values are taken from Robinson and Stokes[48], noting that $\ell n\ A_w$ is calculated from $-[n_x/(55.5 - n_x)]\ \ell n\ A_x$.

7. Methods of Computation

All calculations in the computation of the Gibbs-Duhem and Van't Hoff expressions were done using the general linear models (SAS GLM procedure, Barr et al., 1976), which use the principle of least squares to fit a fixed-effect linear model to virtually any type of univariate and multivariate analysis, including simple linear regression, multiple linear regression, analysis of variance, analysis of covarience, and partial correlation analysis.

The statistical analysis system language, interfaced with PL/1, was utilized for all routines (GLM 127, GLM 131). The sub-routines used in our computations were (1) proc scatter, (2) proc print, (3) proc GLM, (4) proc sort.

The stepwise regression procedure (STE 251, Barr et al., 1976) which was applied to our analysis of thermodynamic parameters includes five techniques to find that variable of a collection of independent variables which is most likely to be included in a regression model. This method of computation was extremely useful for data screening, permitting some insight into the relative strengths of the relationship between proposed independent variables and dependent variables (largest R^2 statistic).

The five-part procedure includes: 1) a forward selection technique which finds first the single-variable model, 2) a backward elimination technique which is first performed for a model including all the independent variables, 3) stepwise modification of the forward selection, 4) the maximum R^2 improvement and 5) minimum R^2 improvement. The fourth and fifth procedures produce models which fit the data equally well in our computation.

RESULTS

1. Bovine liver L-glutamate dehydrogenase association; molecular weight distribution as a function of temperature

In a recent publication, Reisler and Eisenberg[8,13,14] have found by light scattering measurements of different concentrations of bovine L-glutamate dehydrogenase that the equilibrium constant varies as a function of temperature. Using their data and our molecular sieve chromatography data, we have recalculated the distribution of the weight fraction of monomer undergoing isodesmic association as a function of temperature. The correlation coefficient of each value at each concentration was found to be 0.9999.

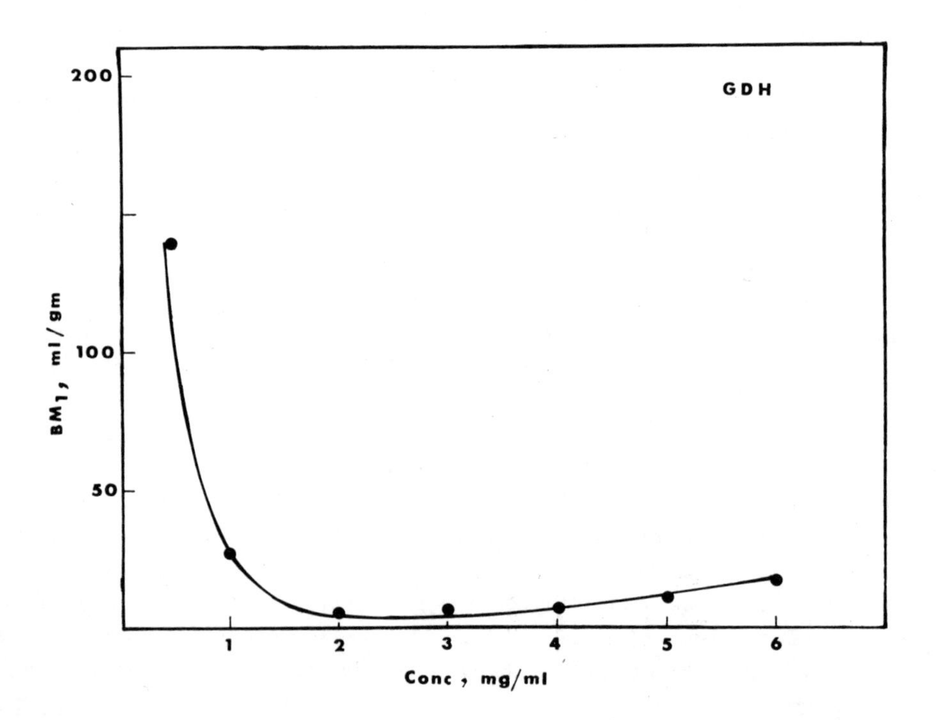

Figure 1. Variation of BM_1 (ml/gm, β = (ml - mole/gm^2) as a function of concentration (GDH in 0.2 M phosphate buffer containing 1 x 10^{-4}M EDTA at 20°C), computed from a plot of the weight fraction of monomer[17] using combined molecular sieve and molecular weight data. Molecular weight data were taken from Reisler and Eisenberg[14]. The evaluation of confidence limits of each BM_1 value were within the limit of a correlation coefficient of 0.9999.

The non-ideality term BM_1 is also evaluated from molecular weight and partition data at each concentration. It is apparent that distribution of the weight fraction monomer at a given concentration first decreases, then increases. Thus it appears that variation of the equilibrium constant is influenced not only by concentration, but also by non-ideality and temperature. Typical variation in BM_1 as a function of concentration is shown in Figure 1.

The thermodynamic parameters of bovine liver L-glutamate dehydrogenase given in Tables IV and V indicate that at the enthalpic temperature ($T_{\phi h}$) of 303 K, the enthalpy of the solution becomes zero. At higher temperatures, the enthalpy has a negative value, as shown in Figure 2.

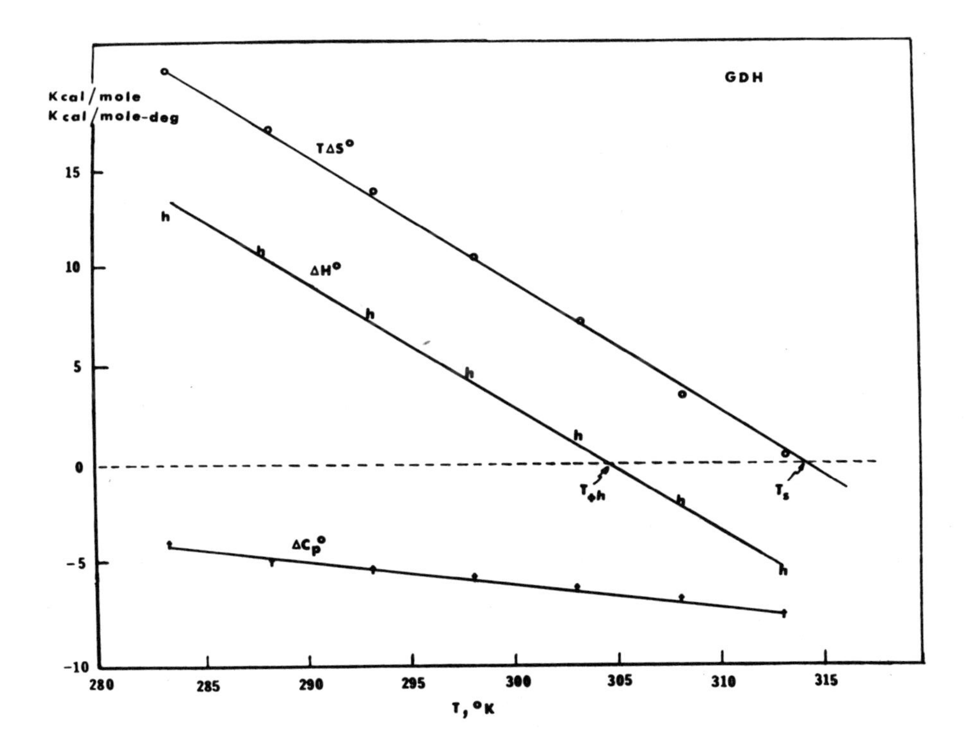

Figure 2. Thermodynamic plot of bovine liver L-glutamate dehydrogenase (GDH) association as a function of temperature, with thermodynamic temperatures indicated on the plot. $\Delta S° \times 10^{-3}$ (cal/mole-degree), $\Delta C_p^\circ \times 10$ (Kcal/mole-deg). Each data point was evaluated with interpolation of F-statistics. The sum of the square residuals is close to zero.

The entropy of the solution reaches zero at 313 K, while the harmonious temperature (T_H), at which point the Gibbs free energy change is at a maximum, occurs at 302 K. Values for the thermodynamic temperatures of bovine liver L-glutamate dehydrogenase and other associating systems which we examined are summarized in Table V.

Initially, then, this isodesmic association is characterized by a positive enthalpy change and a larger positive entropy change, while the Gibbs free energy change of the reaction is negative. As the solution approaches 303 K, the enthalpy

TABLE IV

THERMODYNAMIC PARAMETERS OF BOVINE LIVER L-GLUTAMATE DEHYDROGENASE AS A FUNCTION OF TEMPERATURE (NON-IDEAL CASE)

NOTE: $\Delta G° = A + BT + CT^2 + DT^3$, where A = -50.02, B = 0.80, C = -39×10^{-3}, D = 55×10^{-6}, and C/D = 709. Mean residue square = 0.9997 and mean square of error= 5.2×10^{-5}. The sum of the square of the deviation = 0.006.

Temp, in °K	K (ml/mg)	$\Delta G°$ (Kcal/mole)	$\Delta H°$ (Kcal/mole)	$\Delta S°$ (Kcal/mole-deg)	$\Delta C_p°$ (Kcal/mole-deg)
283	1.29	-7.27	12.55	0.0701	-0.429
288	1.62	-7.53	10.21	0.0616	-0.484
293	1.95	-7.77	7.59	0.0524	-0.541
298	2.10	-7.95	4.68	0.0424	-0.600
303	2.05	-8.06	1.48	0.0315	-0.660
308	1.82	-8.12	-2.04	0.0198	-0.722
318	1.50	-8.14	-5.87	0.0073	-0.785

changes are small, while the major contributor to the Gibbs free energy of interaction is the $T\Delta S°$ term. As may be seen in Table IV, the maximum Gibbs free energy value is reached at T_H, 302 K. Over the temperature range studied, $\Delta C_p°$ varied from -430 to -785 cal/mole-degree, noting that T_H should be equivalent to T_S.

The large changes in the BM_1 value (seen in Figure 1) and the distribution of the weight fraction of monomer (from Table V*) strongly suggest that the effects of non-ideal behavior of the solute are much greater at low temperature than at high temperature.

2. Glucagon Association: The mean residue ellipticity as a function of temperature.

The thermodynamic parameters of glucagon association, which exists in an equilibrium between monomer and trimer, have been reevaluated from the data of Formisano et al.[11], as shown in Tables IV and VI. After evaluating the equilibrium constants as a function of temperature and the second virial coefficient, we regenerated a theoretical curve of the mean residue ellipticity as a function of concentration.

We found that the mean residue ellipticity of monomer varied from -4550 ± 30 at 285 K ($T_{\phi h}$) to -5710 ± 32 at 318 K, and for the trimer from -6570 ± 54 at 285 K ($T_{\phi h}$) to -10,530 ± 206 degrees cm^2/decimole at 318 K. The second virial coefficient is seen to vary from -1030 to -250 deg-cm^2-ml/decimole-mg over the temperature range studied, as shown in Figure 3. Although the variation in the second virial coefficient as a function of temperature was quite pronounced between the ideal and non-ideal case, the overall variation in the thermodynamic parameters of glucagon at a concentration of 3.132 mg/ml was only 5.6 percent. As may be seen in Tables IV and VI, the enthalpy change, $\Delta H°$, is small and positive at 283 K, but becomes negative and continues to increase as a function of temperature over the range studied.

*Table V [illegible]

TABLE VI

THERMODYNAMIC PARAMETERS OF GLUCAGON ASSOCIATION (NON-IDEAL CASE). ALL CALCULATIONS WERE MADE AT A GLUCAGON CONCENTRATION OF 3.132 mg/ml.

NOTE: A = 216, B = 1.79, C = -4.4×10^{-3}, D = 3.0×10^{-6} and C/D = 1467. The mean residue square = 0.9929, mean square of error = 0.047. The sum of the square deviation = 0.008.

Temp, °K	$K(ml/mg)^2$	$\beta(\frac{deg\text{-}cm^2\text{-}ml}{decimole\text{-}mg})$	ΔG° (Kcal/mole)	ΔH° (Kcal/mole)	ΔS° (Kcal/mole-deg)	ΔC_p° (Kcal/mole-deg)
283	22.86	-1029	-6.34	0.04	0.0263	-1.0440
288	22.54	-1092	-6.44	-5.16	0.0081	-1.0364
293	25.05	-1050	-6.64	-10.31	-0.0097	-1.0280
298	7.135	-1020	-5.99	-15.43	-0.0279	-1.0186
303	6.886	-1004	-5.87	-20.50	-0.0439	-1.0083
308	6.046	-770	-5.70	-25.52	-0.0603	-0.9972
313	2.218	-479	-5.56	-30.48	-0.0762	-0.9851
318	0.598	-257	-4.82	-35.37	-0.0918	-0.9721

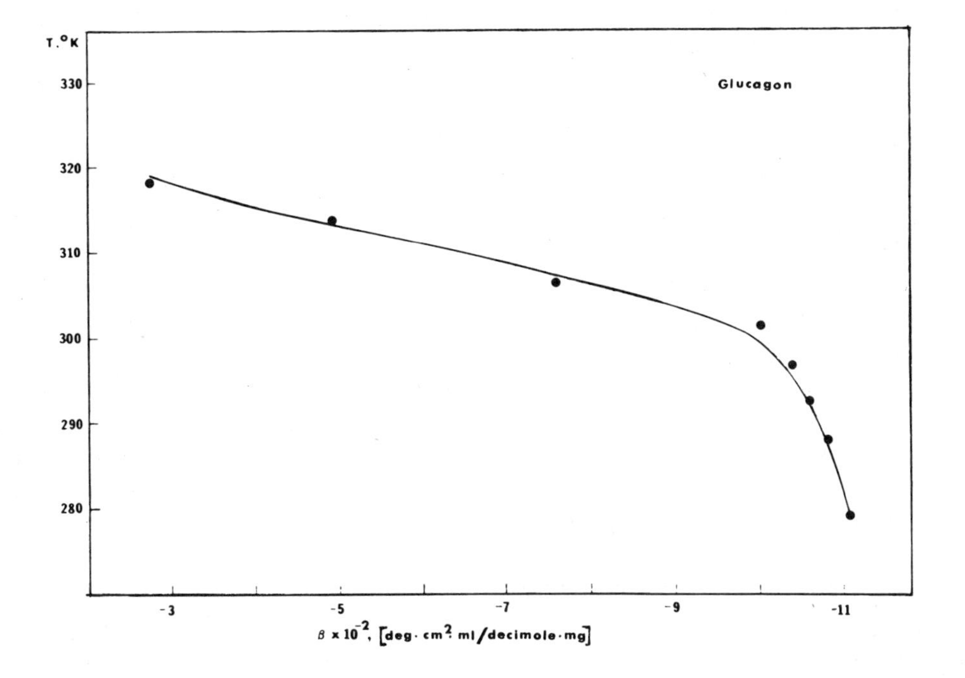

Figure 3. Second virial coefficient as a function of temperature in the glucagon monomer-trimer equilibrium (ascertained by analytical equilibrium sedimentation). β = deg-cm^2-ml/decimole-mg. The data were fitted equally well to a model describing isodesmic association.

The characteristic feature of this monomer-n-mer association appears to be the domination of the free energy change by the entropy term at temperatures below 290 K, where the enthalpy term is positive. At higher temperatures, the association is more distinctly driven by enthalpy. The Gibbs free energy of association is at a maximum at 287 K (T_H).

The heat capacity change, ΔC_p°, is seen to decrease gradually from -1044 cal/mole-degree at 283 K to -972 cal/mole-degree at 318 K. There is a strong correlation between the change in heat capacity and the corresponding alteration in the second virial coefficient, suggesting that there is a significant change in structural conformation over this temperature range upon association of the monomeric units into trimer ($\sim$ 300 K).

3. ApoA-II association; the mean residue ellipticity as a function of temperature.

The thermodynamics of the non-linear Van't Hoff expression of the monomer-dimer association of the reduced carboxymethylated form of apoA-II protein from human high density lipoproteins have been described for the ideal case[10]. Using the experimental data for the effect of temperature on the mean residue ellipticity of apo A-II at 220 nm, we have extended these thermodynamic considerations to the non-ideal case. Although it is usually necessary to change the pressure, pH, salt or solvent composition with more stable proteins to observe the difference in heat capacity (ΔC_p°) change in the equilibration between native and denatured states (two-state model)[5,7,51-58], the free energy of association of Cm apo A-II can be measured over a wide range of temperatures without modifying the solvent, making it an excellent choice for thermodynamic studies.

The thermodynamic parameters of self-associating S-carboxymethylated apo A-II protein in the non-ideal case at a low concentration of 0.137 mg/ml are shown in Tables IV and VI. The mean residue ellipticity of monomer varied from -4320 $\pm$ 50 at 283 K to -7450 $\pm$ 50 at 320 K, and for the dimer from -13,150 at 283 K to -12,850 at 320 K. The second virial coefficient, as seen from Tables IV and VII and Figure 4, varied from -1481 to -1023 deg-cm^2-ml/decimole-mg, decreasing exponentially over the temperature range of 280$\sim$320 K. ΔC_p° varied from -2300 to $\pm$ 23 cal/mole-degree in this temperature range. The experimentally obtained value of ΔC_p° from our computation is -1056 cal/mole-deg at 299 K (T_H). Thermodynamic temperatures of $T_{\phi h}$ and T_s were found to be 298 K and 303 K respectively.

4. Thermodynamic Temperatures

As seen from Table III, we observed a variation of some three degrees Kelvin between values of T_H, the harmonious temperature, and T_S, the entropic temperature, in each of the three associating systems we examined. By definition, however, the two temperatures are equivalent, since $d\Delta G^\circ/dT = 0$ where

$$T_H^2 + \left(\frac{2\ell}{3\ell}\right)T_H + \frac{\alpha}{3\gamma} = 0 \quad \text{or} \quad T_S^2 + \left(\frac{2\ell}{3\gamma}\right)T_S + \frac{\alpha}{3\gamma} = 0$$

$$T_{\phi h}^3 + \frac{\ell}{2\gamma}\, T_{\phi h}^2 + \frac{\Delta H_0^\circ}{2\gamma R} = 0, \text{ where } d(\Delta G^\circ_{(T)}/T)/d(1/T) = 0$$

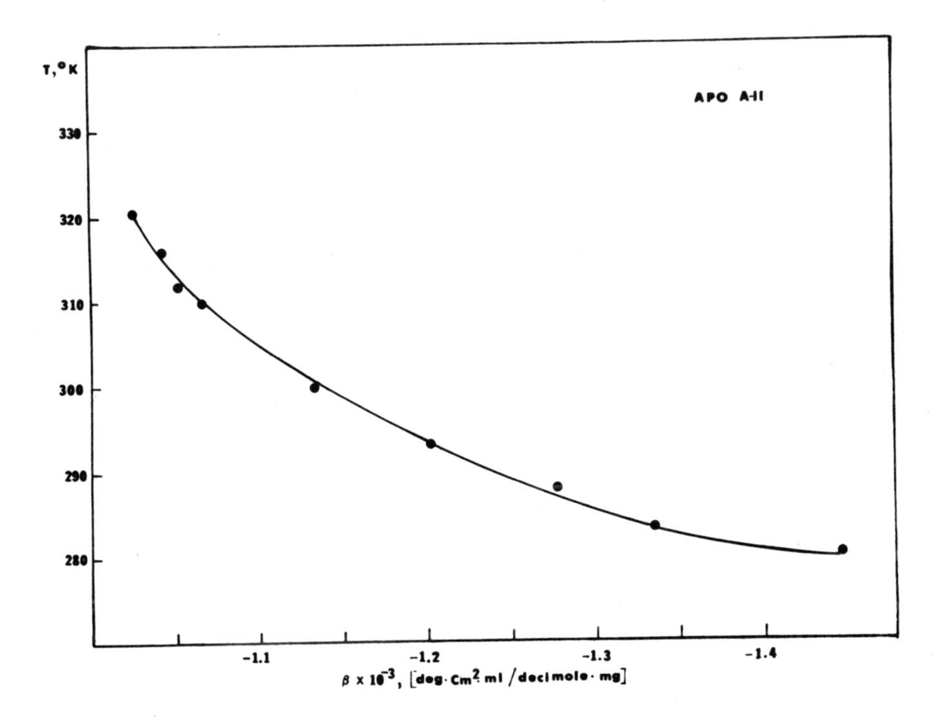

Figure 4. Second virial coefficient of cm apo A-II protein as a function of temperature in 0.01 M phosphate buffer, pH 7.4. These values were then used to recalculate the thermodynamic parameters listed in Table VI.

Gill and Wädso[24] have pointed out that at a temperature we designate as $T_{\phi}h$, the enthalpy of the solution becomes zero; that is, there is zero heat of solution from the liquid phase of hydrocarbon in water. Theoretically, then, $T_{\phi}h \neq T_H$.

In order to accurately determine the temperature of compensation, $\langle T_c \rangle$, where $\langle T_c \rangle = d\Delta H^\circ_T / d\Delta S^\circ_T$, we plotted values of $\Delta H^\circ_{(T)}$ versus $\Delta S^\circ_{(T)}$. In the resulting linear relationship, shown in Figure 5, we observed that, within the allowances of statistical error, the temperatures of compensation for glucagon, apo A-II and bovine liver L-glutamate dehydrogenase are equivalent at 295 ± 2 K, as seen from Table V.

5. Monomer-dimer association of α-chymotrypsin as affected by a Hofmeister series of anions.

The molecular weight as a function of concentration distribution of α-chymotrypsin was examined. Several model systems were used to regenerate the curve representing $(M_1/M_{wapp}C) = [\sum_i iK_iC_1{}^i] + BM_1$ and $1/M_{wapp} = \frac{1}{M_{w(c)}} + 2B_1C + 3B_2C^2$. At each concentration, we found the molecular weight distribution shows a constant increase, with no plateau region observed.

TABLE VII

THERMODYNAMIC PARAMETERS OF S-CARBOXYMETHYLATED apo A-II (cm Apo A-II) ASSOCIATION. Cm Apo A-II CONCENTRATION OF 0.1373 mg/ml (NON-IDEAL CASE).

NOTE: A = 138.9, B = 0.88, C = 1.1×10^{-3}, D = 7.8×10^{-7} and C/D = 1222. Mean residue of square = 0.9995. Mean square of error = 0.047 and the sum of the square deviation = 0.005.

Temp, °K	K(ml/mg)	$\beta(\frac{\text{deg-cm}^2\text{-ml}}{\text{decimole-mg}})$	ΔG° (Kcal/mole)	ΔH° (Kcal/mole)	ΔS° (Kcal/mole-deg)	ΔC_P° (Kcal/mole-deg)
280	1.88	-1482	-5.40	18.40	0.0846	-0.9829
285	3.18	-1336	-5.79	13.48	0.0670	-1.0071
290	4.07	-1280	-6.03	8.38	0.0493	-1.0316
295	5.35	-1204	-6.30	3.16	0.0315	-1.0563
300	5.67	-1121	-6.44	-2.18	0.0135	-1.0812
305	4.49	-1066	-6.41	-7.65	-0.005	-1.1064
310	3.50	-1061	-6.36	-13.24	-0.023	-1.1317
315	2.36	-1050	-6.21	-18.97	-0.041	-1.1574
320	1.38	-1023	-5.97	-24.81	-0.060	-1.1932

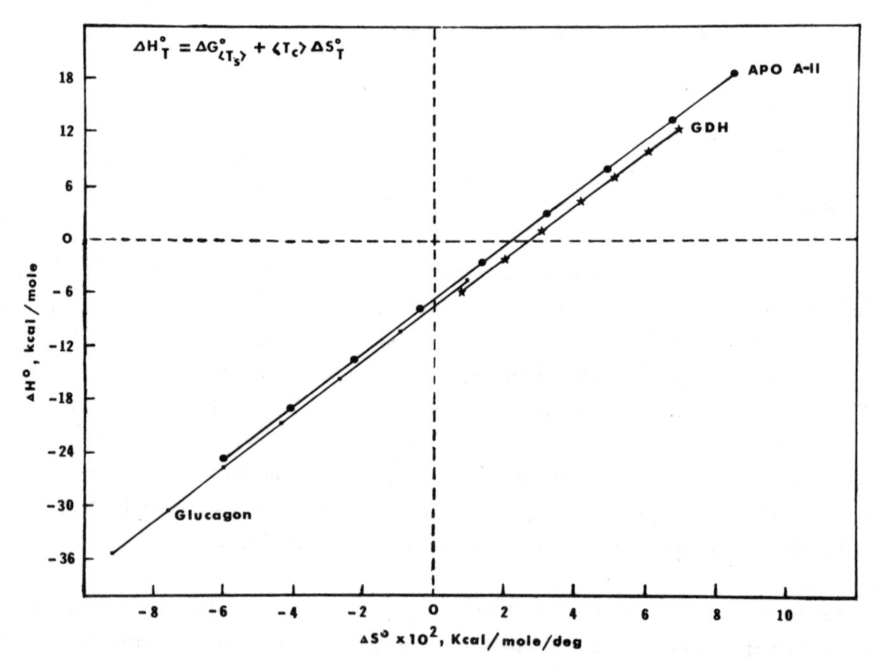

Figure 5. A plot of $\Delta H^\circ_{(T)}$ versus $\Delta S^\circ_{(T)}$ for bovine liver L-glutamate dehydrogenase (GDH), glucagon and cm apo A-II protein, with data points taken from Tables IV, VI and VII. The slope of these lines represents $\langle T_c \rangle$, the compensatory temperature for each associating protein system, as listed in Table V.

We compared the variation in equilibrium constants as a function of concentration for several model systems and found that two possible modes of association -- isodesmic and monomer-dimer -- were equally well fitted to the data, as shown in Table III. In both model systems the value of BM_1 varied as a function of concentration. In the case of isodesmic association BM_1 values were positive, while for monomer-dimer association these values were negative over the concentration range we examined.

Using only the sedimentation methods we have applied, it is difficult to make a choice between the two models. By these procedures the mode of association of α-chymotrypsin cannot be clearly defined. Due to the additional evidence of X-ray crystallographic studies by Birktoft et al.[59], which shows this enzyme to be in the dimeric form, light scattering data[36] and sedimentation coefficient as a function of concentration and pH data[6,35,37] it is probably legitimate to represent the process loosely as monomer-dimer association.

Figure 6 shows a plot of ℓn K versus $\ell n\ A_x$. The lines are theoretical and were determined by a simple GLM least square to fit a fixed-effect linear model, assuming that α-chymotrypsin undergoes monomer-dimer association at pH 4.3. The equilibrium

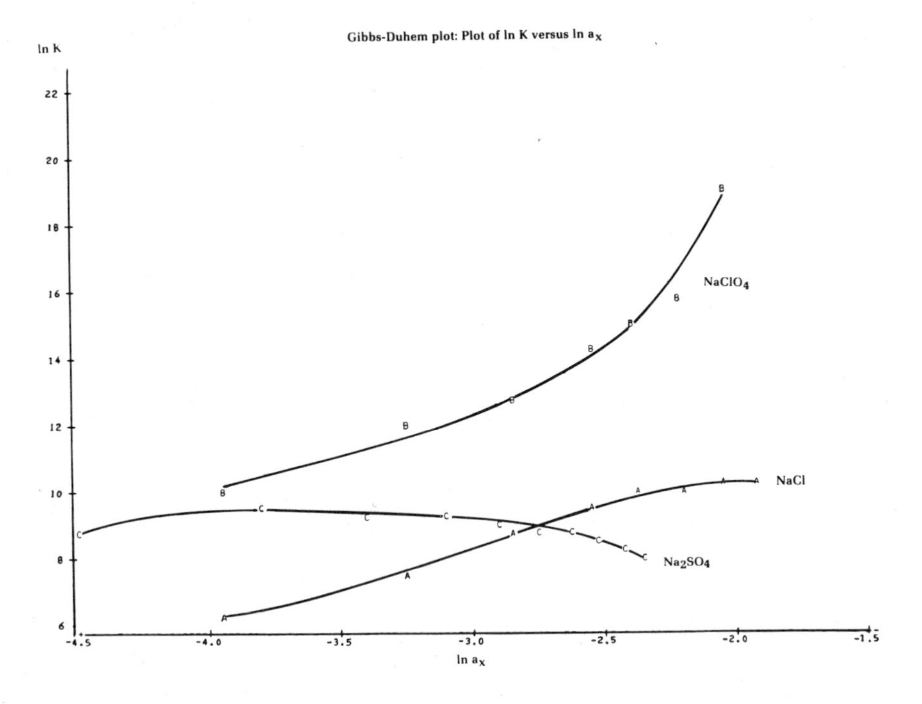

Figure 6. Variation of ℓn Keq as a function of concentration of a Hofmeister series of anions at 25°C. The activity coefficients of various electrolytes were extracted from Robinson and Stokes (1959). Equilibrium constants as a function of electrolytes were determined by sedimentation equilibrium measurements and compared with S°_{20w} of Aune et al.[6].

constants as a function of temperature were recalculated from the data of Aune et al.[6], while $\ell n\, A_X$ values were taken from the tables published by Robinson and Stokes[48]. This figure clearly shows that the salt activity for NaCl, Na_2SO_4 and $NaClO_4$, as a function of the equilibrium constants, is not a linear function at 25°C.

TABLE VIII

THE EFFECT OF VARIOUS SALTS ON PREFERENTIAL BINDING ($\Delta\bar{\nu}_{pref}$) AND SOLVATION PARAMETERS ($\Delta\bar{\nu}_X$ and $\Delta\bar{\nu}_W$) OF α-CHYMOTRYPSIN DIMERIZATION.

NOTE: ℓn K versus $\ell n\, A_X$ values are recalculated based on Aune et al.[6], and extensively compared with our own sedimentation data. $\ell n\, A_X$ values are taken from Robinson and Stokes (1959). $\ell n\, A_W$ was reevaluated from $\ell n\, A_W = -(\frac{n_X}{n_W})\ell n\, A_X$.

The Gibbs-Duhem expression is obtained by combining two osmatic coefficients: $\phi = -(55.51)/\nu m_X \ell n A_W \simeq [1 - \frac{e^2|Z_1Z_2|}{6\varepsilon KT}\kappa\sigma_{(x)}]$ where $\sigma_{(x)} = \frac{3}{x^3}\int_0^x (\frac{x}{1+x})^2\, dx$ and $x = \kappa a$. κ, kappa, is a quantity proportional to the square root of ionic strength, with dimension of 1/length, a is the mean diameter of the ions, K is the Boltzmann constant and ε is the dielectric constant.

$$\phi = +\frac{1000}{\nu W_X m_X}\ell n\, A_X$$, where W_X is the molecular weight of the Hofmeister salt.

Thus, solvent activity, A_X, may be obtained by integration of the Debye-Hückel-Onsager equation, an expression used for evaluation of the electrostatic free energy contribution.

$\Delta\bar{\nu}_{pref}$	NaCl	$NaClO_4$	Na_2SO_4
$(d\,\ell n\, K/d\,\ell n\, A_X)$	+ 2.40	+ 4.64	- 0.18
$\Delta\bar{\nu}_X$	+ 3.6	- 2.30	+ 2.0
$\Delta\bar{\nu}_W$	- 416	+ 2422	- 338
R^2	0.9781	0.9372	0.7897
mean square of error	0.068	0.806	0.010

The effect of various salts on the preferential binding $(\Delta\bar{\nu}_{pref}) = (\Delta\bar{\nu}_X - \frac{n_X}{n_W}\Delta\bar{\nu}_W)$ and solvation parameters $(\Delta\bar{\nu}_W)$ of α-chymotrypsin dimerization are shown in Table VIII. The values of the residue square (R^2) and F-tests show that the data of $\ell n K$ versus $\ell n\, A_X$ for NaCl and $NaClO_4$ fit a linear model within a 95% confidence limit. The fit of the data for Na_2SO_4 was not this close.

The slopes of $(d\,\ell n\, K/d\,\ell n\, A_X) = \Delta\bar{\nu}_{pref}$ at pH 4.3 are 2.4 in NaCl, 4.64 in $NaClO_4$ and - 0.18 in Na_2SO_4. Aune et al.[6], noted that the values of $d\,\ell n\, K/d\,\ell n\, A_X$ at pH 4.1 were about 1.0 in NaCl. Our analysis shows, however, that the change in salt binding to protein for NaCl is in fact greater than 2. The negative slope which we observed for Na_2SO_4 would seem to indicate that this salt acts as a dissociating agent, an assumption which is further supported by the poor fit of data to a linear model.

The values of $\Delta\bar{\nu}_x$ and $\Delta\bar{\nu}_w$ in NaCl solution, shown in Table VIII, are approximately 4 and -400, indicating that some 4 moles of NaCl are bound per mole of dimer while 400 moles of H_2O are expelled during the dimerization of α-chymotrypsin. On the other hand, the $NaClO_4$ solution has a positive value of $\Delta\bar{\nu}_w$ (+ 2400) while $\Delta\bar{\nu}_x$ is negative (- 2.3), indicating that as the concentration of ClO_4^- ions increases, a hydrated perchlorate, perhaps $(ClO_4^-)nH_2O$, is formed which enhances the dimerization of α-chymotrypsin by perturbing the solvation layer of this protein, similar to the action of the L_i^+ ion. The values of $\Delta\bar{\nu}_x$ and $\Delta\bar{\nu}_w$ of Na_2SO_4 are +2 and -338. Since $d\,\ell n\ K/d\,\ell n\ A_x$ is negative in this case, this would indicate a "salting-in" phenomenon which would favor a dissociation reaction. These results contradict the previous assumption that this anion, at the extreme of the Hofmeister series, would salt out hydrophobic groups most effectively[34,60].

Figure 7 shows a plot of the standard free energy change as a function of $\ell n\ A_x$ for α-chymotrypsin dimerization in a 0.178 M NaCl-0.01 M sodium acetate buffer, pH 4.3. The curve shows the influence of a given electrolyte on the free energy change, with each data point representing a different temperature, from calculations based on equation (16). It is distinctly parabolic, with an inflection point at 288 K where $\ell n\ A_x = -2.0159$. At this point $(d\Delta G°/d\,\ell n\ A_x) = \infty$ and the salt activity reaches a maximum (see section 7).

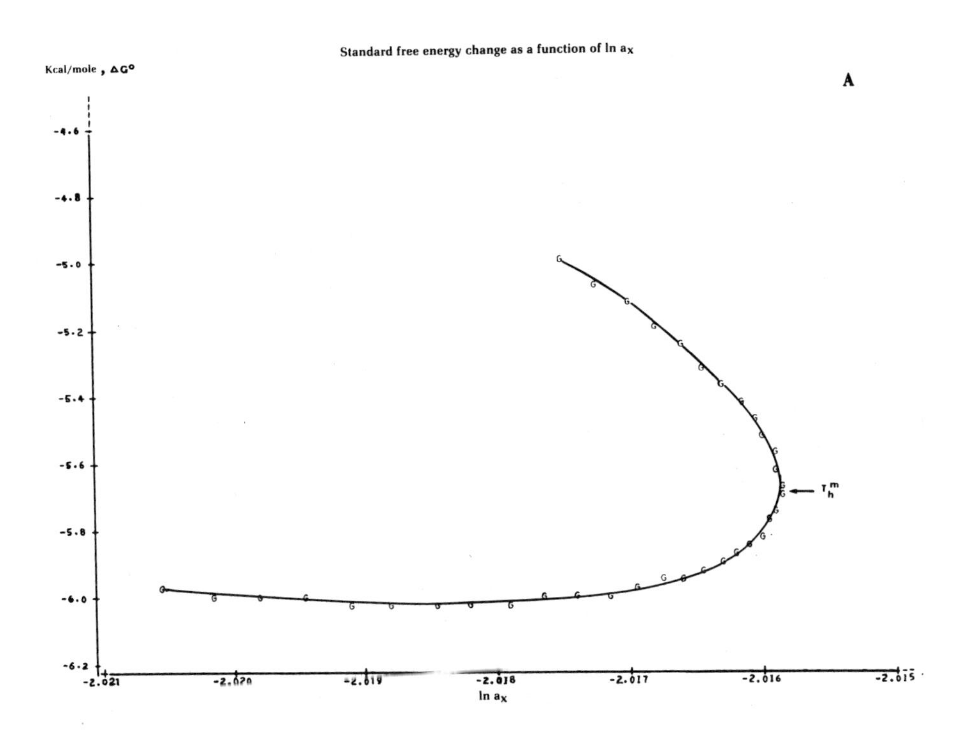

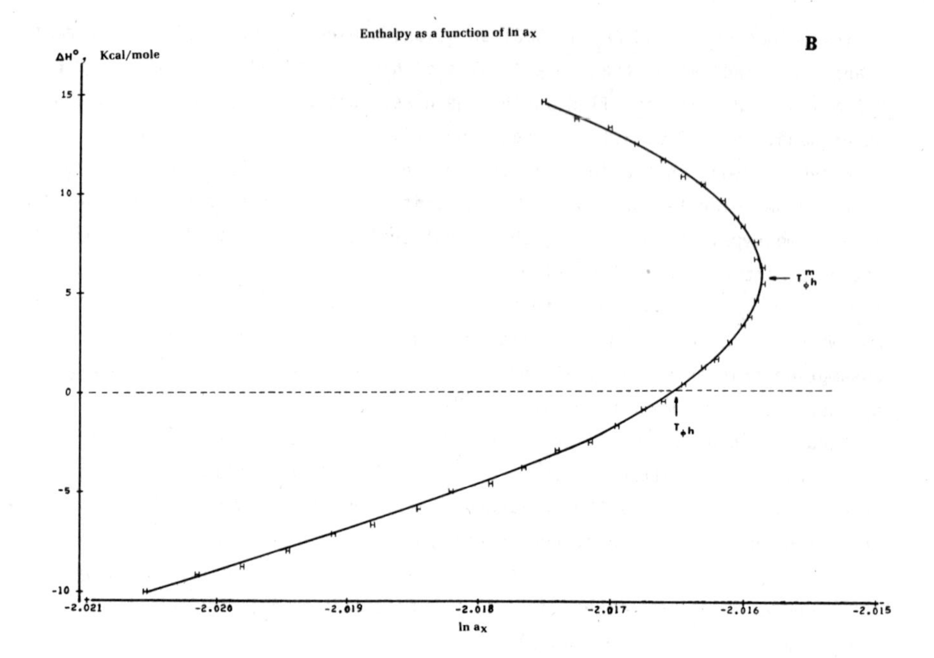
Enthalpy as a function of ln a_X
B
ΔH°, Kcal/mole
15
10
5
0
-5
-10
-2.021
-2.020
-2.019
-2.018
-2.017
-2.016
-2.015
ln a_X

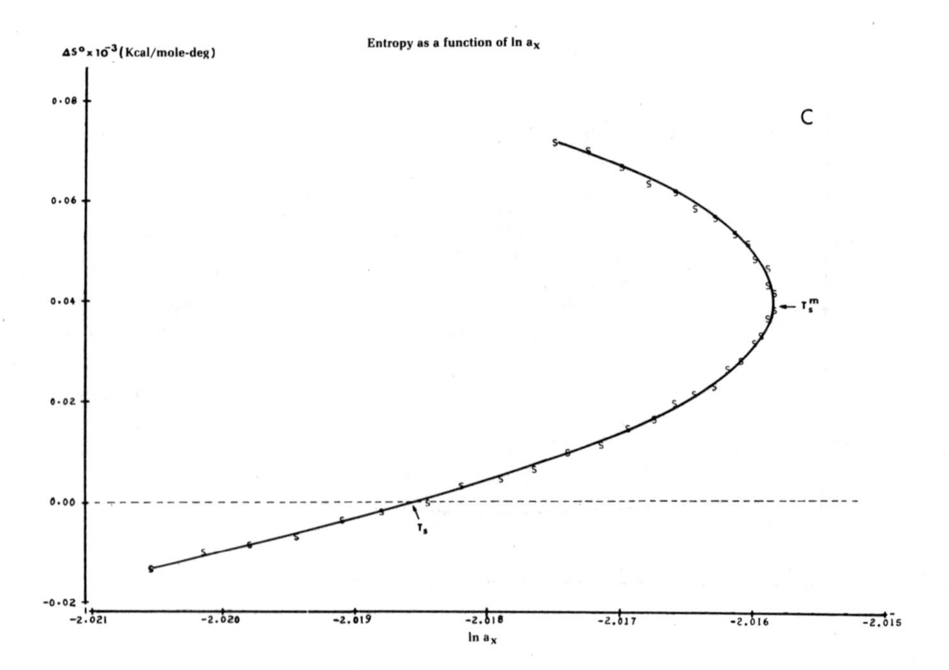
Entropy as a function of ln a_X
ΔS° x 10⁻³ (Kcal/mole-deg)
C
0.08
0.06
0.04
0.02
0.00
-0.02
-2.021
-2.020
-2.019
-2.018
-2.017
-2.016
-2.015
ln a_X

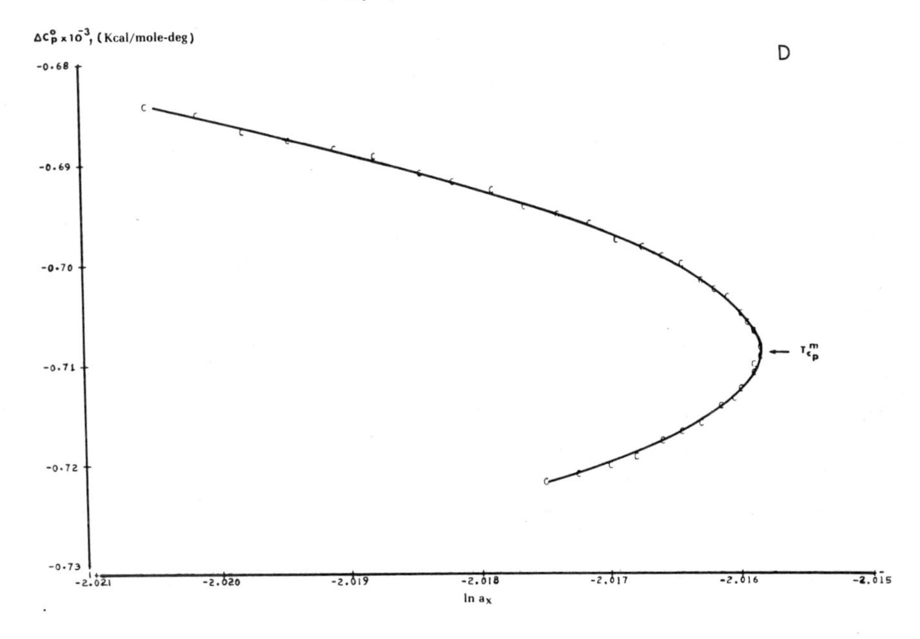

Figure 7A. Dependence of $\Delta G°$, the Gibbs free energy change, as a function of $\ell n\ A_X$ (α-chymotrypsin dimerization). The salt activity coefficient of sodium chloride between 0° and 50°C was based on Robinson and Harned's table[47], where $\log \gamma \pm =$ $[I - A/T - B \log T]$. Conditions: xM NaCl - 0.01 M Na-acetate, pH 4.3. 7B. $\Delta H°$ versus $\ell n\ A_X$. 7C. $\Delta S°$ versus $\ell n\ A_X$. 7D. ΔC_p^o versus $\ell n\ A_X$.

At higher temperatures, the activity of the salt seems to have less influence than the standard Gibbs free energy change as a function of temperature on the dimerization process[1].

A summary plot of thermodynamic data as a function of $\ell n\ A_X$, taken from Table IX, is shown in Figure 8. The curves of $\Delta H°$, $T\Delta S°$, ΔC_p^o and $\Delta G°$ plotted as a function of $\ell n\ A_X$ are all parabolic as shown in Figures 7A, B, C, and D. The temperature at which $\Delta S°$ reaches zero is 303 K at $\ell n\ A_X = -2.0185$, while $\Delta H°$ becomes zero at 294 K ($T_{\phi h}$) at $\ell n\ A_X = -2.0166$. As seen from Figure 8, a common unifying temperature exists at 288 K, where $\ell n\ A_X = -2.0159$. On the various curves (at the inflection point), we designate these points as $T_S{}^m$, where $\Delta S°_{gross}$ is at a maximum; $T_{\phi h}{}^m$, where $\Delta H°_{gross}$ reaches a maximum; and $T_H{}^m$, where $\Delta G°_{gross}$ is a minimum in terms of salt activity. It follows that $T_H{}^m \neq T_H$, where $\Delta G°_{motive} = \Delta G°_{gross} - \Delta G°_{comp}$ since T_H must be defined as that temperature at which $(d\Delta G°/dI) = 0$.[1]

TABLE IX

THERMODYNAMIC PARAMETERS OF α-CHYMOTRYPSIN DIMERIZATION IN 0.178 M NaCl, 0.01 M SODIUM ACETATE BUFFER, pH 4.3.

NOTE: $\beta' = 9.75 \times 10^{-2}$ (Kcal/mole-deg), $\gamma' = -5.42 \times 10^{-4}$ (Kcal/mole-deg^2), $R\Delta\bar{\nu}_{pref}$ = -0.618, R-square = 0.9999; PR>F, 0.0001; sum of squares of error = 5.811×10^{-5}, mean square of error = 1.453×10^{-5}; standard deviation = 3.811×10^{-3}.

Temp	ℓn K	ℓn A_x	ΔG° (Kcal/mole)	ΔH° (Kcal/mole)	ΔS° (Kcal/mole-deg)	ΔC_p^o (Kcal/mole-deg)
273	9.20	-2.0175	-4.99	14.64	0.0719	-0.722
274	9.30	-2.0172	-5.10	13.92	0.0692	-0.721
275	9.39	-2.0170	-5.13	13.20	0.0666	-0.720
280	9.76	-2.0162	-5.43	9.61	0.0537	-0.714
285	10.01	-2.0159	-5.67	6.05	0.0411	-0.709
290	10.14	-2.0161	-5.85	2.52	0.0288	-0.704
295	10.16	-2.0168	-5.96	-0.99	0.0168	-0.698
300	10.08	-2.0179	-6.02	-4.46	0.0052	-0.693
305	9.91	-2.0194	-6.01	-7.91	-0.0062	-0.687
307	9.82	-2.0202	-5.99	-9.28	-0.0107	-0.685
308	9.77	-2.0206	-5.98	-9.97	-0.0129	-0.684

The enthalpic temperature, $T_{\phi h}$, and entropic temperature, T_s, are in good agreement with results from a non-linear Van't Hoff plot, in which the thermodynamic data are plotted as a function of temperature alone, as shown in Figure 9. On this plot, the harmonious temperature, T_H, where ΔG°_{gross} is at a minimum, was determined to be 294 K and $\Delta G^\circ_{motive} \cong \Delta G^\circ_{gross}$.

Both Figures 8 and 9 were generated from the STE 251 program of Barr et al.[43], to permit stepwise regression analysis of forward and backward selection, utilizing R^2 improvement statistics. In both cases $R^2 = 0.9999$ and the probability of an F-test, PR>F, was found to be 0.0001.

It is apparent that the customary Van't Hoff linearization will not provide T_H^m, $T_{\phi h}^m$, T_s^m and $\langle T_c \rangle$, and hence is of only limited utility in understanding the thermodynamics of the dimerization process. Figure 8 offers a much more complete view of the various thermodynamic temperatures.

When the enthalpy and entropy of the dimerization process are plotted as in Figure 10, the linear relationship between the two is readily apparent. The slope, $\langle T_c \rangle = (d\Delta H^\circ/d\Delta S^\circ) = (\Delta H^\circ/\Delta S^\circ)$, was found to be 290 K, the compensating temperature, while ΔG°_{gross} and ΔG°_{motive} were both determined to be -5.85 K cal/mole.

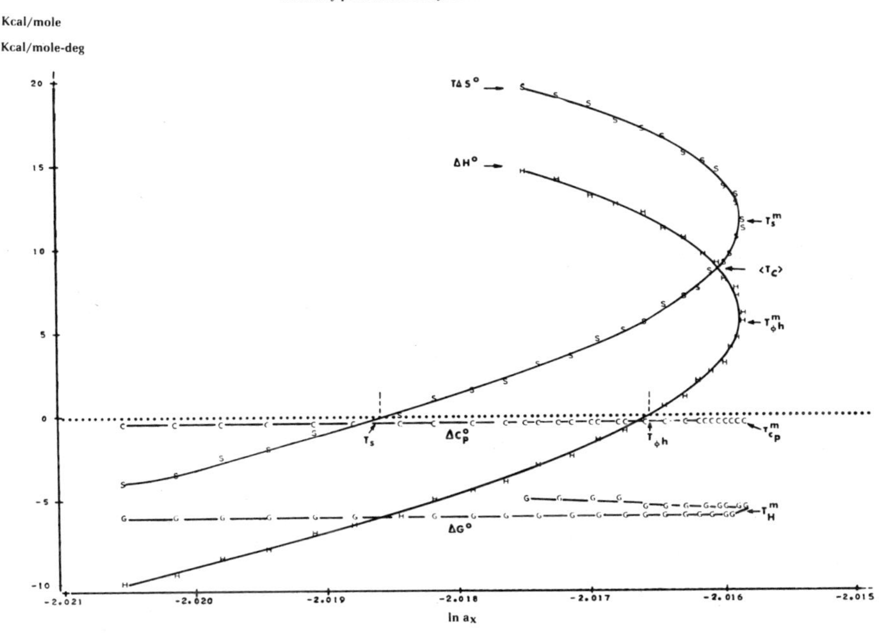

Figure 8. Thermodynamic plot of α-chymotrypsin dimerization as a function of $\ell n\, A_x$ (temperature dependent), with thermodynamic temperatures indicated on the plot. Each data point was evaluated with interpolation of R^2 improvement statistics. Conditions are identical to Figure 7.

As $T_s \Delta S^\circ_{(T)\text{gross}} \rightarrow 0$, that is $T_{exp} \rightarrow T_s$, $\Delta G^\circ_{\langle T_s \rangle}$ becomes zero, as seen in Figure 10. Hence, in the linear compensation process, at $T_{exp} \rightarrow T_s$, the Gibbs free energy term $\Delta G^\circ_{\langle T_s \rangle}$ becomes constant.

A further plot of T_{exp} versus $T_{exp} \Delta S^\circ_{(T)} / \Delta C_{P(T)}^\circ$, shown in Figure 11, also yields a straight line. At T_s, the value of $\langle T_c \rangle$ was 290 K. The compensatory temperature as evaluated from Equation VIII (see Section 6), therefore, is clearly a linear function.

6. Linear Thermodynamic Compensation Defined

It is possible to express the total gross change in the Gibbs free energy which is a result of all the subprocesses ($\Delta G^\circ_{\text{motive}} + \Delta G^\circ_{\text{comp}}$) contributing to any interacting system as follows:

$$\Delta G^\circ_{(T)\text{gross}} = \Delta G^\circ_{(T)\text{motive}} + \Delta G^\circ_{(T)\text{comp}} \qquad \text{(I)}$$

When linear thermodynamic compensation is operating in an associating protein system, where $\Delta G^\circ_{\text{comp}} = 0$,

$$\Delta G^\circ_{(T)\text{gross}} = \Delta H^\circ_{(T)\text{gross}} - T_{exp} \Delta S^\circ_{(T)\text{gross}} \qquad \text{(II)}$$

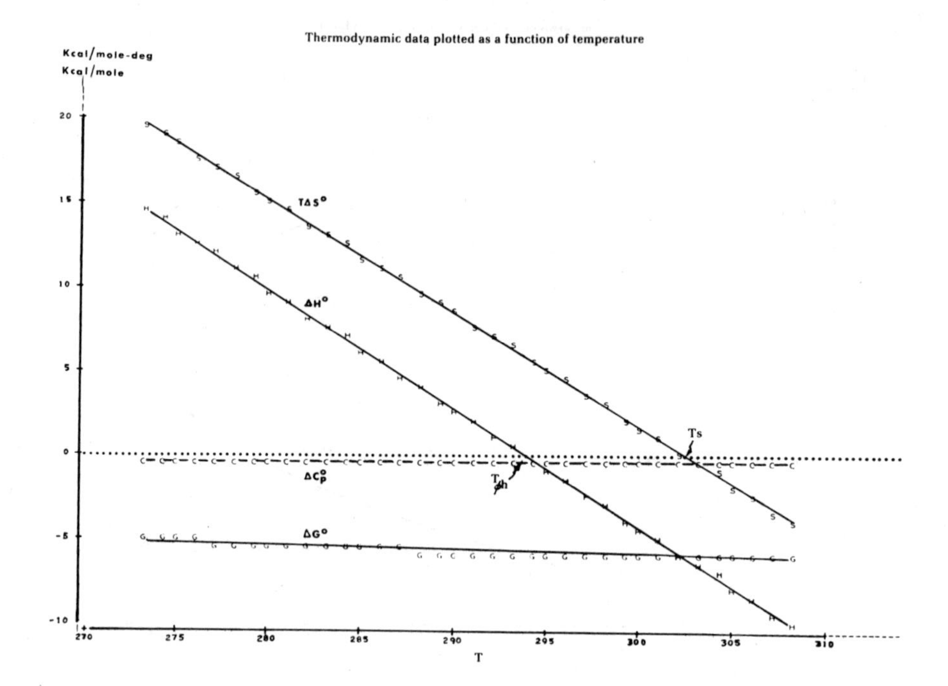

Figure 9. Thermodynamic plot of α-chymotrypsin association as a function of temperature (classic non-linear Van't Hoff plot).

where $\Delta H^{\circ}_{(T)gross} = \Delta G^{\circ}_{<T_S>} + <T_c>\Delta S^{\circ}_{(T)gross}$ (III)

Substitution of Equation (III) into Equation (II) yields

$$\Delta G^{\circ}_{(T)gross} = [\Delta G^{\circ}_{<T_S>} + <T_c>\Delta S^{\circ}_{(T)}] - T_{exp}\Delta S^{\circ}_{(T)gross} \quad \text{(IV)}$$

That is $\Delta G^{\circ}_{(T)gross} = \Delta G^{\circ}_{<T_s>} + [<T_c> - T_{exp}]\Delta S^{\circ}_{(T)gross}$ (V)

Hence, for any system undergoing linear thermodynamic compensation,

$$T_{exp} = <T_c> - [\frac{\Delta G^{\circ}_{(T)gross} - \Delta G^{\circ}_{<T_S>}}{\Delta S^{\circ}_{(T)gross}}] \quad \text{(VI)}$$

where $\lim_{T_{exp}\to T_s} [\frac{\Delta G^{\circ}_{(T)gross} - \Delta G^{\circ}_{<T_S>}}{\Delta S^{\circ}_{(T)gross}}] = \lim_{T_{exp}\to T_s} (\frac{-T_s\Delta S^{\circ}_{(T)gross}}{\Delta C^{\circ}_{P(T)gross}}) = 0,$

since $\Delta C^{\circ}_{P(T)gross} \neq 0$ over the temperature range. Therefore, equation III states that $\lim_{T_{exp}\to T_s} T_{exp} = <T_c>$ and $<T_c> = <T_H> = <T_s>$.

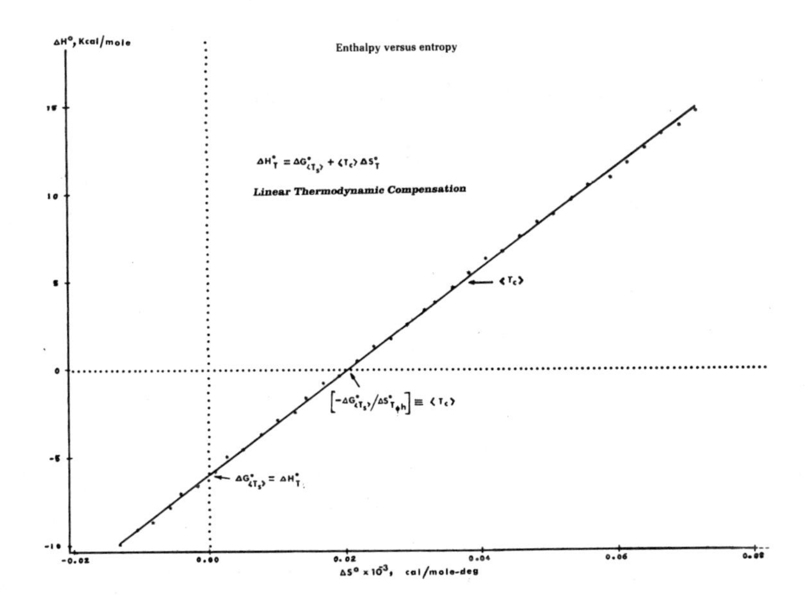

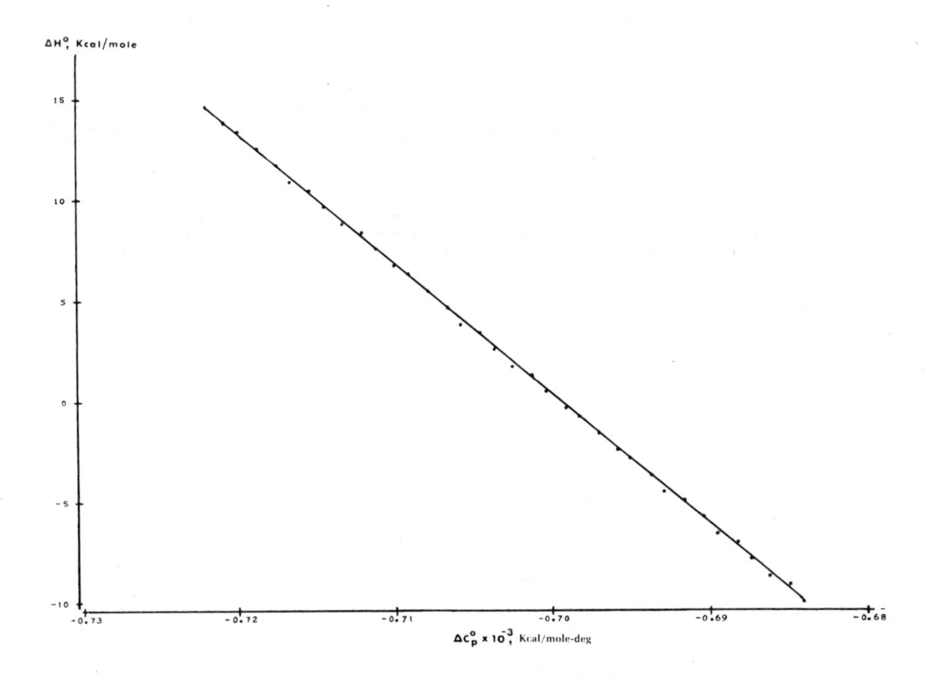

Figure 10. At top, Lumry's plot of $\Delta H^\circ_{(T)}$ versus $\Delta S^\circ_{(T)}$ for α-chymotrypsin dimerization, with data points taken from Table IX. The slope of the line represents $\langle T_C \rangle$. Below, a plot of $\Delta H^\circ_{(T)}$ versus $\Delta C^\circ_{P(T)}$, which is also linear and is used for evaluation of the polynomial coefficients of α-chymotrypsin.

TABLE V

THERMODYNAMIC TEMPERATURE OF INTERACTING PROTEIN SYSTEMS

NOTE: $<T_c>$ obtained from the plot of $\Delta H^\circ_{(T)gross}$ versus $\Delta S^\circ_{(T)gross}$.

Temp, °K	GDH	Glucagon	Cm Apo A-II protein	α-Chymotrypsin
T_H	302	287	299	294
$T_{\phi h}$	303	285	298	302
T_S	313	295	303	303
$\langle T_c \rangle$	293	290	300	290
SEM °K	± 2.0	± 2.0	± 2.0	± 2.0

Compensatory Temperature from the plot of T_{exp} versus $T_{exp}\ \Delta S^\circ_{(T)gross}/\Delta C^\circ_{P(T)gross}$.

Cm Apo A-II Protein	303	monomer ⇄ dimer
Glucagon	290	monomer ⇄ trimer
GDH	313	Isodesmic association
α-Chymotrypsin	290	monomer ⇄ dimer (at low pH)
SEM °K	± 0.1	

NOTE: Residue square = 0.9998, PR>F, 0.0001, sum of squares of error = 0.06 ∿0.29. Standard deviation = 0.11 ∿0.22. The concept of $<T_c>$ and $<T_c>_{app}$ is discussed in greater detail in the text.

Taking the derivative of equation (III) with respect to T_{exp} and setting the result equal to zero gives

$$\frac{d\Delta G^\circ}{dT_{exp}} = 0$$

$$<T_c>\Delta C^\circ_{P(T)gross} - T_{exp}\Delta S^\circ_{(T)gross} - T_{exp}\Delta C^\circ_{P(T)gross} = 0 \qquad \text{(VII)}$$

Thus equation V becomes

$$<T_c> = T_{exp} + [\frac{T_{exp}\Delta S^\circ_{(T)gross}}{\Delta C^\circ_{P(T)gross}}]$$

$$T_{exp} = <T_c> - [\frac{T_{exp}\Delta S^\circ_{(T)gross}}{\Delta C^\circ_{P(T)gross}}] \qquad \text{(VIII)}$$

and as $T_{exp} \rightarrow T_s$, $\Delta S^\circ_{(T)gross} \rightarrow 0$. Therefore $<T_c> = <T_s> = <T_H>$, shown graphically in Figure 11. Experimentally, we have shown that a plot of ΔH°_{gross} versus ΔS°_{gross} is a linear function, with a slope of $<T_c>$, as defined in Equation (VIII).

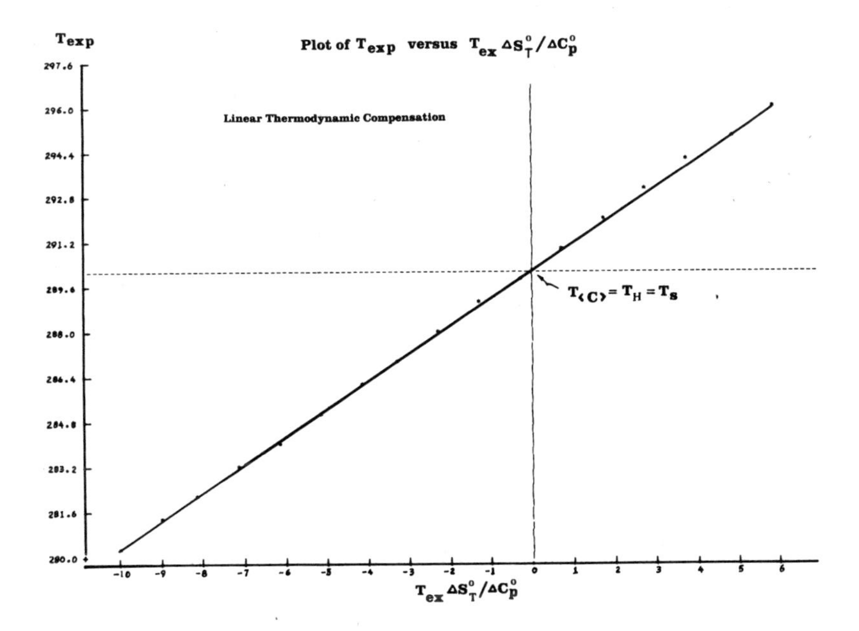

Figure 11. Plot of T_{exp} versus $T_{exp}\Delta S^{\circ}_{(T)}/\Delta C^{\circ}_{P(T)}$ for α-chymotrypsin dimerization, with data points taken from Table IX (see also Equation VIII of section 6).

Furthermore, at $\Delta S^{\circ}_{(T)gross} = 0$, $\Delta G^{\circ}_{<T_S>} = \Delta H^{\circ}_{(T)gross}$ and at $\Delta H^{\circ}_{(T)gross} = 0$, $<T_c> = (-\Delta G^{\circ}_{<T_S>}/\Delta S^{\circ}_{T_{\phi h}})$. This linear function confirms the principle of linear thermodynamic compensation operating in each of the associating systems we examined.

In order to prove that $<T_c> = T_{exp}$, it is necessary to take the derivative of $\Delta G^{\circ}_{(T)gross}$ with respect to $\Delta S^{\circ}_{(T)gross}$ from Equation IV which yields

$$\frac{d\Delta G^{\circ}_{(T)gross}}{d\Delta S^{\circ}_{(T)gross}} = \frac{d\Delta G^{\circ}_{<T_S>}}{d\Delta S^{\circ}_{(T)gross}} + <T_c> - T_{exp} - \Delta S^{\circ}_{(T)gross}\frac{dT_{exp}}{d\Delta S^{\circ}_{(T)gross}} \quad (IX)$$

$$-\frac{\Delta S^{\circ}_{(T)gross}}{\left[\dfrac{\Delta C^{\circ}_{P(T)gross}}{T_{exp}}\right]} = <T_c> - T_{exp} - T_{exp}\frac{\Delta S^{\circ}_{(T)gross}}{\Delta C^{\circ}_{P(T)gross}} \quad (X)$$

Hence, from Equation X, it may be seen that $<T_c> = T_{exp}$.

7. Thermodynamic temperatures defined, T^{m}_{j} where $j = H$, ϕ^{h}, s and ΔC°_{p} at the inflection point of the parabolic curves of Figure 8, where $\ln A_x = -2.0159$.

Since it is possible to express $\Delta G^\circ_{(T)gross}$ as follows,

$$\Delta G^\circ_{(T)gross} = \ell T \ln A_x + \Delta H^\circ_o + \alpha' T + \beta' T^2$$

its derivative with respect to $\ln A_x$ is equal to ∞ , i.e. $(d\Delta G^\circ_{(T)gross})/(d \ln A_x)=\infty$ and at this point the salt activity reaches a maximum.

In order to determine the temperature at which salt activity is at a maximum, the following relationship must be considered for 1:1 electrolytes:

$$\frac{d \ln A_x}{dT} = \frac{d\left[I - \frac{A}{T} - B \ln T + \ln C_x\right]}{dT} = 0$$

We have observed that the salt activity of NaCl in 0.1M to 0.3M solution reaches a maximum at 286 K. In α-chymotrypsin, T^m_j was found to be 288 K where j = H, ϕh, s and ΔC^o_p .

8. Van't Hoff and Gibbs-Duhem expressions in the thermodynamic compensation process.

When linear thermodynamic compensation is operating in an associating system as influenced by both a Hofmeister series of anions and temperature, it is necessary to consider both the Van't Hoff and Gibbs-Duhem expressions in evaluating the compensation temperature, $\langle T_c \rangle$.

DISCUSSION

1. Thermodynamic analysis of the non-linear Van't Hoff expression

The positive dependence of the enthalpy term, ΔH°, on temperature suggests that the accompanying change in heat capacity, ΔC°_p, is directly related to a reorganization of the water molecules surrounding the non-polar groups of the associating species. In such hydrophobic interaction, the change in ΔH° and ΔS° for the solution of hydrocarbons and non-polar groups will be negative, as the water freezes to ice[21].

Thermodynamic analysis of these apolar groups suggests that the chief contributor to the change in heat capacity is the exposure of hydrogen atoms in hydrocarbon chains (CH_2) to water[25]. Hydrogen bonding appears to be a minimal factor[61,62], since it is favored by non-polar solvents but is extremely weak in water and other polar solvents[63]. Hence, the positive, unfavorable enthalpy term at low temperature is a direct consequence of negative heat capacity changes.

Lumry and Rajender[2] have described the compensating temperature, $\langle T_c \rangle$, as a constant where $\Delta G^\circ_{comp} = 0$. We have found in these studies, however, that the compensatory temperature is most accurately defined as $\langle T_c \rangle = T_{exp} + T_{exp}\left(\frac{\Delta S^\circ_{(T)gross}}{\Delta C^\circ_{P(T)gross}}\right)$. $\langle T_c \rangle$ is evaluated from the linear plot of $\Delta H^\circ_{(T)}$ versus $\Delta S^\circ_{(T)}$ shown in Figure 5.

It is clear from our thermodynamic analysis of the non-linear Van't Hoff plot for these three self-associating protein systems that the variation of the equilibrium constants as a function of temperature depends not only on the concentration, but also upon the value of the second virial coefficient, which is also a function of concentration and temperature, and is indicative of alteration in the molecular asymmetry, accounting for the geometric exclusion volume effect and molecular size[8,13,64,65,66].

We have shown that the non-ideality term, BM_1, contributes significantly to the molecular volume changes in self-associating protein systems[67]. Our evaluation of thermodynamic parameters from the non-linear Van't Hoff expression has established the fact that β or BM_1 plays an important role in the change in heat capacity of the sample protein systems as a function of temperature. We have yet to determine, however, why an abrupt change is observed in the second virial coefficient of glucagon at T_s, where the entropy of the solution approaches zero, suggesting a major structural alteration upon association.

Any such changes in the folding of the tertiary structure which accompany self-association will involve a change in the heat capacity term, $C_p \cong \sigma_H^2/kT^2$. Here σ_H^2 is the variation of H_{comp}, the compensatory enthalpy of the system, a heat quantity which describes fluctuations in internal energy, exclusion volume and molecular asymmetry of the interacting species[3].

2. The Van't Hoff and Gibbs-Duhem expression in thermodynamic compensation process.

Evaluation of the $\Delta\bar{\nu}_{pref}$ terms of the Gibbs-Duhem expression has shown that a Hofmeister series of anions can modify the association equilibria or cause conformational changes in the association reaction either by preferential binding to one of the equilibrating species of the monomer or by changing the activity of water[12,32,33].

$$HPO_4^{-2} > SO_4^{-2} > F^- > Cl^- > Br^- > ClO_4^- > CNS^- \qquad 34$$

$\longleftarrow$ salting out (hydrophobic interaction) — salting in (dissociation) $\longrightarrow$

Our computations and data from Aune et al.[6], both seem to indicate that the generally accepted ordering of the Hofmeister series of anions may not be valid. The values of $(d\ \ell n\ K/d\ \ell n\ A_x)$ are suggestive of a hydrophobic interaction which takes part in the association of the α-chymotrypsin monomer. Our results also suggest, however, that the added electrolytes also influence their own solvation sphere -- particularly in the case of the ClO_4^- anion which also perturbs the solvation layer of the protein monomer and further enhances the association. On the other hand, our data indicate the SO_4^{-2} anion promotes dissociation, which would contradict its accepted order in the Hofmeister series.

At low pH, then, we must assume that the influence of charged groups in the protein[35] as they interact with various electrolytes cannot be ignored in any consideration of association parameters. In the pH range we examined, pH 4.1 to 4.4, the influence of $\Delta\bar{\nu}_{H^+}$ was too small to be considered in evaluating the electrostatic Gibbs free energy change.

The theoretical formulation of the electrostatic interaction between two charged-sphere particles[68-70] in terms of the second virial coefficient[71] is well established. The mode of action of the electrolytes at the binding site remains unclear, however, as does the relationship of $\Delta\bar{\nu}_w$ and $\Delta\bar{\nu}_x$ to the binding site. Certainly any further studies should define both $\Delta\bar{\nu}_x$, $\Delta\bar{\nu}_{H^+}$ and $\Delta\bar{\nu}_{H_2O}$ as a function of temperature for the full Hofmeister series.

3. Compensatory temperature, $<T_c>$, defined.

At $<T_c>$ in the linear compensation process, as defined by Equation IV where $d\Delta G^\circ_{(T)gross}/dT = 0$,

$$T_{exp} = <T_c> - \left[\frac{T_{exp}\Delta S^\circ_{(T)gross}}{\Delta C^\circ_{P(T)gross}}\right]$$

Hence we would assume that the values for $<T_c>$, T_s and T_H would be equivalent, within the allowance for statistical error, in each of the systems we examined. Our results presented in Table V, however, show that although there is relatively good agreement in the values of $<T_c>$, the variance in the other values is much greater than we might expect.

Why is this the case? Are we in fact determining the compensatory temperature, $<T_c>$, in plotting ΔH°_{gross} versus ΔS°_{gross}, or does the slope of the resulting line only represent $<T_c>_{app}$? Two possibilities present themselves.

First, we might assume that the variation in T_H, T_s and $<T_c>$, which was at most ten degrees Kelvin or within the 3 percent error limit, is indicative of the fact that the value of 295 ± 2.0 K which we have designated $<T_c>$ is in fact $<T_c>_{app}$, a temperature close to the compensatory temperature, but still influenced by sensitive changes in entropy and the heat capacity of system, as in Equation IV.

This equation can be modified to consider such slight variations, as follows:

$$<T_c> \frac{\Delta C^\circ_{P(T)gross}}{T_{exp}} - \Delta S^\circ_{(T)gross} - T_{exp}\frac{\Delta C^\circ_{P(T)gross}}{T_{exp}} = 0$$

where $<T_c>_{app} = <T_c> + (<T_c>_{app} - T_s)$. Substitution into the following equation,

$$<T_c>_{app} = T_{exp}\left(\frac{\Delta S^\circ_{(T)gross} + \Delta C^\circ_{P(T)gross}}{\Delta C^\circ_{P(T)gross}}\right) \qquad \text{gives}$$

$$T_{exp} = <T_c> + (<T_c>_{app} - T_s) - T_{exp}\left(\frac{\Delta S^\circ_{(T)gross}}{\Delta C^\circ_{P(T)gross}}\right)$$

A second possibility is that our choice of the polynomial function for ΔG° as a function of temperature in evaluating our thermodynamic parameters was not the best possible selection for the precise evaluation of that unique temperature which we have designated $<T_c>$ for any interacting system.

If, as is true in our case, $\Delta G^\circ_{(T)}$ is experimentally determined from a plot of Keq as a function of temperature from equilibrium measurements, based on l'Hopital's rule the resulting polynomials derived from $\Delta G^\circ_{(T)}$, where $\frac{d\Delta H^\circ_{(T)}}{d\Delta S^\circ_{(T)}} = \frac{\Delta H^\circ_{(T)}}{\Delta S^\circ_{(T)}}$, will always result in a linear function of temperature.

If, however, the polynomials for $\Delta H^\circ_{(T)}$ and $\Delta S^\circ_{(T)}$ are independently, experimentally determined as a function of temperature, the resulting plot will always be non-linear. Thus, by our procedure, it is essential to precisely evaluate Keq as a function of temperature to give the best possible fit of the data. This, in turn, will affect the accuracy of the polynomial chosen to represent $\Delta G^\circ_{(T)}$, from which a plot of $\Delta H^\circ_{(T)}/\Delta S^\circ_{(T)}$ is derived.

In either case, the argument for the existence of a unique compensatory temperature for interacting protein systems undergoing the thermodynamic compensation process remains a sound one. We have defined $\Delta S^\circ_{(T)}/\Delta C^\circ_{P(T)} = \langle \Delta T_c' \rangle / \langle T_{exp} \rangle$, where $\langle \Delta T_c' \rangle = (\langle T_c \rangle - T_{exp})$. It is to be hoped that further refinement of our experimental techniques, which will result in more precise experimental data, will permit the precise determination of $\langle T_c \rangle$ in future examinations of any self-associating protein system.

CONCLUSIONS

1. Our data demonstrate the existence of a linear thermodynamic compensation process operating in any associating biological system.

2. We have demonstrated the existence of a compensatory temperature at which the enthalpy and entropy of the interacting system are in balance, for maximum effectiveness of biological function.

3. At $\langle T_C \rangle$, the standard Gibbs free energy of association is at a minimum in the thermodynamic sense. Hence ΔH° and $T\Delta S^\circ$ are balanced, and the protein structure retains its maximum stability.

Our definition of the compensatory temperature suggests a number of interesting hypotheses which will require further examination. For example, we might speculate that all systems in vivo have a similar, unique compensatory temperature -- for example, 37 degrees C. in the human body -- at which entropy and enthalpy are maintained in balance, permitting all individual body processes to operate at a maximum level of effectiveness.

We would speculate that even if the primary and secondary structural sequences of a protein are altered, its compensatory temperature in vitro will remain unchanged as long as it undergoes self-association (tertiary-tertiary interaction) and retains its biological function.

Our results suggest that this unique compensatory temperature may very well be at or near 295 K in any self-associating system examined in vitro.

Ultimately, the questions we must ask ourselves are: Does temperature determine the structure of proteins and their interaction? Or do protein structure and interaction determine the compensatory temperature?

In terms of evolutionary events in biological systems, it would seem that the latter case is more likely.

ACKNOWLEDGEMENTS

This work was supported by National Science Foundation Grant PCM 76-04367 and in part by General Research Support, College of Arts and Sciences, University of Florida.

The author is grateful to Professor Rufus Lumry of the University of Minnesota for his advice and council in the examination of the process of thermodynamic compensation.

APPENDIX 1

THERMODYNAMIC TEMPERATURES DEFINED BASED ON A NON-LINEAR VAN'T HOFF PLOT.

1. $\langle T_c \rangle$ --- The temperature of compensation, given by the slope of a plot of $\Delta H^\circ_{(T)}$ versus $\Delta S^\circ_{(T)}$. $\langle T_c \rangle = d\Delta H^\circ_{(T)}/d\Delta S^\circ_{(T)}$ and $\langle T_c \rangle = T_{exp} + (\frac{T_{exp}\Delta S^\circ_{(T)}}{\Delta C^\circ_{P(T)}})$.

2. T_H --- The harmonious temperature, where the Gibbs free energy change (ΔG°) is at a minimum, i.e., $d\Delta G^\circ_{(T)}/dT = 0$ where $T_H^{\,2} + \frac{2\ell}{3\gamma} T_H + \frac{\alpha}{3\gamma} = 0$.

3. $T_{\phi h}$ --- The enthalpic temperature ($T_{\phi h}$) at which the enthalpy of the solution becomes zero, where $T_{\phi h}^{3} + \frac{\ell}{2\gamma} T_{\phi h}^{2} + \frac{\Delta H^\circ_0}{2\gamma R} = 0$, i.e. $d(\Delta G^\circ/T)/d(1/T) = 0$

4. T_s --- The temperature at which the entropy of the solution approaches zero, where $T_s^2 + \frac{2\ell}{3\gamma} T_s + \frac{\alpha}{3\gamma} = 0$, thus $T_H = T_s$.

5. T_{exp} --- Experimental temperature, determined by the conditions of a particular experiment, which can be equivalent to any of the thermodynamic temperatures, at random.

APPENDIX 2

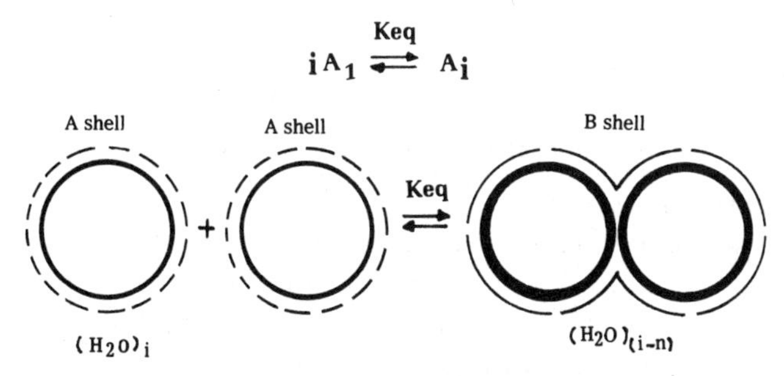

Diffusion control process

1. Perturbation of local dielectric properties.
2. Decrease in dielectric properties of the particle will result in a negative heat capacity change.

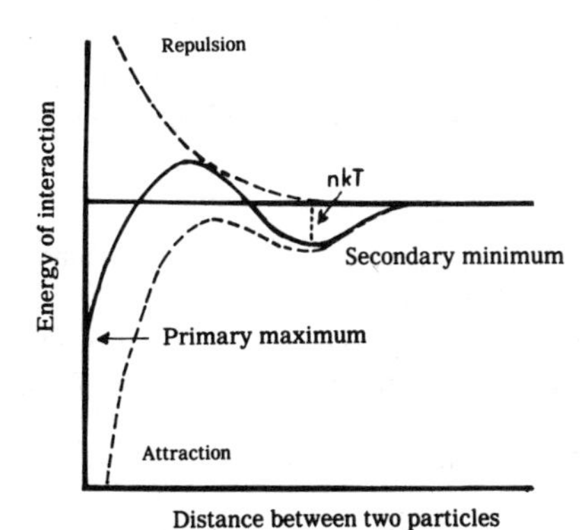

Protein interaction is in general a diffusion control process. Local perturbation of the dielectric properties of a monomer in its A shell (solvation layer) results in a decrease in the dielectric constant. This will cause a negative heat capacity change characteristic of the hydrophobic effect, which may perhaps be explained in terms of Wien's secondary effect.

In terms of the energy of interaction, it appears that the attractive force is not a purely hyperbolic function. The energy barriers of the secondary minimum and secondary maximum, which are a function of the reaction coordinates, must be small to accommodate equilibrium between the interacting species. If, however, the energy level nkT is large, the system will be predominantly in the monomeric state with no interaction. If this energy barrier is overcome by perturbation of equilibrium by varying temperature, pH, ligand binding or the presence of Hofmeister anions, dimerization will result.

REFERENCES

1. Chun, P. W. and Saffen, E. E., Jr. (1978) unpublished results.
2. Lumry, R. and Rajender, S. (1970), Biopolymers, 9,1125.
3. Lumry, R. (1978) Personal communication, "Practical Significance of Enthalpy-entropy Compensation in the Study of Couple Processes"; Frank, H. S. and Lumry, R., unpublished result.
4. Chun, P. W., Saffen, E. E., Jr., and Oeswein, J. Q. (1978), unpublished results.
5. Biltonen, R. and Lumry, R. (1969), J. Am. Chem. Soc., 91,4256.
6. Aune, K. C., Goldsmith, L. C. and Timasheff, S. N. (1971), Biochemistry, 10,1967.
7. Shiao, D. F., Lumry, R. and Fahey, J. (1971), J. Am. Chem. Soc., 93,2024.
8. Reisler, E. and Eisenberg, H. (1971), Biochemistry, 10,2549.
9. Edelhoch, H. and Osborne, J. C., Jr. (1976), Adv. Prot. Chem., 30,183.
10. Osborne, J. C., Jr., Palumbo, G., Brewer, H. B., Jr. and Edelhoch, H. (1976), Biochemistry, 15,317.
11. Formisano, S., Johnson, M. L. and Edelhoch, H. (1977), Pro. Natl. Acad. Sci., U.S., 74,3340.
12. Formisano, S., Johnson, M. L. and Edelhoch, H. (1978), Biochemistry, 17,1468.
13. Eisenberg, H. and Tomkins, G. M. (1968), J. Mol. Biol., 31,37.
14. Eisenberg, H., Joseph, R. and Reisler, E. (1976), Adv. Prot. Chem., 30,101.
15. Chun, P. W. and Kim, S. J. (1969), Biochemistry, 8,1633.
16. Chun, P. W. and Kim, S. J. (1970), Biochemistry, 9,1957.
17. Chun, P. W., Kim, S. J., Williams, G. D., Cope, W. T., Tang, L. H., and Adams, E. T., Jr. (1972), Biopolymers, 11,197.
18. Gratzer, W. B. and Beaven, G. H. (1969), J. Biol. Chem., 244,6075.
19. Gratzer, W. B., Creeth, J. M. and Beaven, G. H. (1972), Eur. J. Biochem., 31, 505.
20. Sasaki, K., Dockerill, S., Adamiak, D. A., Tickle, I. J. and Blundell, T. (1975), Nature, 257, 751.
21. Frank, H. S. and Evans, M. S. (1945), J. Chem. Phys., 13, 507.
22. Tanford, C. (1970), Adv. Prot. Chem., Vol. 24, 1; Edited by Anfinsen, C. B., Edsall, J. T. and Richards, F. M., Acad. Press, New York.
23. Anfinsen, C. B. and Scheraga, H. A. (1975), Adv. Prot. Chem., Vol. 29, p. 205, Edited by Anfinsen, C. B., Edsall, J. T. and Richards, F. M., Acad. Press, New York.
24. Gill, S. J. and Wadsö, L. (1976), Proc. Natl. Acad. Sci., 73, 2955.
25. Gill, S. J., Nichols, N. F. and Wadsö, L. (1976), J. Chem. Thermo., 8, 445.
26. Von Dreele, P. H., Lotan, N., Ananthanarayanan, V., Andretta, R. H., Poland, D., and Scheraga, H. A. (1971), Macromolecules, 4, 408.
27. Alter, J. E., Taylor, G. T., and Scheraga, H. A. (1972), Macromolecules, 5, 739.
28. Maxfield, F. R., Alter, J. E., Taylor, G. T., and Scheraga, H. A. (1975), Macromolecules, 8, 479.
29. Edsall, J. T. (1935), J. Am. Chem. Soc., 57, 1506.

30. Kauzmann, W. (1959), Adv. Prot. Chem., Vol. 14, 1; Edited by Anfinsen, C. B., Edsall, J. T. and Richards, F. M., Acad. Press, New York.

31. Barclay, I. M. and Butler, J. A. V. (1938), Trans. Faraday Soc., 34, 1445.

32. Klotz, I. M. (1966), Arch. Biochem. Biophys., 116, 92.

33. Tanford, C. (1969), J. Mol. Biol., 39, 539.

34. Von Hippel, P. H. and Schleich, T. (1969), In: "Structure and Stability of Biological Macromolecules," edited by Timasheff, S. N. and Fasman, G. D., p. 417, Dekker, New York.

35. Aune, K. C. and Timasheff, S. N. (1971), Biochemistry, 10, 1609.

36. Steiner, R. F. (1954), Arch. Biochem. Biophys., 53, 457.

37. Egan, R., Michael, H. O., Schluster, R. and Jandorf, B. J. (1957), Arch. Biochem. Biophys., 66, 366.

38. Gilbert, G. A. (1955), Discuss. Faraday Soc., 20, 32.

39. Massay, V., Harrington, W. F. and Hartley, B. S. (1955), Discuss. Faraday Soc., 20, 24.

40. Neurath, H. and Dreyer, W. J. (1955), Discuss. Faraday Soc., 20, 32.

41. Rao, M. S. N. and Kegeles, G. (1958), J. Am. Chem. Soc., 80, 5724.

42. Bethune, J. L. and Kegeles, G. (1961), J. Phys. Chem., 65, 1761.

43. Barr, A. J., Goodnight, J. H., Sall, J. P. and Helwig, J. T. (1976), "A User's Guide to SAS, 76," Statistical Analysis System, Univ. of Florida, Circa Computing Facilities.

44. Brumbaugh, E. E., Saffen, E. E., Jr., and Chun, P. W. (1978), Biophys. Chem., in press.

45. Yphantis, D. A. (1964), Biochemistry, 3, 297.

46. Wyman, J., Jr. (1964), Adv. Protein. Chem., 19, 223.

47. Robinson, R. A. and Harned, H. S. (1941), Chem. Rev., 28, 419.

48. Robinson, R. A. and Stokes, R. H. (1959), In: "Electrolytic Solutions," Butterworths, London.

49. Bates, R. G. (1973), In: "Determination of pH in Theory and Practice," pp. 72-73, John Wiley and Sons, New York.

50. Ten Eyck, L. F. and Kauzmann, H. S. (1967), Proc. Natl. Acad. Sci., U.S.A., 58, 888.

51. Hagler, A. T., Scheraga, H. A. and Nemethy, G. (1972), Ann. N. Y. Acad. Sci., 204, 51.

52. Ben-Naim, A. and Yaacobi, M. (1974), J. Phys. Chem., 78, 170.

53. Biltonen, R. and Lumry, R. (1969), J. Am. Chem. Soc., 91, 4256., ibid (1971), 98, 224.

54. Brandts, J. F. and Hunt, L. (1967), J. Am. Chem. Soc., 89, 4826.

55. Brandts, J. F. (1964a), J. Am. Chem. Soc., 86, 4291.

56. Brandts, J. F. (1964b), J. Am. Chem. Soc., 86, 4302.

57. Zipp, A. and Kauzmann, W. (1973), Biochemistry, 12, 4217.

58. Hawley, S. A. (1971), Biochemistry, 10, 2436.

59. Birktoft, J. J., Mathews, B. W. and Blow, D. M. (1969), Biochem. Biophys. Res. Comm., 36, 131.

60. Schrier, E. E. and Schrier, E. B. (1967), J. Phys. Chem., 71, 1851.

61. Schrier, E. E., Pottle, M. and Scheraga, H. A. (1964), J. Am. Chem. Soc., 86, 3444.

62. Kresheck, G. C. and Scheraga, H. A. (1965), J. Phys. Chem., 69, 1704.

63. Klotz, I. M. and Franzen, J. S. (1962), J. Am. Chem. Soc., 84, 3461.

64. Zimm, B. H. (1946), J. Chem. Phys., 14, 164.

65. Ishihara, A. and Hayashida, T. (1950), J. Phys. Soc. Jap., 6, 40.

66. Chun, P. W., Kim, S. J., Stanley, C. and Ackers, G. K. (1969), Biochemistry, 8, 1625.

67. Chun, P. W. and Yoon, Y. J. (1977), Biopolymers, 16, 2579.

68. Stigter, D. and Hill, T. L. (1959), J. Phys. Chem. 63, 551.

69. McMillan, W. G. and Mayer, J. E. (1954), J. Chem. Phys., 13, 276.

70. Verwey, E. F. and Overbeck, J. Th. G. (1948), In: "Theory of the Stability of Lyophobic Colloids," Elsevier Publishing Co., Amsterdam.

71. Tung, M. S. and Steiner, R. F. (1974), Eur. J. Biochem., 44, 52.

72. Harned, H. S. and Owen, B. B. (1943), In: "The Physical Chemistry of Electrolytic Solutions," Chapter 15, Reinhold Publishing Corp., New York.

Catsimpoolas, ed. Physical Aspects of Protein Interactions

ENTROPY-DRIVEN POLYMERIZATION OF PROTEINS: Tobacco Mosaic Virus Protein and Other Proteins of Biological Importance

MAX A. LAUFFER
Andrew Mellon Professor of Biophysics, University of Pittsburgh, Pittsburgh, Pennsylvania 15260

ABSTRACT

The coat protein of tobacco mosaic virus polymerizes at pH values above 7 to form double-discs and at somewhat lower values to form helical rods. Our laboratory has established that polymerization is endothermic and, therefore, entropy-driven and can be carried out reversibly. The positive enthalpy and entropy of polymerization have been evaluated from the variation of the equilibrium constant with temperature; the positive enthalpy has been verified calorimetrically. Polymerization is favored by an increase in ionic strength and by a decrease in pH; it is accompanied by the binding of hydrogen ions during early stages. Reducing the pH lowers the thermodynamic liability resulting from the necessity to bind hydrogen ions. Polymerization is accompanied by an increase in volume. This and the increase in entropy which drives the polymerization come from the release of water molecules. Structure-breaking chemicals inhibit polymerization; structure-makers favor it. This polymerization resembles in many respects the entropy-driven polymerization of tubulin, sickle cell hemoglobin, collagen, actin, myosin, flagellin, phycocyanin, pyruvate carboxylase, various synthetic polypeptides and other biologically important molecules. It is a reasonable inference that the source of the entropy increase for all of them is the release of water molecules.

INTRODUCTION

A. Theory of Entropy-Driven Processes

Reversible chemical reactions and biological processes carried out at constant temperature and constant pressure, involving no non-mechanical work, exhibit no change in Gibbs free energy. While reversible processes are dynamic, there is no net change except in response to infinitessimal stresses imposed upon the system. Spontaneous processes, in contrast, invariably exhibit a decrease in Gibbs free energy under the above conditions. $\Delta G<0$. For such processes,

$$\Delta G = \Delta H - T\Delta S \qquad \text{Eq. 1}$$

where ΔG is the change in Gibbs free energy, ΔH is the change in heat content or enthalpy, T is the absolute temperature and ΔS is the change in entropy. For most familiar processes in biology and chemistry, ΔG is negative because ΔH is negative; because heat is given off in the reaction. There are exceptions, however. In some spontaneous processes, heat is absorbed; the processes are endothermic. From Eq. 1 it follows logically that when ΔH is positive, ΔG can be negative only when ΔS is positive and large enough that $T\Delta S$ overbalances ΔH. Regardless of the ultimate source of the entropy increase, processes which proceed spontaneously because of such large increase in entropy are defined as entropy-driven processes.

When all reactants and all products of a reaction are at unit activity, the Gibbs free energy change, called the standard free energy change and designated by ΔG^o, is related to the equilibrium constant by the equation,

$$\Delta G^o = -RT \ln K_a \qquad \text{Eq. 2}$$

where K_a is the equilibrium constant expressed in terms of activity. By substituting Eq. 1 into Eq. 2, one obtains,

$$\ln K_a = \Delta S^o/R - (\Delta H^o/R)(1/T) \qquad \text{Eq. 3}$$

For processes in which ΔH^o and ΔS^o are constant over the temperature range studied, the graph of $\ln K_a$ versus $(1/T)$ is a straight line. If ΔH^o and ΔS^o are both positive, as in entropy-driven processes, the slope of the line, $d \ln K_a/d(1/T)$ or $-\Delta H^o/R$, is negative, as illustrated in Figure 1. For many biological and biochemical processes, particularly those which can be studied over a wide temperature range, ΔH^o, and therefore ΔS^o, vary with temperature. In these cases, $\partial\Delta H^o/\partial T$ and also $T\partial\Delta S^o/\partial T$ are equal to the difference between the heat capacities at constant pressure, ΔC_p, of products and reactants.[1] The linear graph of Figure 1 represents the case in which ΔC_p is either zero or too small to be measured, given the error of the experiment, over the temperature range of the study.

While most familiar processes in chemistry and biology have negative values of ΔH, are exothermic and enthalpy-driven, entropy-driven processes are not at all uncommon. The most familiar is the melting of ice at appropriate temperatures. Entropy-driven processes involving material of biological origin have been known for half a century. Szegvari[2] discovered in 1924 that a viscous sol at room temperature of nitrocellulose dissolved in amyl acetate and benzine is transformed into a rigid gel at higher temperature. The process is reversed by lowering the temperature. Heymann[3] found in 1935 that colloidal aqueous solutions of methyl cellulose undergo a sol-gel transformation when the temperature is raised and return to the sol state when temperature is lowered again. To explain these processes, Freundlich[4] postulated that bound solvent molecules are released when the solute molecules join together to form a gel network on heating. Heymann observed, further, that colloidal aqueous methy cellulose increased in volume upon gelation. This has turned out to be behavior common to many entropy-driven processes and is to be expected if closely packed bound solvent molecules are released to form less dense liquid. At constant temperature,

$$\partial \Delta \overline{G} / \partial P = \Delta \overline{V} \qquad \text{Eq. 4}$$

Here P is pressure, $\Delta \overline{G}$ is change in partial molar free energy and $\Delta \overline{V}$ is change in partial molar volume. When Eq. 2 is substituted into Eq. 4 and the result is integrated, Eq. 5 follows.

$$\ln K_a = \ln K_o - (P - P_o)\Delta \overline{V}/RT \qquad \text{Eq. 5}$$

K_o is the value of K_a at atmospheric pressure, P_o. When $\ln K_a$ is plotted against P, a straight line with slope equal to $-\Delta \overline{V}/RT$ is obtained, as illustrated in Figure 2 for the case of an entropy-driven process with a positive value of $\Delta \overline{V}$.

The source of the entropy increase which drives these processes must be discussed. Assembling dissolved molecules into large ordered structures of itself involves a decrease in entropy. Obviously,

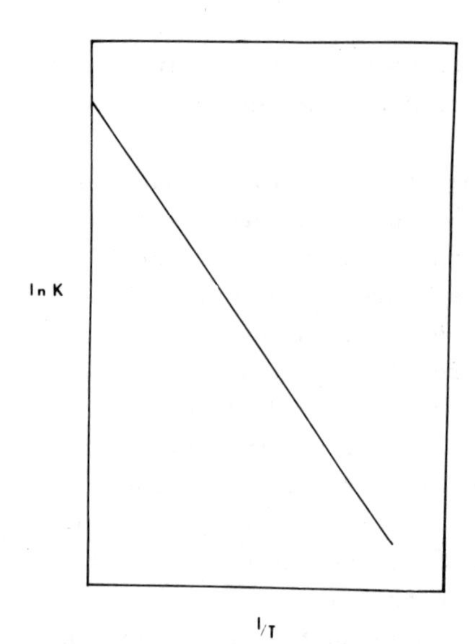

Fig. 1. Logarithm of the equilibrium constant, K, versus reciprocal of absolute temperature. T. From Lauffer[7]. Reprinted by permission of the copyright owner, Springer-Verlag.

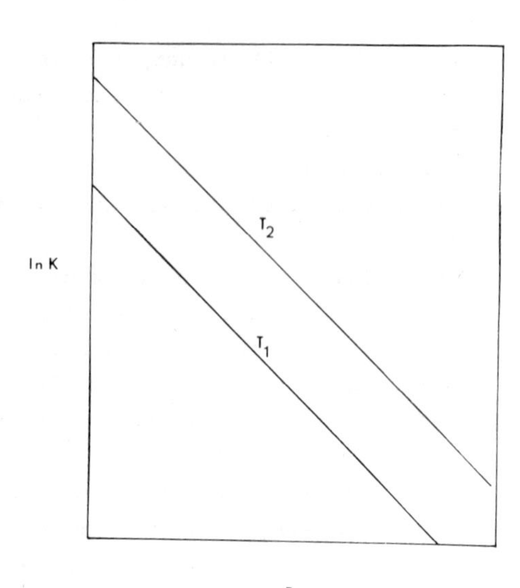

Fig. 2. Logarithm of the equilibrium constant, K, versus pressure, P, at two temperatures. From Lauffer[7]. Reprinted by permission of the copyright owner, Springer-Verlag.

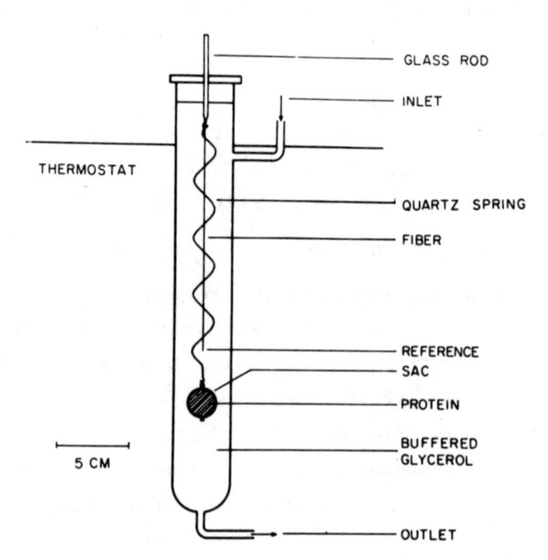

Fig. 3. Diagram of spring-balance. From Stevens and Lauffer[5]. Reprinted by permission of the copyright owner, The American Chemical Society.

something must be released during the process to provide the necessary disordering. It was long ago proposed[3,4] that solvent molecules are released. That this actually happens was demonstrated for the entropy-driven polymerization of tobacco mosaic virus (TMV) protein. Stevens and Lauffer[5] measured the water released during TMV protein polymerization with a quartz spring-balance, illustrated diagrammatically in Figure 3. The apparatus consisted of a cellophane dialysis sac attached to a quartz spring. Protein dissolved in water, glycerol and buffer was put inside the sac. The sac was suspended in a water-glycerol buffer solution. The idea inherent in this method is that "free" water, glycerol, and electrolyte are able to diffuse through the membrane so that the weight of each is cancelled by buoyancy. Protein and "bound" water, however, are not able to pass the membrane and, therefore, are weighed, corrected for buoyancy. If "bound" water is released upon polymerization, it will come to equilibrium and will no longer be weighed. The purpose of the glycerol in the system is to increase the density of the solution; otherwise, the weight of "bound" water would be cancelled by buoyancy.

The theory of the method is in reality more complicated than described above. The complications have been considered on a number of occasions [5,6,7,8]. This brief, simplified discussion of the method and the results is included here because of the crucial dependence of our understanding of entropy-driven processes upon these experiments on TMV protein. In the pH range 6-7, TMV protein is unpolymerized at room temperature and at pH 5.5 it is polymerized both in the cold and at room temperature[9]. In our first attempt to measure the water released when TMV protein polymerizes, 0.10 g of TMV protein dissolved in approximately 0.1 μ sodium chloride solution buffered to pH 7.50, containing 0.25 g glycerol per ml, was placed inside the sac. The outside tube contained a water-glycerol-salt solution of the same pH, ionic strength and glycerol content. The extension of the spring was then measured after thermal equilibrium had been reached at 4.0°C. When the spring reached a stable extension, the external solution was replaced by one buffered to pH 5.5. After sufficient time to permit equilibration of the electrolyte across the membrane, with the resultant drop in pH and polymerization of the protein, the extension of the spring was again measured. Finally, to demonstrate reversibility, the original solution at pH 7.5 was placed in the external tube, and sufficient time was allowed for equilibration. The results are shown in Figure 4. The spring extension was completely reversible. The change in weight was calculated from the difference between the original spring extension and the stable value achieved after equilibration at pH 5.5. Another experiment was carried out with TMV protein in buffer solution containing no glycerol. In order to be able to subtract water release associated with hydrogen ion binding, identical experiments were performed on TMV, which binds very nearly the same amount of hydrogen ion between pH 7.5 and 5.5 as does TMV protein.

If "bound" water is released during polymerization, it is to be expected that there will be a slight increase in volume of the system. This will be expressed as an increase in the partial specific volume of the protein, which, in turn, will change the buoyant weight of the protein. It is, therefore, necessary to measure this and correct for it in interpreting the spring-balance experiment. Stevens and Lauffer[5] obtained a value of +0.0074 ml/g for the increase in partial specific volume when the polymerization was carried out in a Linderstrom-Lang dilatometer. With this value the results of the spring-balance experiment were interpreted to show that 0.027 g of water are released per gram of protein upon polymerization. Since the molecular weight of the protein monomeric unit is about 17.5×10^3, this amounts to about 26 water molecules per protein monomer. Jaenicke and Lauffer[10] repeated the spring-balance experiment but brought about polymerization by raising the temperature. An almost identical value for water release was obtained.

Release of water molecules upon polymerization can now be considered a demonstrated fact rather than a theoretical speculation. It is sufficient to account for the entropy increase needed to drive the polymerization of TMV protein. As will be discussed later, hydrogen ions are bound during polymerization, but this contributes a small negative entropy change. All experiments reported thus far indicate that no other ions are bound or released when TMV protein polymerizes at pH values near 6.5. Therefore, water released is the only known source of the entropy increases for this entropy-driven polymerization.

The term, "hydrophobic bond", is frequently used to describe the type of union formed by entropy-driven reactions. This term is misleading because these unions are quite different from chemical bonds, even weak chemical bonds. The principal reason for the stability of such unions is that the surfaces in contact will be forced to interact with water when they separate. Interaction with water is somewhat analogous to the freezing of water on the surface. Just as there is a temperature above which water will not form ice because it would require an increase in free energy, there is a temperature above which water cannot interact with exposed protein surfaces without an increase in free energy. At such temperatures, the protein surface must remain in contact with an adjacent protein surface, that is, maintain union with it, because this is the only way to avoid interaction with water. Therefore the union will be stable. However at a temperature a few degrees lower, there is a decrease in free energy when the surface interacts with water just as there is a decrease in free energy at temperatures below the freezing point for the formation of ice. At such temperatures, the union between protein surfaces will break to permit the thermodynamically favorable interaction of the surface with water. Figure 5 illustrates this situation. On the right, two protein surfaces are separated. Both surfaces are in contact with water and interact with water as indicated by the curved lines. The interaction with water can be the type involving hydrocarbon side chains as discussed by Kauzmann[11], electrostriction about a charged center or any of the other usual interactions involving water. The surfaces will remain apart as long as the net interaction of the groups on the exposed surfaces with water is thermodynamically favorable, that is, involves a negative free energy change. On the left, the protein surfaces are shown in close contact, a condition under which it is unnecessary for them to interact with water. Since the diagram shows that like charges oppose each other, it will be necessary for the charges to be neutralized in the process involving the joining of the surfaces. The condition found on the left will be stable at higher temperatures because water interaction would be thermodynamically unfavorable at such temperatures. In an almost literal sense,

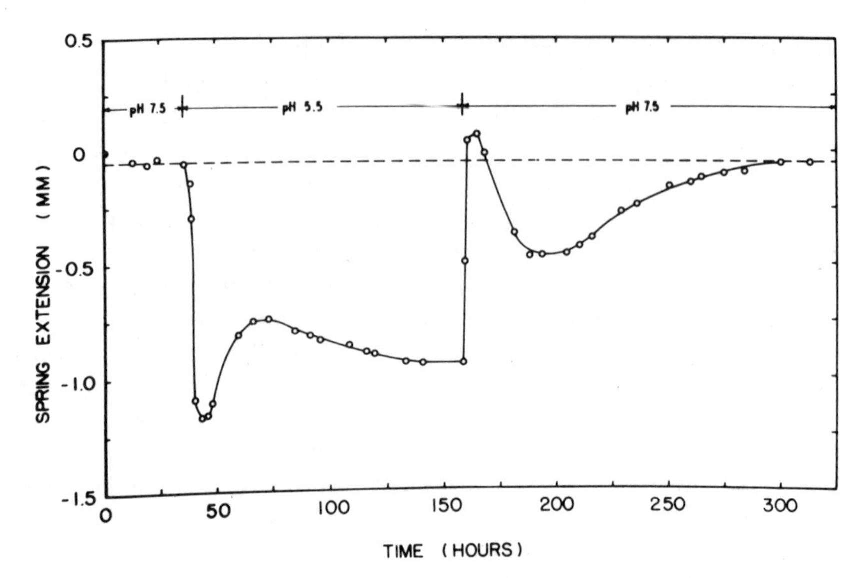

Fig. 4. Time course of equilibration of TMV protein in the spring-balance. From Stevens and Lauffer[5]. Reprinted by permission of the copyright owner, The American Chemical Society.

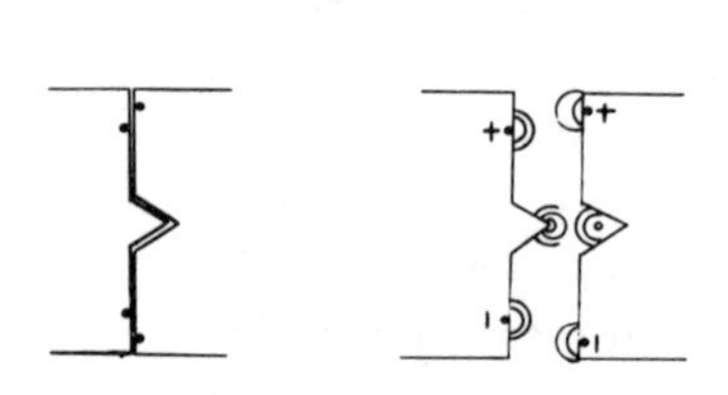

Fig. 5. Diagramatic illustration of water interaction with a protein surface. Left-entropic union. Right-surfaces interacting with water, denoted by curved lines.

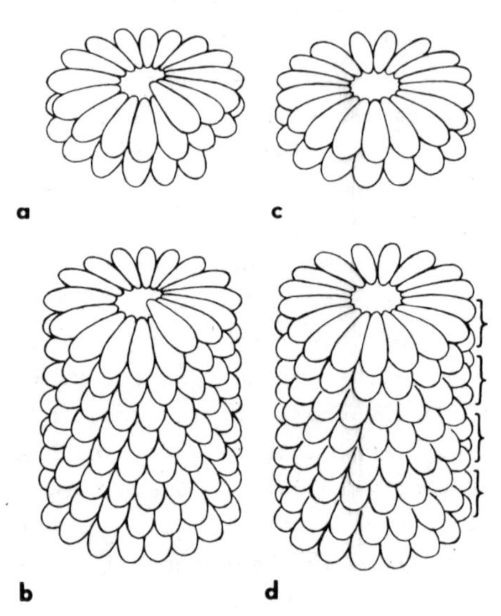

Fig. 6. Models of four polymers of TMV protein. (a) the hypothetical double spiral (b) the helical rod; (c) double disc; (d) stacked double discs. From Lauffer[7]. Reprinted by permission of the copyright owner, Springer-Verlag.

water pushes these protein surfaces into close contact with each other when the temperature is raised to the point that interaction of the surfaces with water is thermodynamically disadvantageous. The author has proposed that the term "entropic union" would be much more appropriate than "hydrophobic bond"[7].

Compared to ordinary chemical bonds, which have negative free energies of formation in the neighborhood of 100 kilocalories per mole, entropic unions are extremely weak. As will be discussed more fully later, the polymerization of tobacco mosaic virus protein involves an entropy increase of about 100 calories per degree per mole of trimer with a molecular weight of about 52.5×10^3. At pH values slightly above 6, this protein changes from completely unpolymerized to highly polymerized over a temperature range of 5^oC. The increase in $T\Delta S$ is about 500 calories per mole. Thus, the strength of the union is only of the order of 0.5 kilocalories per mole, close to the value of RT.

While water interaction is the dominant factor involved in entropic unions in aqueous solution, there must be other forces which come into play when the polymerized protein is in the dried state, for these polymers are stable under such conditions. One would expect at the very least that there are van der Waals interaction between the closely fitting protein surfaces, interactions which are completely overpowered by the water interaction at sufficiently low temperatures in solution but which could become dominant in the dried state.

B. Survey of Entropy-Driven Processes in Biology

Biologically important entropy-driven processes include the polymerization of tobacco mosaic virus protein, the polymerization of sickle cell anemia hemoglobin, the formation of collagen fibers from soluble collagen, the polymerization of globular actin into fibers, the polymerization of tubulin, the crystallization of some proteins, the cold inactivation of some enzymes, formation of the spindle apparatus during cell division, sol-gel transformations in protoplasm and many others. Many of these have been reviewed previously by the author[7,12]

The present review deals largely with material published or found in the literature by the author since his latest review[7] was written, but brief reference will be made to previously considered material as background.

TMV PROTEIN

A. Historical Survey

Tobacco mosaic virus (TMV) is a hollow cylinder composed of 95% protein and 5% ribonucleic acid (RNA) with a molecular weight of approximately 40 x 10^6 daltons. When virus is subjected to dilute alkali[13] or to 67% acetic acid[14], protein can be isolated from the RNA. Because it was first obtained by alkaline degregation, it is called A-protein[13]. The chemical subunit has a molecular weight of 17.53 x 10^3. The amino acid sequence has been determined and is available in the Atlas of Protein Sequences and Structures[15]. Of special significance for present purposes is the number of carboxyl and amino groups. There are 8 aspartic acid residues, 7 glutamic acid residues, 1 terminal carboxyl group, 2 lysine residues, 11 arginine residues and no terminal amino group, since it is acetylated. There is no histidine. Therefore at pH values in the range 7-8, the net charge would be -3 if all groups ionized normally. At pH values near 7 and at concentrations near 1 mg/ml, TMV protein seems to be a trimer of the ultimate protein subunit[16]. The author and associates[9] found that A-protein polymerizes endothermically and reversibly in 0.1 ionic strength (μ) phosphate buffer at pH 6.5. It is therefore an entropy-driven process in which the free energy decrease is derived primarily from release of water molecules.

TMV RNA can also be obtained in purified form by extracting the virus with phenol. Fraenkel-Conrat and Williams[17] and Lippincott and Commoner[18] found that copolymerization of the RNA and the protein leads to the formation of rod like particles apparently indistinguishable from the virus, possessing biological activity. Calorimetric studies carried out in the author's laboratory show that this, too, is an endothermic reaction[19]. The A-protein can be dissociated into monomers by diluting to a concentration of 0.1 mg/ml at low temperature[20]. In acidic media, TMV-A-protein polymerizes to form particles with very high sedimentation coefficients[21] and with a helical arrangement very much like that of TMV itself[22]. A-protein, which has a sedimentation coefficient of 4S, can be converted in the cold and at room temperature into high molecular weight polymers by reducing the pH from 7.7 to 5.0 or at pH values between about 6.0 and 6.8 in 0.1 μ ionic strength phosphate buffer by raising the temperature from $4^{O}C$ to or slightly above room temperature. Upon raising the temperature, 4S material disappears and 20S appears at intermediate temperatures and material with much higher sedimentation coefficients appears at $30^{O}C$[9]. Caspar[23] identified 20S material as the double disc, the structure of which was initially identified by Markham et al.[24] and more recently by Finch et al.[25] and Champness et al.[26]. Double discs can further polymerize to form stacked double discs[24]. This takes place at pH values above 7 and is favored by high ionic strength[27,28]. Schematic representations of the various polymers of TMV protein are represented in Figure 6. In the upper right hand corner is the double disc and below it the stacked double disc. The lower left illustrates the helical form of polymer obtained at low pH values. These three structures have been well identified. The upper left is a hypothetical particle termed a "double spiral" by the author[7] and a "lock washer" by Durham et al.[29] There is disagreement concerning the role of this hypothetical structure. The author has proposed[7] that it is the structure formed at pH values not far above 6.0 when 4S material is converted to 20S material by raising the temperature. It should be pointed out that the sedimentation coefficient of this hypothetical double spiral should be indistinguishable from that of a double disc. Durham et al.[29] regard it as only a transient when double discs polymerize to form the helical rods illustrated on the far left. More will be said about this later.

Evidence was first reported in 1962 that polymerization of A-protein involves the binding of H^{+} ions[30]. It was evident from the earliest titration experiments[23,31] that two hydrogen ions are bound per protein monomer when A-protein is polymerized by reducing pH from above 7 to about 5.5. Furthermore, when temperature is raised at pH 6.5, hydrogen ions are bound during polymerization. Also, hydrogen ions are released when polymerized protein in unbuffered media is depolymerized by lowering the temperature from approximately 30° to $4^{O}C$, as shown by a significant drop in pH[31]. Many titrations experiments have been published more recently; this will be discussed in some detail later.

B. Known Variables

The polymerization of TMV protein involves loss of freedom of the reactants, release of water molecules, binding of H^+ ions, condensation of charged particles and possibly other as yet unidentified contributions to the free energy[7]. This statement can be summarized by Eq. 6.

$$\Delta G^o_{net} = \Delta G^o_{S.T.} + \Delta G^o_{H_2O} + \Delta G^o_{H^+} + \Delta W_{el} + \Delta G^o_x \equiv$$

$$\Delta G^o + \Delta W_{el} = -RT \ln K'_a \qquad \text{Eq. 6}$$

The subscripts, net, S.T., H_2O, H^+ and x refer to the total process, the loss of freedom, the release of water, the binding of H^+ ions, and unidentified sources, respectively. ΔW_{el} is the difference between the electrical work of charging the product and that of charging the reacting units. When Eq. 1 is substituted for ΔG^o,

$$\ln K'_a = \frac{\Delta S^o}{R} - \frac{\Delta H^o + \Delta W_{el}}{RT} \qquad \text{Eq. 7}$$

Let the polymerization of TMV protein be represented by the reaction: $nA + mH^+ \rightarrow P$. This is a completely general formulation. Nothing is implied about the nature of A or P; only the ratio, m/n, of the number of H^+ ions bound to the number of polymerizing units must have a specific value. The equilibrium constant, K'_a for this reaction is $a_p/a_A^n\, a_{H^+}^m$, where a denotes activity. The apparent equilibrium constant at constant pH, K_a is a_p/a_A^n. By dividing one by the other and taking the logarithms, one obtains:

$$\log K_a = \log K'_a - m\,\text{pH} \qquad \text{Eq. 8}$$

The appropriate counterpart on the concentration basis of K_a is K_c defined as $[P]/[A]^n$. Khalil and Lauffer[32] proposed that the dominant reason for a_A differing from [A] in electrolyte solutions is the salting-out effect. In salting-out theory, $\log \gamma = K'_s \mu$, where γ is the activity coefficient and K'_s is the salting-out constant appropriate for expressing electrolyte concentration in terms of ionic strength, μ. Since salting-out should affect reactants but not products, the activity coefficient for P can be taken as unity. By this reasoning $K_c = \gamma^n K_a$.

$$\log K_c = n \log \gamma + \log K_a = n K_s'\mu + \log K_a \qquad \text{Eq. 9}$$

When Eq. 8 is substituted into Eq. 9, log is converted to ln, and Eq. 7 is substituted in the result, Eq. 10 is obtained.

$$\ln K_c = \frac{\Delta S^o}{R} + 2.303\ n\ K_s'\mu - 2.303\ m\ pH - \frac{\Delta H^o + \Delta W_{el}}{RT} \qquad \text{Eq.10}$$

Following the practice previously adopted in our laboratory[32,33], let a particular value of K_c be chosen and designated as the characteristic equilibrium constant, K_c^*. Assume only that the constant K_c^* represents the same distribution of reactants and product, regardless of ionic strength, pH, ΔW_{el}, etc. The value of K_c represented by K_c^* will be achieved at a particular value of T defined as the characteristic temperature, T^*.

The parameters ΔS^o, ΔH^o and ΔW_{el} in Eq. 9 refer to the formation of one mole of P. If both sides of Eq. 10 are divided by n and if K_c^* and T^* are substituted for K_c and T, it can be solved for $1/T^*$.

$$\frac{1}{T^*} = \frac{\Delta S^o/n - \frac{R}{n} \ln K_c^* - 2.303\ R\frac{m}{n}\ pH + 2.303\ RK_s'\mu}{\Delta H^o/n + \Delta W_{el}/n} \equiv \frac{\Delta S^*}{\Delta H^* + \Delta W_{el}^*} \qquad \text{Eq.11}$$

All terms in the numerator have the dimensions of entropy and all in the denominator have the dimensions of energy. ΔS^*, ΔH^* and ΔW_{el}^* are parameters appropriate to the disappearance of one mole of A.

When $\Delta W_{el}^*/(\Delta H^*) << 1$, Eq. 11 can be approximated accurately by Eq. 12.

$$\frac{1}{T^*} = \frac{\Delta S^*}{\Delta H^*}\left(1 - \frac{\Delta W_{el}^*}{\Delta H^*}\right) = \frac{\Delta S^*}{\Delta H^*} - \frac{\Delta S^*}{\Delta H^*}\ \frac{\Delta W_{el}^*}{\Delta H^*} \qquad \text{Eq. 12}$$

Equations 11 and 12 provide a structure for the interpretation of the effect of ionic strength, pH and electrical work (which depends on ionic strength and pH) on the characteristic temperature, T^*, at which an arbitrarily chosen degree of polymerization occurs. With the exception of two assumptions, namely that activity coefficients are salting-out terms and that K_c^* represents an invariant distribution of unpolymerized and polymerized protein, the reasoning involved in the derivation of these equations is purely thermodynamic.

C. Hydrogen Ion Binding

Several hydrogen ion titration studies have been published[34-42] since the original ones referred to previously[23,31]. The results are in general agreement except in the region between about pH 6 and 7, the region of most critical interest because here hydrogen ion binding is associated with polymerization. The results of the experiments of Shalaby and Lauffer[41], carried out in 0.1 μ KCl at a protein concentration of 5 mg/ml, are shown in Figure 7. The titration is completely reversible at all temperatures studied from 4°C to 20°C. The open symbols represent data obtained by titrating with HCl from about pH 8.0 to about pH 5.3 and the closed symbols represent titration in the reverse direction with KOH. The number of H^+ ions bound in going from pH 8 to pH 5.3 is significantly greater, 0.08 ions per protein monomer, at 4° than at 20°. This can be explained in terms of the enthalpy of H^+ ion binding by carboxyl groups. However, some investigators, Scheele and Schuster[39], Durham and Hendry[40] and Durham et al.[42], for example, observed hysteresis loops at 20°C. In the region around pH 6.2, the loop exhibits the maximum spread, about 0.4 H^+ ions per mole of monomer in the experiments of Scheele and Schuster. That is, H^+ ion binding at a given pH is less when the protein is titrated from pH 7 with HCl than when titrated from pH 5 with KOH. The higher values obtained in the KOH titration coincide with those shown in Figure 7. The only known differences between the experiments of Scheele and Schuster and those of Shalaby and Lauffer were in the rates of titration and in the starting pH on the alkaline side. Shalaby and Lauffer began at pH 8 and spent fourteen hours titrating to pH 5.3 and back. Scheele and Schuster began at pH 7.0 and spent seven hours titrating to pH 5.0 and back; some other investigators titrated much more rapidly than this. Shalaby and Lauffer duplicated the hysteresis loop of Scheele and Schuster when they did the titration at the same faster rate and began at pH 7.0. In this experiment, without removing from the titration vessel, the same protein was then titrated reversibly from pH 8 to 5.3 and back at the slower rate. Apparently, equilibrium is achieved with greater difficulty when protein is assembling and binding H^+ ions than when it is disassociating and releasing H^+ ions. The key to avoiding hysteresis in TMV protein titration seems to be extremely slow addition of acid.

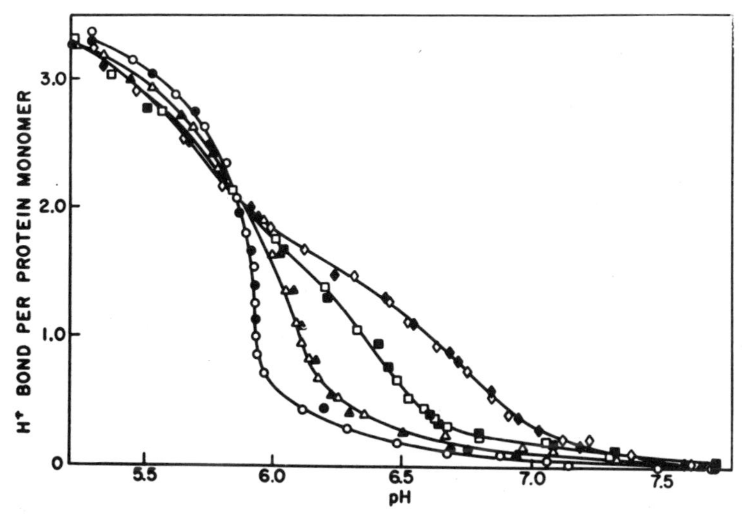

Fig. 7. Acid-base titration of TMV protein in 0.1 M KCl. Open symbols-HCl; closed symbols-KOH; circles-4°C; triangles-10°C; squares-15°C; diamonds-20°C. From Shalaby and Lauffer[41]. Reprinted by permission of the copyright owner, The Journal of Molecular Biology.

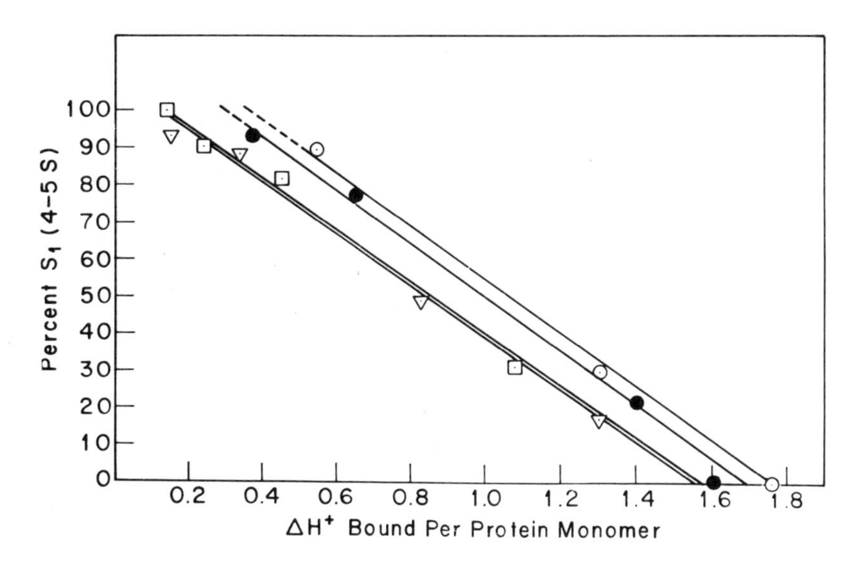

Fig. 8. Hydrogen ion binding by TMV protein versus fraction in 4-5 S state. O - pH 6.05; ● - pH 6.19; ∇ - pH 6.42; ☐ - pH 6.56. From Shalaby and Lauffer[41]. Reprinted by permission of the copyright owner, The Journal of Molecular Biology.

To correlate H^+ ion binding with polymerization, sedimentation velocity experiments were carried out at 4, 10, 15 and 20°C, at protein concentrations of 5 mg/ml in 0.08 μ KCl buffered with 0.02 μ potassium phosphate to pH values of 6.05, 6.19, 6.42 and 6.56. The results are presented in Table 1 along with the corresponding H^+ ion binding taken from Figure 7. The percentages listed are assumed to be subject to an error of ±5.

TABLE 1
SEDIMENTATION VELOCITY-HYDROGEN ION BINDING DATA FOR TMV PROTEIN POLYMERIZATION

Temp. °C	pH	6.05	6.19	6.42	6.56
4	s_1 and % s_1	4.4, 89	4.5, 93	4.9, 93	4.2, 100
	s_2 and % s_2	22, 11	20, 7	11, 7	
	H^+bd per monomer	0.54	0.37	0.15	0.14
10	s_1 and % s_1	5.0, 30	5.0, 78	5.3, 88	4.4, 90
	s_2 and % s_2	125, 69	24, 25	21, 11	18.3, 12
	H^+bd per monomer	1.30	0.65	0.34	0.24
15	s_1 and % s_1	125, 96	4.3, 22	4.3, 49	4.9, 81
	s_2 and % s_2		32, 12	23, 51	22, 20
	s_3 and % s_3		170, 64		
	H^+bd per monomer	1.68	1.40	0.83	0.45
20	s_1 and % s_1	122, 87	35, 13	4.1, 17	4.6, 31
	s_2 and % s_2	173, 11	151, 82	28, 20	24, 48
	s_3 and % s_3			129, 64	28, 20
	H^+bd per monomer	1.76	1.59	1.30	1.08

A plot of the number of H^+ ions bound per monomer against the percentage of 4S material remaining under the different conditions is shown in Figure 8. The different lines represent different pH values and the points on each line represent different temperatures. There is, within the experimental error postulated for the estimation of concentration, a linear relationship between the disappearance of 4S material and the number of H^+ ions bound. At the four pH values studied, the lines are parallel, indicating a constant value, 1.4 H^+ ions per monomer. Some H^+ ions are bound before any polymerization takes place. This initial binding is greater at the lower pH values, where the sum of initial binding and binding observed for the polymerization (1.4 per monomer) gives a value close to 2 H^+ ions per monomer, the value agreed upon for the polymerization. The disappearance of 4S material leads to the formation of either or both 20S material and higher polymers of 100S and above. H^+ ion binding takes place in the 4S to 20S transformation; further transformation to yield higher polymers does not involve H^+ ion binding.

In Figure 9 the same variables as in Figure 8 are plotted, H^+ ion binding versus percentage of 4S component. Each of the lines represents a single temperature and the points on each line represent different pH values. Figure 9 represents H^+ ion binding when polymerization is induced by changing pH. All the conclusions drawn from Figure 8 are applicable in this case, that is, a linear relation between H^+ ion binding and the disappearance of 4S material at all temperatures studied and the amount bound per monomer is the same at all temperatures, 1.5 H^+ ions per monomer. The uptake of H^+ ions upon A-protein polymerization was also observed by Vogel and Jaenicke[38] and by Durham et al.[42].

As analyzed above, A-protein, which is predominantly the cyclic trimer, polymerization is accompanied by the binding of 3 x 1.5 or 4.5 moles of H^+ ion per mole. There is another way of looking at these data. Equation 13 expresses the concentration of A-protein remaining as a function of pH.

$$\log [A] = \frac{1}{n} \log \frac{[P]}{\gamma^n K_a'} + \frac{m}{n} \text{pH} \qquad \text{Eq. 13}$$

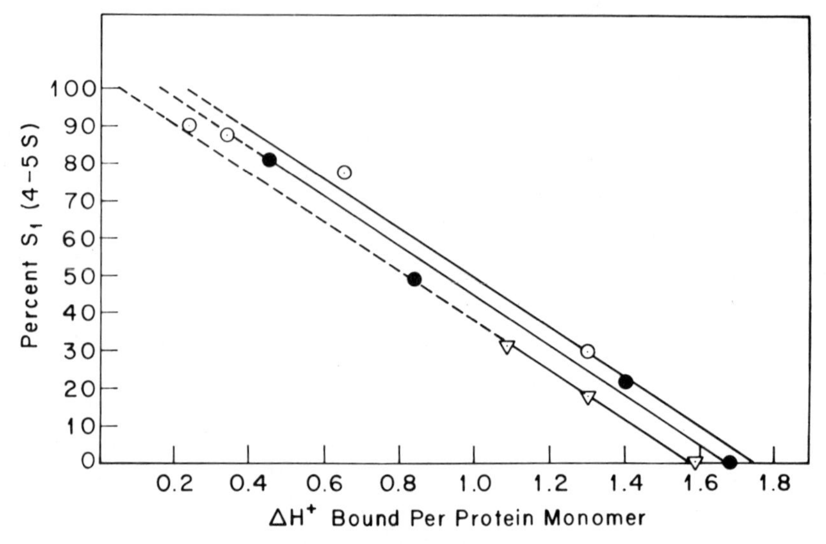

Fig. 9. Hydrogen ion binding by TMV protein versus fraction in 4-5 S state. O - 10°C; ● - 15°C; ▽ - 20°C. From Shalaby and Lauffer[41]. Reprinted by permission of the copyright owner, The Journal of Molecular Biology.

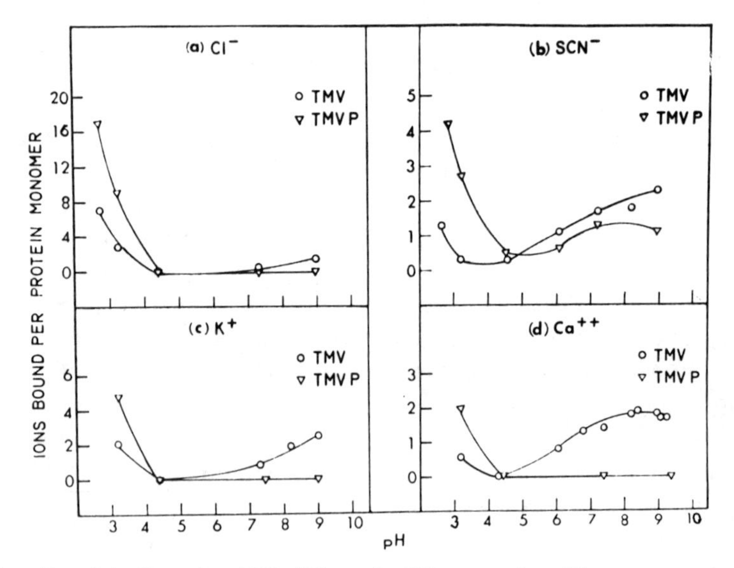

Fig. 10. Ion binding by TMV (O) and TMV protein (▽). From Shalaby et. al.[45]. Reprinted by permission of the copyright owner, The American Chemical Society.

This equation can be derived by substituting Eq. 8 into Eq. 9, replacing K_c by $[P]/[A]^n$, and solving for log [A]. The first term on the right of the equality sign is almost constant for all useful values of [A] when n is greater than 10. Thus, log [A] is practically a linear function of pH with a slope of m/n, the number of H^+ ions bound per trimer. Some error is introduced by assuming that [A] is given by the area under the 4S peak, because of the possibility that species differing somewhat from trimer are included in the area. However, log [A] is not greatly in error. When the data of Table 1 were analyzed in this manner, a value a little greater than 1 was obtained for m/n instead of 4.5 as anticipated from the preceding analysis. Stevens and Loga[43] performed elegant experiments to determine the variation of [A] with pH and found a value very close to 1.5 for m/n. The conclusion is that, even though upon polymerization 1½ H^+ ions are bound per monomer or 4½ per trimer, only ½ per monomer or 1½ per trimer participate critically in the polymerization reaction; the rest are bound as a result of the changes brought about by polymerization.

There is, thus, general agreement among several investigators that H^+ ions are bound when 4S material is converted into 20S material at pH values between 6 and 7 and released when 20S material is depolymerized. In apparent contradiction to this observation is the finding of Shalaby and Lauffer[41] that at pH 7.18 and 20°C, 20S material at a concentration of 31 mg/ml releases no H^+ ions at all when depolymerized to 4S material at 1 mg/ml by diluting with 0.1 M KCl titrated to pH 7.18. In this experiment, the 20S material was identified by sedimentation analysis. Because of the high concentration, the sedimenting boundary had a refractive index gradient too high to be recorded; the concentration was determined by subtracting the concentration of a slower sedimenting component from the loading concentration. The observed sedimentation coefficient was 16S. When it was corrected to zero concentration by the solution viscosity method of Lauffer[44], a value of 19.4S was obtained, clearly identifying this as 20S material. When this material was diluted at 20°C with 0.1 M KCl titrated to pH 7.18, after a delay of four hours a single boundary with a measured sedimentation coefficient of 4.18S was obtained. When corrected in the usual manner this yielded a value of 4.3S, clearly the 4S component. There was no change in pH. This result was confirmed in three additional similar experiments. Furthermore, it was found in additional experiments that when this 4S material at pH 7.18 undergoes temperature change back and forth between 20 and 4°C, there is no change in pH. Thus, when 20S material in unbuffered 0.1 M KCl at pH 7.18 dissociates into 4S material upon dilution, no H^+ ions are released.

When the unbuffered material at 20°C, pH 7.18 at a concentration of 31 mg/ml was lowered in temperature to 4°C, sixty percent of it was converted to 6S material. There was a change in pH from 7.18 to 6.45, but this amounts to the release of only 1.2 x 10^{-4} moles of H^+ ion per mole of protein monomer. The material at 4° was diluted to a concentration of 1 mg/ml with 0.1 M KCl titrated to pH 6.45. Depolymerization to 4S was complete and there was no change in pH, again showing that the 20S material remaining in this experiment releases no H^+ ions when depolymerized to 4S. But one can calculate from the data of Table 1 for pH 6.42 that 1.55 moles of H^+ ions are released when 20S material is depolymerized to 4S material by reducing the temperature from 15° to 4°C. Clearly there is more than one kind of 20S material - at least two. In our laboratory, the 20S material at pH 7.18 which released no H^+ ions upon depolymerization by dilution is identified as the double disc. This is primarily because of our early observation[27] that double discs and stacked double discs are observed in the electron microscope when grids are prepared from relatively concentrated TMV protein at pH 7.5 and the subsequent confirmation by Durham et al.[29]. The 20S material which, in our experiments and in those of others, binds H^+ ions in formation and releases them upon depolymerization must be something else. In our view, it is the two-turn helix, the double spiral. However, other investigators do not distinguish between this and the double disc.

It is clear by contrasting the results of our dilution experiment at pH 7.18 with our titration results at 20°C in Figure 7 that either form of 20S material can exist in the pH range 6.45 to 7.18. What determines the form obtained? If polymerization of 4S material takes place in water or neutral salt solution, where there is no buffering capacity and the protein is in great molar excess over the H^+ ion concentration, polymerization must take place without binding H^+ ions. The form of 20S material we identify as the double disc is formed. If the polymerization takes place in a buffer present in great molar excess, the buffer acid can supply, without appreciable change in pH, all the H^+ ions needed to produce upon polymerization the form of 20S material which has bound H^+ ions. Similarly, the same 20S form will be produced when all the H^+ ions needed are supplied by addition of acid in a titration experiment[41] or by an autotitrator[42].

An interesting feature of the H^+ ion titration curve of TMV protein is that 17 H^+ ions are bound per protein monomer between pH 8 and pH 1[34] while the amino acid analysis reveals only 16 carboxyl groups. The isoionic point is at pH 4.3-4.6. At this pH range the protein should be nearly isoelectric at the concentrations used in the experiments of Scheele and Lauffer[34]. Four H^+ ions are dissociated between the isoionic point and the plateau between pH 7 and 9, giving a net charge of -4 per protein monomer. The total number of positively ionizing groups revealed by the amino acid analysis is 13. If all of these were protonated at pH 7-8, the net charge should be -3. Both of these observations indicate that one positively ionizing group is not ionized at pH 7-8 and ionizes or titrates below this pH value. It must be in a position not readily available to H^+ ions. The titration is reversible from pH 11 to pH 2.7. Between pH 2.7 and pH 1, six H^+ ions are bound. There must be a drastic structural change at very low pH.

D. Binding of Ions Other Than H^+

Ion binding by TMV and TMV protein was measured by Shalaby et al.[45] at 22°C by using cation and anion-exchange membrane electrodes. The results are shown in Figure 10. While TMV binds Ca^{++} and K^+ ions to the extent of two per protein subunit and Cl^- ion to the extent of 1.4 per protein subunit at pH 9, TMV protein does not bind any of these ions at the pH values studied between the isoionic point, approximately pH 4.3-4.6, and pH 9. However, TMV protein binds all three ions at pH values below the isoionic point. Seventeen Cl^- ions are bound per protein subunit at pH 2.7; much smaller numbers of K^+ or Ca^{++} ions are bound at this pH value. The CNS^- ion was bound, but this ion is not usually present in TMV protein solutions. Banerjee and Lauffer[46] measured K^+ and Cl^- binding with specific ion electrodes at pH 6.5 on unpolymerized TMV protein at 2-6°C and on polymerized protein at 22-26°C. No binding of either ion was detected for either polymerized or unpolymerized protein. The error of measurement was such that the binding of one ion per protein subunit could have been detected. Durham and Hendry[40] observed no H^+ ion displacement from TMV protein at pH 6.8 and room temperature upon addition of $CaCl_2$ or $MgCl^2$, but as much as 0.4 H^+ ions displaced per protein subunit at pH values close to 6. They interpreted H^+ ion displacement in terms of cation binding. Several times larger displacements were reported for TMV, indicating substantial Ca^{++} + Mg^{++} ion binding, in agreement with direct potentiometric observations. In contrast, potentiometric studies in the author's laboratory reveal no Ca^{++} binding by polymerized TMV protein[47]. The weight of the evidence is that binding of ions other than the H^+ ion plays no direct role in temperature-reversible polymerization of TMV protein.

Calcium ions do, however, affect the polymerization of TMV protein indirectly or in a secondary manner. Durham and Hendry[40] found that Ca^{++} and Mg^{++} ions shift the pH at which depolymerization occurs about 0.3 pH units. Observations made in the author's laboratory indicate that Ca^{++} ions facilitate polymerization[47]. These observations can perhaps be understood in terms of the salting-out effect, to be discussed in detail subsequently.

A different kind of Ca^{++} ion effect was observed by McMichael and Lauffer[48]. When TMV protein is stored at high concentrations in the cold at low ionic strength at pH 5 - 5.5, it is highly polymerized. It is still in the polymerized state when it is brought to room temperature and mixed with acetate buffer at pH 6.5 containing additional electrolyte to give an ionic strength of 0.1. When the temperature is dropped to 5°C, the protein depolymerizes rapidly except when the added electrolyte is $CaCl_2$. In that case, depolymerization requires up to 12 hours. When temperature is subsequently raised, polymerization is rapid and when temperature is lowered a second time, depolymerization is rapid. Ion binding studies show that the protein in the first high temperature state has bound Ca^{++} ion, that this Ca^{++} ion is released during the slow depolymerization step and is not bound again in the rapid polymerization which occurs on raising the temperature. McMichael and Lauffer[48] suggested that the protein is in a tight helical form capable of binding Ca^{++} at pH 5 - 5.5, that it remains in this form long enough to bind Ca^{++} when it is added to the pH 6.5 solution containing Ca^{++} ions at room temperature. This stabilizes the tight structure. When the temperature is reduced, the Ca^{++} is released very slowly, permitting only very slow depolymerization. When temperature is raised, a looser polymer incapable of binding Ca^{++} ion is formed. This suggestion is consistent with observations that there is an increase in the partial specific volume, explained by release of bound water molecules, when the pH of a TMV protein solution at 25°C is reduced from 6.7, where the protein is polymerized, to 5.45 [10]. However, it is difficult to square this interpretation with the observation that TMV protein does not bind Ca^{++} ions at room temperature at pH 4.4 [45]. Further research is required.

E. Electrical Work

Realization that the polymerization of charged particles to form a charged cylinder might involve a change in electrical work of charging prompted the author to derive an equation for the electrical work, W_{el}, of charging a cylindrical model judged to be appropriate for TMV protein polymers[49]. The change in electrical work during polymerization, ΔW_{el}, is the difference between the electrical work of charging the product and the sum of the electrical work of charging all of the reactants consumed. ΔW_{el} has the same dimensions as ΔG and is, in fact, one of the components of ΔG. W_{el} for charging a particle surrounded by ions in solution is obtained by evaluating $\int_0^Q \zeta dQ$, where Q is the final charge and ζ is the electrokinetic potential - the potential between the position of the closest approach of ions in the solution to the surface of the particle and the bulk of the liquid. The model chosen as appropriate for polymerized TMV protein and TMV itself is a cylinder with uniform surface charge density from one end to the other. One consequence of choosing such a model is that the resultant equation is valid for rods of any length, even for very short discs if the charge is on the edge of the disc. The original publication of this theory is marred by type-setting errors in some of the equations. The correct equations for $\bar{\zeta}$, the average electrokinetic potential along the cylindrical surface, and for W_{el} are:

$$\bar{\zeta} = \frac{2\,Q\,\ln(1 + 1/\kappa a_i)}{DL^2}[\sqrt{(R + 1/2\kappa)^2 + L^2} - (R + 1/2\kappa)] \qquad \text{Eq.14}$$

$$W_{el} = \frac{Q^2 \ln(1 + 1/\kappa a_i)}{DL^2}[\sqrt{(R + 1/2\kappa)^2 + L^2} - (R + 1/2\kappa] \qquad \text{Eq.15}$$

Equations similar to but not identical with the approximation of the author's equations for very long thin rods were previously published by Gorin[50] and by Hill[51]. In Eqs. 14 and 15, D is the dielectric constant, R and L are the radius and length of the cylinder, κ is the Debye-Hückel constant and a_i is the radius of closest approach of ions to the cylinder. When $L/(R + 1/2\kappa) < 1/2$, as in a disc, Eq. 15 reduces to:

$$W_{el} = \frac{Q^2 \ln(1 + 1/\kappa a_i)}{2D(R + 1/2\,\kappa)} \mathrel{\hat{=}} \frac{Q^2\,1/\kappa a_i}{2D(R + 1/2\kappa)} \text{ when } 1/\kappa a_i << 1 \qquad \text{Eq.16}$$

This equation is appropriate for the double discs or double spirals formed by TMV protein polymerization. The starting material is a trimer of the protein subunit. No really satisfactory model exists for such a particle but a sphere is perhaps as good as any. Equations for ζ and W_{el} for spheres come directly from Debye-Hückel theory.

$$\zeta = \frac{Q}{Da_i(1 + \kappa a_i)} = \frac{Q\,1/\kappa}{Da_i^2\,(1 + 1/\kappa a_i)} \qquad \text{Eq.17}$$

$$W_{el} = \frac{Q^2\,1/\kappa}{2Da_i^2\,(1 + 1/\kappa a_i} \qquad \text{Eq.18}$$

It should be noted that the equations show that both ζ and W_{el} became proportional to $1/\kappa$, and therefore to $1/\sqrt{\mu}$, for both models when $\kappa a_i >> 1$.

For the purpose of investigating the possible magnitude and sign of ΔW_{el} for the formation of double spirals from trimers, assume that 11 trimers polymerize to form a double spiral. Let the trimers be represented by spheres with a radius of 26.35 x 10^{-8}cm and a radius of closest approach, a_i, of 28.35 x 10^{-8}cm. This radius was chosen because it represents a sphere of the same volume as the ellipsoid of revolution corresponding to a sedimentation coefficient of 4S. Let the double spiral be represented by a short cylinder of radius, R, of 90 and of a_i of 92 x 10^{-8}cm and of length 47.5 x 10^{-8}cm. The charge on each trimer is 3q and on each short cylinder is 33q, where q is the charge on a protein monomer in the polymerized state expressed in electronic units. For the purpose of the calculation, charge must be converted to c.g.s. units by multiplying q by 4.8 x 10^{-10}. It is true that the charge on the trimer in the unpolymerized state is greater than 3q as thus defined, but the binding of H^+ ions is taken care of by ΔG_{H^+} in Eq. 6 and the pH terms in Eqs. 11 and 12. The dielectric constant is assigned a value of 78.54. The Debye-Hückel constant, κ, is proportional to $\sqrt{\mu}$; the proportionality constant appropriate for 25°C was used in the calculations. When these values are substituted into Eq. 16 or 18, whichever is appropriate, when molecules are converted to moles by multiplying by Avogadro's number and when W_{el} is converted from ergs per mole to k cal per mole by dividing by 4.185 x 10^{10}, the results in Table 2 are obtained.

TABLE 2

ELECTRICAL WORK CALCULATIONS FOR POLYMERIZATION OF TMV PROTEIN

μ	$\frac{10^8}{\kappa}$cm	Cylinder W_{el}/q^2 kcal/q^2	Spheres W_{el}/q^2 kcal/q^2	Spheres $11W_{el}/q^2$ kcal/q^2	Per Reaction $\Delta W_{el}/q^2$ kcal/q^2	Per Trimer $\Delta W_{el}/q^2$ kcal/q^2
0.025	19.32	4.173	0.272	2.989	1.184	0.108
0.050	13.68	3.119	0.218	2.399	0.720	0.066
0.075	11.16	2.596	0.189	2.082	0.514	0.047
0.100	9.71	2.306	0.171	1.881	0.425	0.039
0.125	8.62	2.067	0.156	1.719	0.348	0.032
0.150	7.87	1.905	0.146	1.603	0.302	0.027

All of the values in the extreme right column are positive. When q is -4, as at pH 8, ΔW_{el} ranges from 0.43 at 0.15 μ to 1.73 kcal/mole of trimer at 0.025 μ. When q is -2, as at pH 5.9, the values for ΔW_{el} are one fourth of these. These calculations demonstrate that, if the models chosen are appropriate, differences in the electrical work term at different values of μ and pH can have a significant effect on the polymerization of TMV protein, for as was pointed out earlier, the difference in ΔG^* between unpolymerized and polymerized protein is only about 0.5 kcal/mole. Beyond indicating the kind of variation and the rough magnitude of these variations, these calculations should not be taken too seriously; electrokinetic theory rarely yields values in close agreement with experiment.

F. Electrophoretic Mobility

With our present knowledge and understanding, the beautiful electrophoretic mobility data of Kramer and Wittmann[52], shown in Figure 11, can be interpreted more fully than when originally published. The lack of parallelism between the titration curve and the electrophoretic mobility curve and the great difference between the isoelectric point, pH 3.2, and the isoionic point, pH 4.3-4.6, are striking features of the data. The low mobility of unpolymerized protein at pH 6 and above compared to the polymerized protein at pH 6 is easy to explain. Even though unpolymerized protein has greater negative charge per unit of mass than the polymerized protein, it has vastly more surface area per unit mass than does the polymerized protein and, therefore, a lower surface charge density[7].

While electrokinetic theory has not been very successful in predicting the exact magnitude of the charge from electrophoretic mobility, it is completely successful in predicting that the mobility of a fixed particle is proportional to its charge. Thus, if charge is known at one pH value, charge can be calculated at a different pH value from the ratio of the two mobilities. At pH 8, unpolymerized protein has a charge of -4 proton units per monomer as determined by H^+ ion titration[34]. There is no ion binding[45,46]. Therefore, the net charge is -4. From this value the net charge

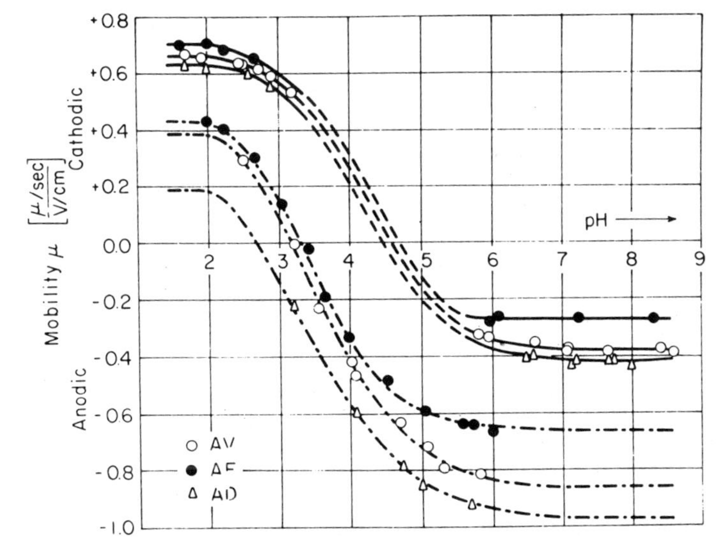

Fig. 11. Electrophoretic mobility of the protein from three strains of TMV. AV-vulgare; AF-flavum; AD-Dahlemense. From Kramer and Wittmann[52]. Reprinted by permission of the copyright owner, Zeitschrift für Naturforschung.

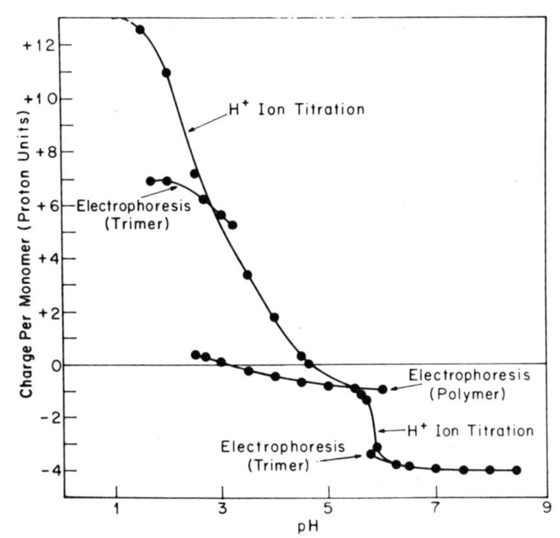

Fig. 12. Charge on TMV protein at 4°C from H^+ ion titration or electrophoresis.

at other pH values for unpolymerized protein (trimer) can be calculated from the ratio of mobilities. The results are displayed graphically in Figure 12. In like manner, the charge on polymerized protein can be calculated from electrophoretic mobility data calibrated at pH 5.5, where the charge is -0.92 proton units per monomer[41,45,46]. This result is also displayed in Figure 12. To contrast electrophoretic mobility data with H^+ ion titration data, the charge per monomer determined from the titration data of Scheele and Lauffer[34] and of Shalaby and Lauffer[41] are also presented in the figure. For unpolymerized protein, the agreement between charge calculated from mobility and determined by titration is good between pH 6 and 8.5. It is fair between pH 2.7 and 3.25. Below pH 2.7 the two estimates of charge are widely different, but this is the region in which H^+ ion titration is not reversible, presumably because of protein denaturation. In contrast, except at the calibration point, pH 5.5, the charge on polymerized protein calculated from mobility data differs widely from the H^+ ion titration charge. At pH 3.2 the charge from mobility is 0 and from titration, +4.4. At this pH value both Cl^- and K^+ ions are bound. The excess of bound Cl^- over K^+ ion is 4.2 per monomer. This is sufficient to explain the difference between the two estimates of the charge.

G. Effect of Ionic Strength

It has been known for more than a decade that an increase in ionic strength facilitates TMV protein polymerization[32,53,54] Khalil and Lauffer[32] found that, for the most part, the effect of ionic strength on polymerization was linear, and, for that reason, they interpreted it as an aspect of the salting-out phenomenon. Later it was shown that Bovine serum albumin also facilitated TMV protein polymerization and that this effect, too, could be interpreted in terms of salting-out[33]. Even though the data of Khalil and Lauffer were predominantly linear, significant deviations from linearity were observed, particularly at high pH and low ionic strength. Additional similar departures from linearity have been observed in the author's laboratory from time to time. This question was therefore reinvestigated in light of our present understanding of the possible role of electrical work[55]. The results are presented in Figure 13. While the data at ionic strengths above 0.075 can be fitted to straight lines, there is no doubt about non linearity when ionic strengths are lowered to 0.025.

If $\Delta S^{*'} \equiv \Delta S^{*} - 2.303\ RK_s'\mu$, Equation 12 can be rewritten in the form:

$$\frac{1}{T^*} = \frac{\Delta S^{*'}}{\Delta H^*} + \frac{2.303RK_s'\mu}{\Delta H^*} - \frac{\Delta S^{*'}}{\Delta H^*}\frac{\Delta^* W^*_{el}}{\Delta H^*} - \frac{2.303RK_s'\mu}{\Delta H^*}\frac{\Delta W^*_{el}}{\Delta H^*} \qquad \text{Eq.19}$$

As was shown in Eq. 16 and 18, W_{el} is a function of $1/\kappa$ and, therefore, of $1/\sqrt{\mu}$ for both cylindrical and spherical models. When $\kappa a_i >> 1$, W_{el} is accurately proportional to $1/\sqrt{\mu}$. When this condition is not met, the proportionality is only crude. To avoid unnecessary complications, let us assume that $\Delta W_{el} = \phi q^2/\sqrt{\mu}$, where ϕ is a constant whose magnitude is determined by Eqs. 16 and 18. This can not be expected to be a highly accurate approximation. This value for ΔW_{el} can be substituted into Eq. 19. Since the last term on the right in Eq. 19 is very much smaller than any of the others, it can be dropped. The result is Eq. 20.

$$\frac{1}{T^*} = \frac{\Delta S^{*'}}{\Delta H^*} + \frac{4.57K_s'\mu}{\Delta H^*} - \frac{\Delta S^{*'}}{(\Delta H^*)^2}\ \frac{\phi}{\sqrt{\mu}} = C + B\mu - A\ 1/\sqrt{\mu}$$

A, B and C in Eq. 20 can be true constants only when ΔH^* and $\Delta S^{*'}$ are independent of T. As will be seen later, there is evidence both for and against the possibility that ΔH^* and, as a consequence, $\Delta S^{*'}$ for TMV protein polymerization might vary somewhat with T and, therefore, with T^*. In spite of this, the ratio, $\Delta S^{*'}/\Delta H^*$, which is equal to C, will be much more nearly constant. A problem remains for A and B. However, T^* varies only about $5^{\circ}C$ or less from its mean value in a typical experiment covering the ionic strength range from 0.025 to 0.15. The variation of ΔH^* and, therefore, of A and B about their mean values should be of the order of magnitude of 10% compared to a 6 fold range in μ and a 2½ fold range $1/\sqrt{\mu}$.

T^* is determined experimentally from data of the sort illustrated in Figure 14 in which optical density, OD, a measure of turbidity, is plotted as a function of T. The data were obtained by the

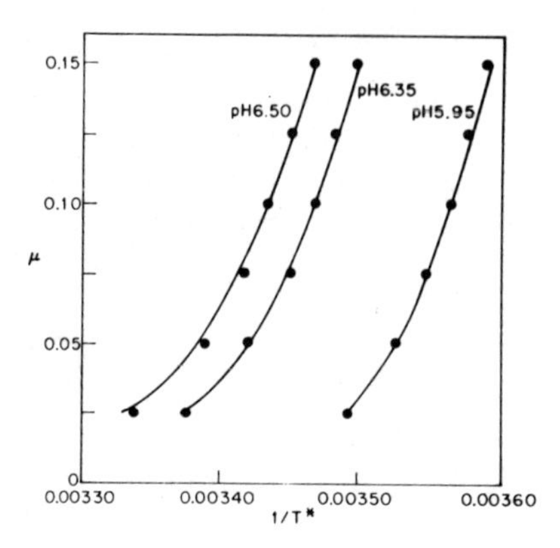

Fig. 13. Effect of ionic strength (μ) on TMV protein polymerization. T^* is the characteristic temperature corresponding to the characteristic value of the equilibrium constant, K_c^*.

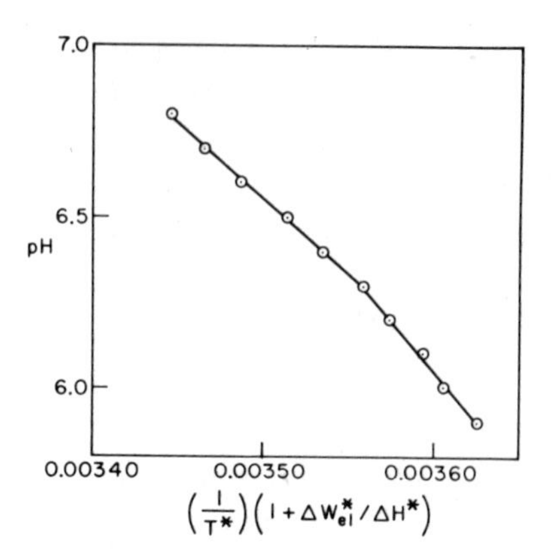

Fig. 15. Effect of pH on the characteristic temperature T^* for polymerization of TMV protein at 0.10μ. See Fig. 13 and text.

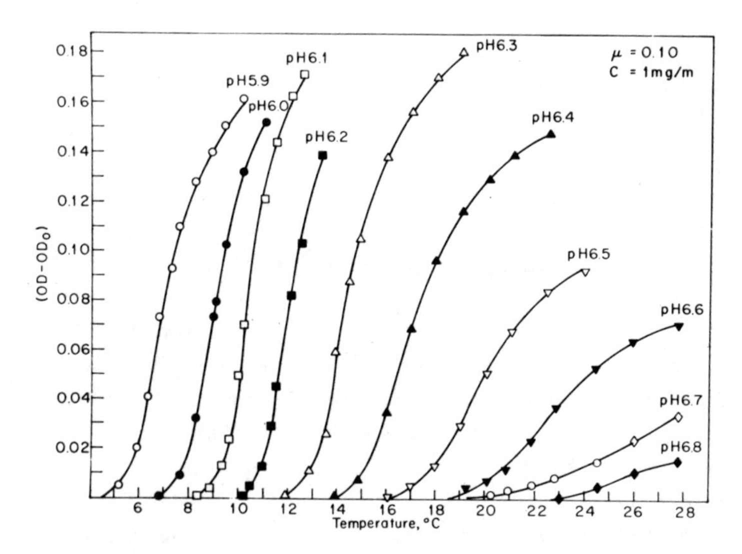

Fig. 14. Optical density ($OD - OD_o$) of TMV protein solution as a function of temperature at various pH values.

method of Smith and Lauffer[53]. A value of $(OD - OD_o)$ of 0.01 was chosen as the characteristic value. This value was chosen because it represents a weight-average molecular weight of about 500,000 and is thus a measure of early stages in the polymerization but yet is high enough so that the temperature at which it occurs, the characteristic temperature, T^*, can be read with reasonable accuracy. The measurements displayed in Figure 14 are completely reversible; data obtained with increasing and also with decreasing temperature are included. The only important assumption involved is that the distribution of products and reactants is the same at $(OD - OD_o)$ of 0.01, therefore representing the characteristic equilibrium constant, K_c^*, regardless of the temperature, pH or ionic strength at which it occurs.

TABLE 3

PARAMETERS EVALUATED FROM IONIC STRENGTH DATA

pH	5.95	6.35	6.50
$A \times 10^5$	1.353	2.027	2.363
$B\ (=4.574K_s'/\Delta H^*) \times 10^4$	4.00	4.00	4.00
$C\ (=\Delta S^{*\prime}/\Delta H^*) \times 10^3$	3.568	3.491	3.469
q (charge/monomer)	2.10	2.60	2.82
$A/C(q)^2 \times 10^4$	8.61	8.59	8.57

As a preliminary analysis, the data of Figure 13 were fitted to Eq. 20. The parameters involved are listed in Table 3. No great accuracy can be claimed for the parameter assigned, partly because of the approximations and uncertainties in the theory and partly because considerable variation is tolerated by the data. The curved lines in Figure 13 represent Eq. 20 with the appropriate parameters found in the table. The closed circles are the experimental data. In fitting these data, the parameter B was kept the same for all three values of pH, because B equals

$4.574 \times 10^4\ K_s'/\Delta H^*$ and the salting-out constant, K_s' is expected to be independent of pH[56]. The values of q listed in the table were derived from the titration curves at $20^{\circ}C$ in Figure 7 and the assigned charge per monomer at pH 8 of -4. Eq. 20 shows that A/Cq^2 equals $\phi/\Delta H^*$. This must be independent of pH if the final term on the right of Eq. 20 really is an electrical work term. Inspection of the bottom line of Table 3 shows that it is remarkably constant; the extreme values differ by only 0.5% even though the extreme values of A differs by 74%. It can be concluded, therefore, that both salting-out and electrical work contribute to the effect of ionic strength on TMV protein polymerization even in the early stages. It was shown previously that electrical work could be a factor limiting the maximum extent of polymerization[49].

If $\Delta S^{*''}$ is defined as $\Delta S^* + 2.303\ R\frac{m}{n}\ pH$, Eq. 11 can be transformed into:

$$\frac{1}{T^*}\left(1 + \frac{\Delta W^*_{el}}{\Delta H^*}\right) = \frac{\Delta S^{*''}}{\Delta H^*} - \frac{2.303\ R\ \frac{m}{n}\ pH}{\Delta H^*} \qquad \text{Eq.21}$$

The data of Figure 14 can be analyzed in terms of Eq. 21. The values of $1/T^*$ are obtained as above. The values of $\Delta W_{el}/(\Delta H^*)$ appropriate to each pH value can be evaluated from the parameters of Table 3 and the value of q obtained from the titration curve at $20^{\circ}C$ in Figure 7. By comparing Eq. 19 with Eq. 20, it is found that $A/\sqrt{\mu} = \frac{\Delta S^{*'}}{\Delta H^*} - \frac{\Delta W^*_{el}}{\Delta H^*} = C\ \frac{\Delta W^*_{el}}{\Delta H^*}$. Thus, $\frac{\Delta W^*_{el}}{\Delta H^*}$ at a particular ionic strength and pH value is given by $A/(C\sqrt{\mu})$. At pH 5.95 and 0.1 μ, $\frac{\Delta W_{el}^*}{\Delta H^*}$ equals 0.012. Since ΔW^*_{el} is proportional to q^2, the value of $\frac{\Delta W^*_{el}}{\Delta H^*}$ can be calculated for any other pH value. The results are presented in Table 4 and $\frac{1}{T^*}\left(1 + \frac{\Delta W^*_{el}}{\Delta H^*}\right)$ is plotted against pH in Figure 15.

TABLE 4

ELECTRICAL WORK CORRECTIONS FOR $1/T^*$

pH	q	$\frac{\Delta W^*_{el}}{\Delta H^*}$	$1/T^*$	$\frac{1}{T^*}(1 + \frac{\Delta W^*_{el}}{\Delta H^*})$
5.9	-2.00	.01088	.0035860	.003625
6.0	-2.16	.01270	.0035605	.0036057
6.1	-2.30	.01439	.0035416	.0035926
6.2	-2.42	.01594	.0035204	.0035765
6.3	-2.54	.01755	.003497	.0035584
6.4	-2.66	.01925	.003469	.0035358
6.5	-2.82	.02164	.003440	.0035144
6.6	-2.98	.02416	.0034053	.0034876
6.7	-3.16	.02717	.0033743	.0034660
6.8	-3.35	.03054	.0033438	.0034459

From Eq. 21, $\frac{d\,\frac{1}{T^*}(1 + \frac{\Delta W^*_{el}}{\Delta H^*})}{dpH} = \frac{-2.303R\frac{m}{n}}{\Delta H^*}$

It can be shown from Eqs. 7 and 8 that this equation is simply the reciprocal of $d\ln(K'_a)^{1/n}/d\frac{1}{T^*}(1 + \frac{\Delta W^*_{el}}{\Delta H^*})$. This corresponds, therefore, to the accepted way of evaluating ΔH^*, except for the small correction, $(1 + \frac{\Delta W^*_{el}}{\Delta H^*})$. When the data as a whole are considered, three values of the derivative can be obtained from the figure, one for pH in the range 6.35-6.8, one for the range 5.90-6.35 and an average for the whole range. The values are 0.000231, -.0001665 and 0.000199, respectively. ΔH^* is $-4.574\ \frac{m}{n}$ divided by the slope. If $\frac{m}{n}$ is 1.5, as found by Stevens and Loga[43], the value for ΔH^* are 29.70, 41.2 and 34.5 kcal/mole of A protein, respectively. All of these values are reasonably close to earlier estimates of the ΔH^* for trimer polymerization, 30 kcal/mole obtained from variation of equilibrium constant with temperature[15] and from direct calorimetry[57]. The break in the slope of the graph between pH 6.3 and 6.4 seems to be significant; at least it can be reproduced. Eq. 21 suggests two possible causes for the break. Either ΔH^* is constant and $\frac{m}{n}$ changes at about pH 6.35 or $\frac{m}{n}$ is constant and ΔH^* changes. There is some evidence against the second possibility and in favor of the first. Professor Julian Sturtevant has informed us that preliminary calorimetric studies on the polymerization of TMV protein supplied by our laboratory failed to detect any change in heat capacity with temperature. In addition, the titration data of Figure 7 and Table 1 show that at pH values below 6.5, considerable H^+ ion is bound by the protein before polymerization takes place. If ΔH^* has a value of 29.7 kcal/mole of trimer, $\frac{m}{n}$ would have a calculated value of 1.08 at pH values below 6.35. Nevertheless, the simplest physico-chemical explanation of this result is that ΔH^* varies with T, and, therefore, in the sort of experiment done here, with pH. Brandts[58] pointed out that the enthalpy of transfer of certain amino acid side chains from aqueous to organic media is temperature dependent. As will be seen later, there is evidence that ΔH^* is temperature dependent for many entropy-driven processes. If ΔH^* is 29.7 kcal/mole at 279°A and 41.2 kcal/mole at 299°A, $\Delta\Delta H^*/\Delta T$, an estimate of ΔC_p, is about -600 cal/mole/degree.

If ΔH^* is 29,700 cal/mole, the value of K_s' can be evaluated from the parameter B of Table 3 to be 2.6. This is considerably smaller than the estimates originally obtained by Khalil and Lauffer[32], but the polymerization process studied by Khalil and Lauffer can be interpreted in terms of polymerization of aggregates as large as the double spiral, in contrast with the polymerization of trimers as in the present work.

From the parameters C of Table 3, values for $\Delta S^{*\prime}$ ranging from 106 at pH 5.95 to 103 e.u. at pH 6.50 are obtained. The parameter, A, of Table 3 is equal to $\frac{\Delta S^{*\prime}}{\Delta H^*} \frac{\Delta W^*_{el}}{\Delta H^*}\sqrt{\mu}$ or C $\frac{\Delta W^*_{el}}{\Delta H^*}\sqrt{\mu}$. From the values of A and C in Table 3 and $\Delta H^* = 29{,}700$ cal/mole, one can calculate ΔW_{el} at the pH values 5.95, 6.35 and 6.50 for various values of μ. For μ of 0.10, the values at the three pH values are .357, .545 and .640 kcal/mole, respectively. The theory outlined in Section E, Table 2, leads to 0.039 kcal/mole of trimer at μ = 0.1 for $\frac{\Delta W^*_{el}}{q^2}$. When the values of q from Figure 7 are introduced, the value of ΔW^*_{el} from theory for the three pH values are 0.172, 0.264 and 0.310 kcal/mole, respectively. Considering the problems inherent in electrokinetic theory and the drastic approxmation made in deriving and applying Eq. 20, from which the former set of values was obtained, the agreement between the two sets is as good as can be expected. Theory predicts and experiments confirm, quantitatively within a factor of 2, that changes in electrical work significantly affect the polymerization of TMV A-protein.

H. Polymerization of Protein from Mutants of TMV

The results of polymerization experiments on the proteins from several mutants of TMV and from other related viruses was reviewed previously[7]. These include cucumber virus 4, the Dalamense strain, the PM2 mutant and the protein from cucumber green mottle mosaic virus. Recent studies have been carried out on the polymerization of the protein from the flavum strain and the E66 mutant of TMV.

1. Protein from the flavum strain

The flavum strain of tobacco mosaic virus was isolated from the yellow spot on a plant infected with the common strain of TMV[59]. Yellow signs of disease are observed on Samsum and Java tobacco. Yields of the virus are 5-fold less than with the common strain. This strain is temperature sensitive in that it will not multiply at temperatures above 20°C. The amino acid sequence was determined by Wittmann et al.[60]. It is identical with that of common TMV except for a single mutation in position 19, where alanine appears instead of aspartic acid. This protein thus has one fewer carboxyl group than the protein from the common strain. It should therefore be less negative at pH 7-8 to the extent of 1 charge per protein monomer. The protein should also be more hydrophobic than the protein from vulgare. Flavum protein aggregates to form rods but not at temperatures as high as 30°C. It is apparently unstable and denatures readily.

The polymerization properties of this protein have recently been studied in the author's laboratory[61]. In the pH range 6-7, flavum protein sediments at a concentration of 1 milligram per ml with an s^{20}_{w} of 2.2 - 2.5, compared to the value of 4S for vulgare protein under similar conditions. Sedimentation equilibrium studies were carried out by the method of Stevens[62]. The results are shown in Figure 16. At pH 6.0, a monomer-tetramer fit best represents the data, but at pH 7.0 the best fit is obtained with condensation polymerization of monomers.

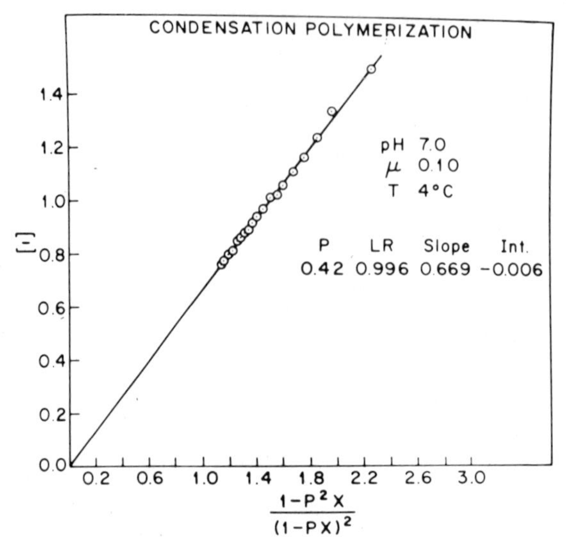

Fig. 16. Sedimentation equilibrium data on flavum TMV protein. Ξ is a function of concentration. X is a reduced positional variable and P is the fraction polymerized. The line is theoretical and the circles are experimental. See Stevens[62].

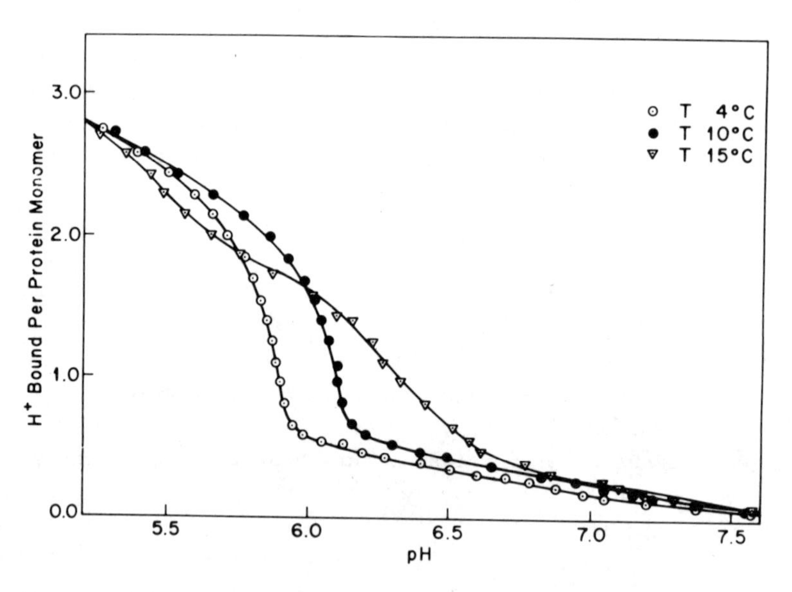

Fig. 17. Acid base titration of protein of the flavum strain of TMV in 0.1 M KCl.

Hydrogen ion titrations were carried out from pH 8 to 5 and back again to pH 8 in 0.1 μ KCl at a protein concentration of 3.5 mg/ml at 4, 10 and 15°C. The results are shown in Figure 17. The titration is reversible in the sense that data obtained in titrating with HCl from high to low pH and back again to high pH with KOH follow the same curve. However, such reversible results were obtained only when the titration was carried out so slowly that the total experiment in both directions took at least 10 hours; otherwise a hystereses loop developed. The isoionic point is near pH 5; thus, the charge at pH 7.5 is -3. Sedimentation velocity measurements in 0.08 μ KCl + 0.02 μ phosphate buffer were performed at 3.5 mg/ml at pH 5.88, 6.08, 6.17, 6.30, 6.45 and 6.65. The percentages of various aggregates present at each pH value were determined from the areas under the schlieren peaks. The number of hydrogen ions bound per monomer was taken from Figure 17. The results are shown in Table 5. As shown in Figure 18, hydrogen ion binding correlates with the disappearance of the smallest particles; it does not depend upon the size of the aggregates formed.

Polymerization studies were carried out by the turbiditic method of Smith and Lauffer[53]. Parallel studies were carried out on TMV protein under identical conditions. The results are displayed in Figure 19, where the solid curves represent the flavum protein and the dashed curves the vulgare protein. These results are completely reversible in the sense that data taken with rising and with falling temperatures fit the same graph. In spite of the lower charge and the higher hydrophobic residue content of the flavum protein, its polymerization closely resembles that of the common protein at 0.2 pH units higher. Stated another way, at a given pH, it requires a temperature several degrees higher to polymerize flavum protein than to polymerize vulgare protein.

When the effect of ionic strength at different pH values was studied, however, results were obtained which do correlate with the higher hydrophobicity and lower net charge. These experiments were conducted exactly the same way as were the experiments on TMV protein shown in Figure 13. The results on the flavum protein

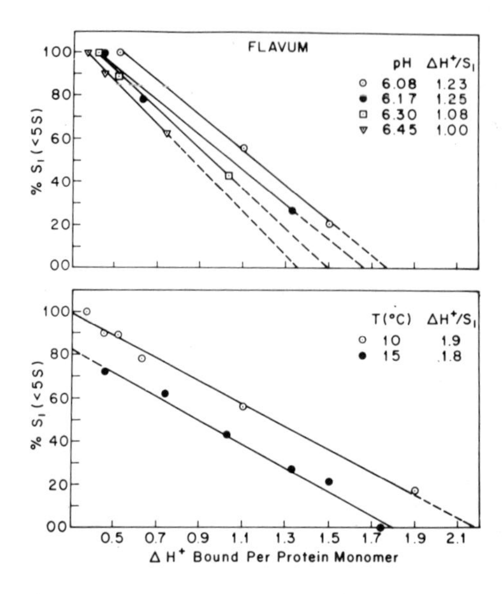

Fig. 18. Hydrogen ion binding by flavum TMV protein versus fraction in <5-S state.

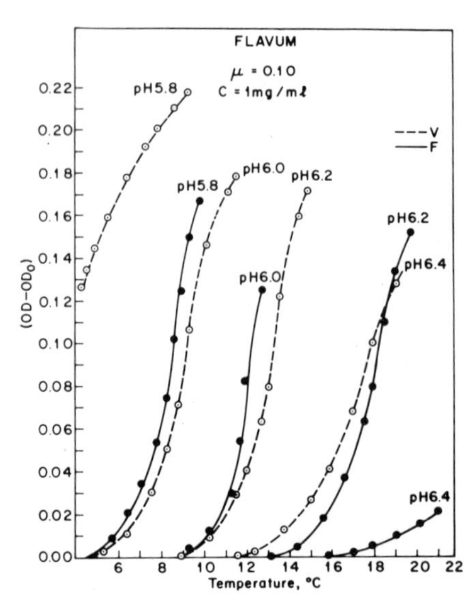

Fig. 19. Optical density ($OD - OD_0$) versus temperature for the polymerization of vulgare (---) and flavum (——) TMV protein at several pH values.

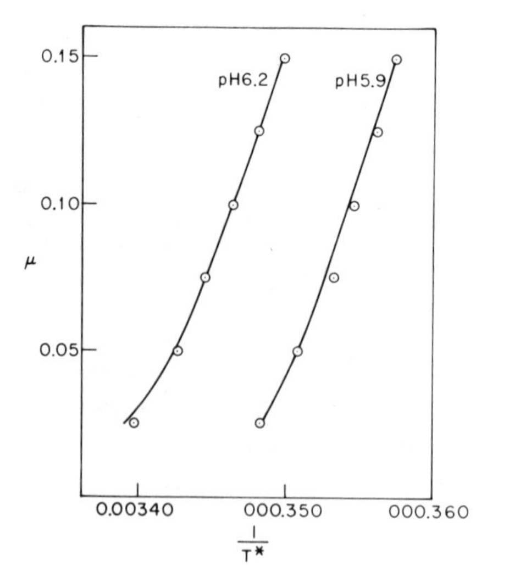

Fig. 20. Effect of ionic strength (μ) on the characteristic temperature, T^* for polymerization of flavum TMV protein. See Fig. 13.

TABLE 5
SEDIMENTATION VELOCITY-HYDROGEN ION BINDING DATA FOR FLAVUM TMV PROTEIN POLYMERIZATION μ = 0.10

T	4					10					15				
pH	S_1	$\%S_1$	S_2	$\%S_2$	H^+_{bd}	S_1	$\%S_1$	S_2	$\%S_2$	H^+_{bd}	S_1	$\%S_1$	S_2	$\%S_2$	H^+_{bd}
5.88	3.7	100	-	-	1.1	2.5	17	173	83	1.9	-	-	153	100	1.74
6.08	3.7	100	-	-	0.52	4.4	56	123	44	1.1	1.9	20	153	80	1.5
6.17	3.9	100	-	-	0.45	4.3	78	35.9	23	0.63	3.5	27	149	73	1.33
6.30	4.0	100	-	-	0.42	4.4	89	29	11	0.52	3.9	43	36.7	57	1.03
6.45	3.7	100	-	-	0.37	4.0	90	21.5	10	0.45	4.4	63	28.9	37	0.74
6.65	4.0	100	-	-	0.30	4.1	100	-	-	0.37	4.4	72	27.4	28	0.46

are shown in Figure 20. The points are experimental data and the curves are graphs of Eq. 20. The parameter B at both pH 5.9 and 6.2 was .00058 compared to a value of .00040 for vulgare. This means that the salting-out constant of the flavum protein is considerably higher than that of type strain protein. This is to be expected of a protein with a higher content of hydrophobic residues. In contrast, the values of the parameter A was 4.8 x 10^{-6} at pH 5.9. Values for TMV protein, presented in Table 4, are from 2 to 3 times higher. The parameter A is related to the change in electrical work upon polymerization. The fact that it is much lower for the flavum protein is consistent with the lower net negative charge expected for this protein.

2. Protein from the E66 mutant

Similar studies were carried out on the E66 mutant of TMV[61]. This mutant has lysine in position 140 instead of asparagine; otherwise its amino acid sequence is identical with that of the protein from the common strain[63]. Like the protein from the flavum strain, E66 protein should also have one fewer net negative charges per monomer at pH 7-8. The acid-base titration curves at 4, 10, and 15°C are shown in Figure 21. Attempts were made to obtain

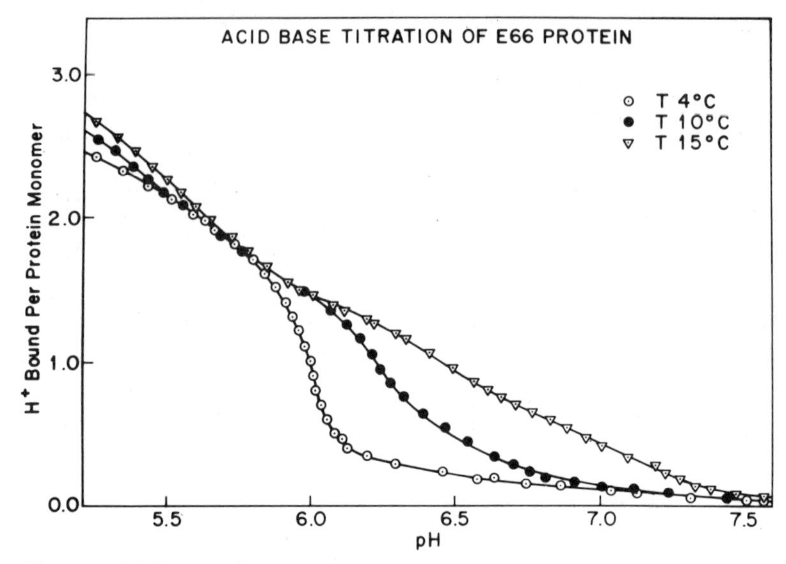

Fig. 21. Acid-base titration of protein of the E66 mutant of TMV in 0.1 M KCl.

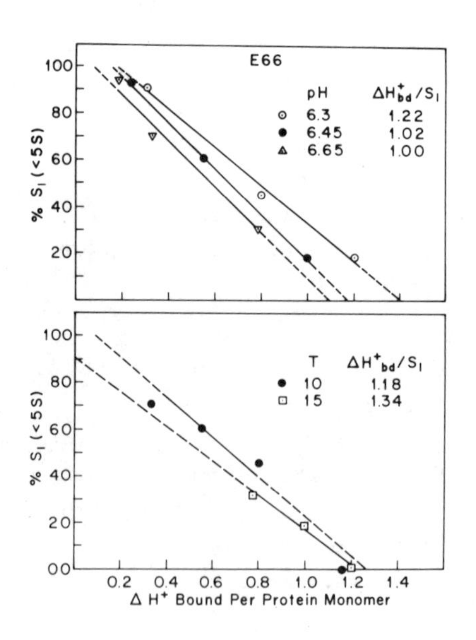

Fig. 22. Hydrogen ion binding by E66 TMV protein versus fraction in <5S state.

titration data at 20°C, but the protein denatured and the titration was irreversible. The results displayed are reversible, but such reversible results are obtained only when the titration in both directions is carried out so slowly that it takes 10 to 14 hours; with faster titration, a hysteresis loop is obtained. It was difficult to obtain reversible titration at 15°C. Rapid titration gives a hysteresis loop, slow titration at 15°C results in denaturation. When the time of the experiment was shortened by titrating down only to pH 5.8 instead of 5.0, it was possible to obtain complete reversibility. For this protein, too, the isoionic point is near pH 5. The titration results then indicate a charge of -3 per monomer at pH 7.5

TABLE 6

SEDIMENTATION VELOCITY-HYDROGEN ION BINDING DATA FOR E66 TMV PROTEIN POLYMERIZATION μ = 0.10

T	4					10					15				
pH	S_1	$\%S_1$	S_2	$\%S_2$	H^+_{bd}	S_1	$\%S_1$	S_2	$\%S_2$	H^+_{bd}	S_1	$\%S_1$	S_2	$\%S_2$	H^+_{bd}
6.08	4.1	86	28.3	14	0.52	-	-	127	100	1.34	-	-	-	-	1.40
6.17	4.1	90	24.4	10	0.40	-	-	109	100	1.16	-	-	109	100	1.32
6.30	3.9	92	23	8	0.30	3.8	45	27.1	55	0.80	45.5	19	74.1	81	1.20
6.45	4.9	94	22.4	6	0.23	4.7	61	24.4	39	0.55	4.9	19	32.3	81	1.00
6.65	4.7	95	21.1	5	0.18	4.6	70	20.9	30	0.33	3.5	31	24.7	69	0.78

These titration results were correlated with the aggregates present at the various pH values and temperatures, just as was done for the flavum strain. The results are shown in Table 6. Figure 22 shows that hydrogen ion binding at the 3 pH values is correlated with the disappearance of the smallest aggregates and is independent of the type of higher polymer formed.

In Figure 23 are shown results of experiments in which polymerization of E66 protein was followed by turbidmetry. The results obtained with the E66 strain are shown as solid lines and those obtained simultaneously under identical conditions for TMV protein are shown as dashed lines. In this case, E66 protein polymerizes at lower temperatures than does TMV protein under comparable conditions. The effect of ionic strength was also determined; the results are shown in Figure 24. Here, too, the curves are graphs of Eq. 20 with parameters A, B, and C adjusted to fit the data. In this case, the parameter B is .00026 compared to .00040 for vulgare TMV. This indicates that the salting-out constant of E66 protein is considerably less than that for TMV, a result consistent with the nature of the amino acid substituted in the mutation. The parameter A has values of 5.2 and 6.2×10^{-6} at pH 6.0 and 6.4 respectively. Again, these numbers are much lower than the parameters for TMV listed in Table 4 for vulgare and are entirely consistent with the lower net negative charge on the E66 protein.

ENTROPY-DRIVEN POLYMERIZATION OF THE COAT PROTEINS FROM OTHER VIRUSES

A. Papaya Mosaic Virus Protein

The polymerization of the coat protein of tobacco rattle virus was reviewed previously[7]. The coat protein of papaya mosaic virus has been isolated by acetic acid degregation[64]. Papaya mosaic virus is a long thin rod, a flexous virus. The coat protein polymerizes endothermically and, therefore, by an entropy-driven process, at pH 4 in 0.01 M citric acid buffer to form polymers with sedimentation coefficients of 13S, 100-120S and 230-250S. The 13S component is probably a double ring of subunits, but the larger particles are long but flexible helices with a pitch of 36 Å, the same as that found in the virus[64].

B. Alfalfa Mosaic Virus Protein

An unusual example of entropy-driven polymerization comes from studies on alfalfa mosaic virus protein carried out by Driedonks et al.[65] Alfalfa mosaic virus is complex in that the total genome consists of three different RNA molecules, each of which

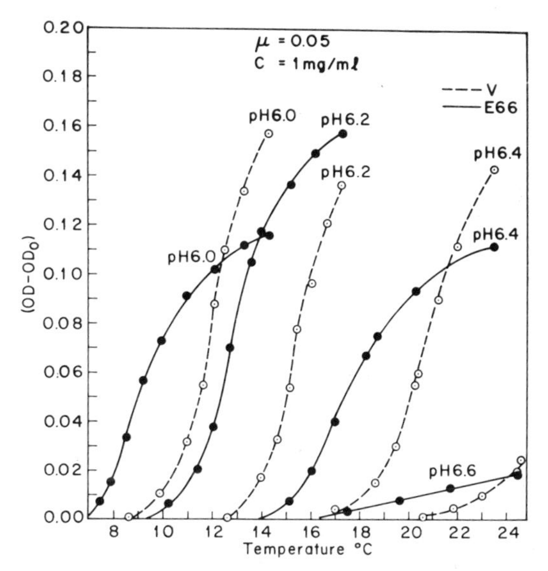

Fig. 23. Optical density (OD - OD_o) versus temperature for the polymerization of vulgare (---) and E66 (—) TMV protein at several pH values.

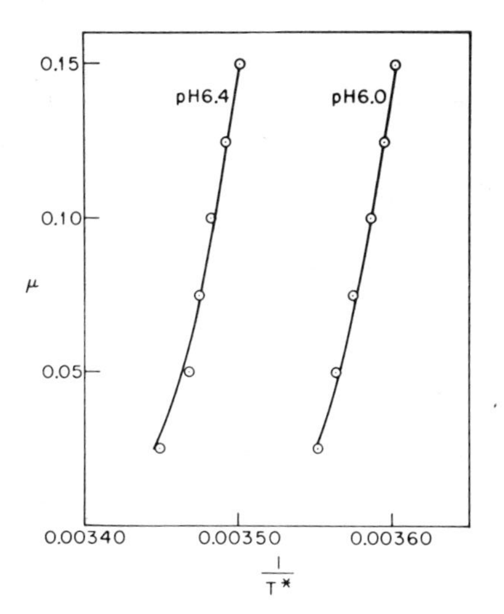

Fig. 24. Effect of ionic strength (μ) on the characteristic temperature, T^* for polymerization of E66 TMV protein. See Fig. 13.

copolymerizes with protein of molecular weight 24,250 to form cylinders of different lengths with rounded ends. The isolated coat protein polymerizes spontaneously, not into rods, but rather into hollow icosahedra, each consisting of thirty dimers of the protein subunit. Such particles, sometimes referred to as spheres, sediment at 30S. The dimers, with molecular weight 48,500, sediment at 3S. Polymerization is entropy-driven. In phosphate buffer at pH 5.5 and ionic strength 0.5, the enthalpy and entropy of inserting one dimer into the lattice are 6.4 kcal/mole and 50 e.u., calculated on the basis of unit mole fraction. In pyrophosphate buffer at $4^{o}C$, the polymerization exhibited a maximum at pH 5.5. The extent of polymerization is considerably greater in buffers of ionic strength 0.50 than when the ionic strength is 0.90. At values of ionic strength below 0.4, the 30S particles associate into larger aggregates. At $4^{o}C$, pH 5.5 and ionic strength 0.50, different anions have widely different effects on the extent of polymerization. The anions, in ascending order of effect upon polymerization, are acetate, oxalate, citrate, phosphate, pyrophosphate, and triphosphate. The equilibrium constant for the polymerization in triphosphate is 40 times that in acetate.

AGGREGATION OF VIRUS PARTICLES

A. Tobacco Mosaic Virus

Brief reference was made previously[7] to unpublished experiments of Anderer and Lauffer who found positive heats of crystallization for tobacco mosaic virus[7]. Additional experiments have been reported by Lauffer and Shalaby[33]. A solution containing 2.56 milligrams per ml of TMV in 0.60 M $(NH_4)_2SO_4$ and 0.077 M PO_4 and pH 6.6 had the appearance of a typical TMV solution at $0^{o}C$. When the temperature was raised somewhat above room temperature, the solution became strongly opalescent and birefringent, characteristic of the earliest stages of the crystallization of TMV. When the temperature was reduced again to $0^{o}C$, the strong opalescence and birefringence disappeared. However, when the experiment was repeated in 0.74 M $(NH_4)_2SO_4$ and 0.095 M PO_4, non reversible opalescence and birefringence developed. These aggregates were not thixotropic. When TMV at concentrations ranging between 1 milligram per ml and 9.9 milligrams per ml was placed in 0.30 - 0.45 M $(NH_4)_2SO_4$ and buffered to pH 5.6 with either 0.15 μ PO_4 or 0.2 μ acetate-chloride buffer, paracrystals were formed which disappeared on vigorous shaking, thus exhibiting thixotropic behavior. The experiments demonstrate that at pH 6.6 the earliest stages of the crystallization of TMV are both reversible and endothermic or entropy-driven and that at pH 5.6, crystallization is thixotropic.

B. Carnation Ringspot Virus

Recent experiments show that the aggregation of carnation ringspot virus is entropy-driven[66,67]. Carnation ringspot virus has an isometric virion which sediments at 132S. Four different strains have been studied. One strain forms aggregates of 12 particles, irreversibly. The other three are opalescent at 4°C and become turbid at higher temperature. The aggregation is reversible; the preparations become opalescent again when cooled to 4°C. There is considerable difference among the strains with respect to the temperature at which turbidity is developed, ranging from 25 to 40°C. Increasing virus concentration reduces the temperature required for aggregation.

PROTEINS AND ENZYMES

A. Hemoglobin

That the formation of dimers, tetramers or higher polymers of many proteins are entropy-driven processes was reviewed previously[7]. Additional examples are cited here. The ΔH^o and ΔS^o of the association of carboxy hemoglobin A_o dimers to form tetramers were found to be +13.2 kcal/mole and +71 e.u. at pH 8.6. Values at pH 6.0 were -8.2 kcal and -4.4 ± 6.3 e.u.[68]. Thus, this association is definitely entropy-driven at pH 8.6 and might be at pH 6.0. The binding of carbon monoxide to hemoglobins M Milwaukee-I and Saskoon at 25°C and pH 7.0 was investigated by Nakamura et al.[69]. The binding of the first CO exhibits negative values of ΔH_o and ΔS_o. However for the binding of the second CO, ΔH^o was about 5 kcal/mole and ΔS^o was about +50 e.u.

Isolated α^{SH} chains and β^{SH} chains from human hemoglobin are capable of forming self polymers as well as the tetrameric hemoglobin molecule. The thermodynamics of the various polymerizations was studied in the laboratory of Ackers. Isolated α chains form α dimers and the β chains form β tetramers. The van't Hoff enthalpies for the two processes are +4.3 and +23.5 kcal/mole and the entropies are +40.6 e.u. and +177.5 e.u. These studies were made on fully oxygenated chains[70]. In contrast, the formation of α-β dimers is exothermic. The calorimetrically determined enthalpy is -15.7 kcal/mole[71]. When the α-β dimers form the hemoglobin tetramer, the reaction is endothermic and therefore entropy-driven when the dimers are oxygenated[72]. The van't Hoff enthalpy is +3.8 kcal/mole and the entropy is +48.4 e.u. However, when the dimers are unoxygenated, the enthalpy is -28.9 kcal/mole and the entropy change is -41.8 entropy units. Since the formation of oxygenated α-β dimers is strongly exothermic, the overall formation of the hemoglobin tetramer from its 2 α and 2 β chains is exothermic.

B. Lipoproteins

The protein, apo A-II is a component of the high density lipoprotein complex. The protein from man is a disulfide dimer, of molecular weight 17,380, of two identical chains of 77 residues each. This protein is capable of self association to form a dimer with molecular weight 34,760. When the protein is reduced with sulfhydryl reagents and then carboxymethylated to protect the exposed sulfhydryl groups, apo A II exists in a monomeric form with molecular weight 8690. This monomer possesses very little organized structure; it is essentially a random coil. It polymerizes to form a dimer with molecular weight 17,380 which has appreciable secondary and tertiary structure. The negative free energy of dimerization varies in an interesting manner with temperature[73]. It has its greatest negative value at about 28°C and decreases upon both lowering and raising the temperature. In 0.01 M phosphate buffer at pH 7.4, the enthalpy of dimerization decreases from about +20 kcal/mole at 7°C to about -28 kcal/mole at 47°. Paralleling this, the entropy decreases from about +100 e.u. at the lower temperature to about -70 e.u. at 47°C. There is a decrease in heat capacity upon dimerization of 1250 cal/mole degree at 25°C. Folding of random coils accompanies dimerization. One would expect a decrease in entropy for the folding and an increase in entropy for the association if it results in released solvent. At low temperatures, the increase in entropy exceeds the decrease and the process is endothermic and entropy-driven, but at temperatures above 28°C, the reverse is true. The change in heat capacity shows that the unfolded monomer has a higher heat capacity than the structured dimer. Similar decreases in heat capacity have been observed for protein denaturation, which resembles the unfolding acconpanying dissociation of the dimer[73]. APO A-II dimerization behaves similarly[74].

C. α-chymotrypsin

Substrate analogues, proflavin, L-AcTryp and D-AcTryp bind to native α-chymotrypsin at 25°C in 0.05 M sodium phosphate buffer at pH 7.8 and 0.1 M NaCl with negative values of ΔH^o and ΔS^o. However, when serine 195 is chemically modified to dehydroalanine or when histidine 57 is methylated, the bindings have very small positive or negative values of ΔH^o but positive values of ΔS^o [75].

Aune et al.[76] studied the dimerization of α-chymotrypsin. ΔH^o varied with temperature. It was +14.51 kcal/mole at 0°C and gradually decreased to -10.13 kcal/mole at 35°C. The unitary entropy change varied from +79 e.u. at 0°C to -5 e.u. at 35°C. ΔC_p was about -700 cal/mole degree. Thus, at least at low temperatures the dimerization of α-chymotrypsin is an entropy-driven process.

D. Glutamate Dehydrogenase

Rather similar results were obtained for bovine liver glutamate dehydrogenase by Reisler and Isenberg[77]. The linear association was studied in 0.2 m sodium phosphate buffer at pH 7 containing 10^{-4} M EDTA. Measurements were made in the temperature range 10-40°C. ΔS^o was approximately +60 e.u. at 10°C and decreased smoothly to 0 at 40°C. ΔH^o decreased from approximately 9 kcal/mol at 10° to about -8 kcal/mole at 40°C. Generally similar results were obtained by Gauper et al.[78].

E. Casein

It was previously pointed out[7] that the precipitation of β casein is an endothermic reaction. More recent information indicates that the complex formation between $\alpha_s 1$ and k-casein polymers in an endothermic, entropy-driven process[79].

F. Binding of Fatty Acids by Albumin

Teresi and Luck[80] measured by equilibrium dialysis the binding of the fatty acid ions, acetate, propionate, caproate, heptanoate and caprylate to serum albumin in 0.2 μ PO_4 buffer at pH 7.6. Values for ΔS^o for binding to the strongest binding centers ranged from +1.42 e.u. for caprylate to +12 e.u. for propionate, but the ΔH^o value ranged from -4.7 kcal/mole to 0 for the same sequence of ions. Lower values of both ΔS^o and ΔH^o were reported for the weaker binding centers. Acetate ion exhibited a value of 0 for ΔH^o for both centers and +8.7 and +4.6, respectively, for the entropy. The workers concluded that the primary binding is of the carboxyl ion with positive groups on the protein. Two kinds of evidence were cited to support this conclusion. The stabilizing effect on thermal denaturation of serum albumin exhibited by the caprylate ion is similar to that of raising the pH as high as 10.8 [81]. The binding of organic anions with chromophores, as determined by spectrometric data, is abolished by acetylation of ε-amino groups and by reacting the guanidium groups with formaldehyde[82]. The fact that ΔS^o increased with increasing hydrocarbon chain length, except for acetate, led the workers to conclude that this resulted from nonpolar interactions with the protein, what we would now call an entropy-driven process. Results of the same general character for propionic, butyric, caproic, caprylic and heptanoic acids were reported recently[83].

COLD INACTIVATION OF ENZYMES

It has already been seen that several enzymes undergo endothermic or entropy-driven dimerization or polymerization. Many enzymes are inactive in the unpolymerized form. In 1949 Hofstee[84] reported that the enzymic activity of crystallized urease was increased by moderate heating and reversibly decreased by storage in the cold.

Since that time the phenomenon of cold inactivation has been reported for many enzymes. Among them are glutamic acid dehydrogenase[85], glutamic acid decarboxylase[86], adenosine triphosphatase[87,88], carbamyl phosphate synthethase[89], glucose-6-phosphate dehydrogenase[90], D -β-hydroxybutyric acid dehydrogenase[91], glycogen phosphorylase[92], a nitrogen fixing enzyme[93], acetyl Co-A carboxylase[94], arginosuccinase[95], 17β-hydroxysteroid dehydrogenase[96] and pyruvate carboxylase[97]. For most [84,85,90-97], evidence is presented that the enzyme is reactivated upon rewarming to room temperature or slightly higher temperatures. For some[90,94-97], ultracentrifugation evidence is presented to show that the active form of the enzyme has a higher sedimentation coefficient than the inactive form found in the cold.

A definitive study of pyruvate carboxylase was reported by Irias et al.[97]. The enzyme studied was isolated from chicken liver. It catalyzes the reaction of pyruvate with CO_2 and ATP in the presence of Mg^{++} and acetyl CoA to form oxalacetate, ATP and phosphate. This enzyme is rapidly inactivated in the cold and partially reactivated upon rewarming to $23^{o}C$. The inactive form is correlated with material sedimenting at 6.75S and the active form with material sedimenting at 14.8S. The reaction was interpreted as an inactive monomer-active tetramer transition. The equilibrium constant is shifted in the direction of the active polymer by high concentrations of enzyme, high concentration of KCl (0.6 M), 2.5 M methyl alcohol, 1.5 M sucrose, 80mM PO_4, oxalacetate and D_2O. Ethanol and acetone decrease the stability of the enzyme at room temperature. There is little effect of pH in the range 6.65-7.42. Even though oxalacetate provided protection, pyruvate did not. In the presence of 5mM ATP, cold inactivation can occur without depolymerization of the tetramer.

Sufficient evidence has been amassed to indicate that the activation of those enzymes which are inactive in the cold is normally associated with entropy-driven polymerization to form the active enzyme. Especially in the case of pyruvate carboxylase, the variables, with the exception of pH, which affect polymerization parallel in general those which affect the polymerization of TMV protein. The evidence is highly suggestive that release of water molecules upon polymerization is the source of the entropy increase with this enzyme, just as with TMV protein. The entropy-driven polymerization-depolymerization affords another possible mechanism for the control of enzymatic activity.

ANTIGEN-ANTIBODY REACTIONS

Two antigen-antibody reactions were discussed previously[7]. In both cases, the enthalpy had small negative or small positive values and the entropy change was strongly positive. Hardie and van Regenmortel[98] studied the binding of univalent TMV-specific Fab fragments prepared from hyperimmune rabbit serum. Their data show that the equilibrium constant decreases from 1.12×10^6 at 4^oC to 0.91×10^6 at 45^oC. Even though the equilibrium constant decreases on increasing temperature, the standard free energy change increases regularly from -32.07 k joules or -7.67 kcal at 4^oC to -36.27 k joules or -8.67 kcal at 45^oC. As will be pointed out in the next paragraph, there is danger inherent in interpreting results in which ΔG^o varies so little. However, in this case the data seem to follow a trend without evidence of substantial error. From these results one can calculate that the enthalpy of the polymerization is -0.896 kcal/mole of antibody bound and the entropy change in polymerization is +24.25 e.u. expressed in terms of calories. Over the whole temperature range, $T\Delta S$ greatly exceeds ΔH in magnitude, so that, even though ΔH is negative, the major portion of the negative ΔG^o is derived from the increase in entropy. This reaction is, therefore, primarily entropy-driven. All of these calculations are based on the usual thermodynamic equations. It is, therefore, a bit disturbing that attempts to reverse the reaction by dilution or by temperature change were unsuccessful.

Szewczuk and Mukkur[99] studied the binding of ε-DNP-L-lysine to bovine colostral anti-DNP IgG 1 and also to rabbit anti-DNP IgG. The free energies of both associations changed only slightly over the temperature range -3 to 67^oC. Eq. 1 shows that when ΔG^o is invariant with temperature, ΔS^o is zero. However, computer analysis of the slight variations with temperature yielded values of ΔH^o and ΔS^o for both systems which ranged from strongly negative at the lowest temperature to positive or strongly positive at the highest temperature for both systems, suggesting enthalpy-entropy compensation. Regardless of the sophistication of the analysis, this result stems from data for which ΔG^o varies only slightly. Furthermore the errors reported for the ΔG^o values are relatively high. In the case of the binding of the colostral antibody, the standard deviation indicated for the ΔG^o data is of such magnitude that no value of ΔG^o at any temperature from -3.0 to 67.0^o differs from the mean of the values at all temperatures by as much as 1 standard deviation. In the case of the binding of the rabbit antibody, the variation with temperature of ΔG^o is sufficiently greater than the indicated standard deviations to suggest a trend.

BIOLUMINESCENCE

In the bioluminescent system from the boring mollusc, *Pholas dactylus*, both the luciferase, with a molecular weight of 310,000 and the luciferin, with a molecular weight of 34,000 are glycoproteins. Henry and Monny[100] showed that the formation of the complex between oxyluciferin and luciferase is entropy-driven; ΔH^o is +3.5 kcal/mole and ΔS^o is +47.3 e.u.

MICROTUBULES

The evidence that *in vitro* polymerization of tubulin isolated from rat brain is entropy-driven was discussed previously[7]. In the intervening years, considerable controversy has arisen over the question whether tubulin isolated from the calf brain can be polymerized into microtubules and whether or not this is an entropy-driven process. The issue has been clarified in the laboratory of Timasheff. Tubulin prepared by the method of Weissenberg can be polymerized. It is reversed at 20°C, inhibited by calcium ion and favored by glycerol[101]. The polymerization process involves the binding of 1 magnesium ion and 1 proton and presumably the release of water. The enthalpy and entropy are strongly positive at 23°C, but both decrease on increasing the temperature to 42°C. At that temperature the enthalpy is slightly negative but the entropy is still slightly positive. The heat capacity decreases 1500 calories per degree mole[102]. Vinblastine induces stable tubulin dimers to form tetramers. The binding of vinblastine to tubulin involves a positive enthalpy change of 5.8 kcal/mole and a positive unitary entropy change[103].

ACTIN-MYOSIN INTERACTION

Fragments of evidence indicating that the interaction of actin and myosin, the most important reaction in muscle contraction, is an entropy-driven process were reviewed previously[7]. A new study by Highsmith[104] strengthens the case substantially. The association constant for the binding of myosin subfragment S-1 to actin was determined as a function of ionic strength and temperature. In 0.15 M KCl at pH 7.0, ΔH^{o} was found to be +184 kcal/mole and ΔS^{o} +1170 e.u. The binding was not only endothermic, but also it is inhibited by high ionic strength. This interaction is extremely important to the process of muscle contraction because the S-1 fraction is the head of the myosin molecule.

ACKNOWLEDGEMENTS

This is a publication from the Biophysical Laboratory of the Department of Biological Sciences, University of Pittsburgh. Work was supported by a U.S. Public Health Service Grant (GM 21619).

REFERENCES

1. Tanford, C. (1970) Adv. Prot. Chem. 24, 1-95.
2. Szegvari, A. (1924) Kolloid Z. 34, 34-37.
3. Heyman, E. (1935) Trans. Faraday Soc. 31, 846-864.
4. Freundlich, H. (1937) J. Phys. Chem. 41, 901-910.
5. Stevens, C.L. and Lauffer, M.A. (1965) Biochemistry 4, 31-37.
6. Lauffer, M.A. (1964) Biochemistry 3, 731-736.
7. Lauffer, M.A. (1975) "ENTROPY-DRIVEN PROCESSES IN BIOLOGY: Polymerization of Tobacco Mosaic Virus Protein and Similar Reactions. Molecular Biology, Biochemistry and Biophysics, Vol. 20, Springer-Verlag, Berlin-Heidelberg, New York.
8. Lauffer, M.A. and Stevens, C.L. (1968) Adv. Virus Res. 13,1-63.
9. Lauffer, M.A., Ansevin, A.T., Cartwright, T.E., Brinton, C.C., Jr. (1958) Nature 181, 1338-1339.
10. Jaenicke, R. and Lauffer, M.A. (1969) Biochemistry 8, 3083-3092.
11. Kauzmann, W. (1959) Advan. Protein Chem. 14, 1-63.

12. Lauffer, M.A. (1964) Protein-Protein Interaction; Endothermic Polymerization and Biological Processes. In: Proteins and Their Reactions (H.W. Schultz, A.F. Anglemier, eds.), Chap. 5, Westport/Conn.: Avi Publ. Co.

13. Schramm, G. (1943) Naturwissenschaften, 31, 94-96.

14. Fraenkel-Conrat, H. (1957) Virology 4, 1-4.

15. Atlas of Protein Sequence and Structure (1972) Vol. 5, (M.O. Dayhoff, Ed.), p. D-285.

16. Banerjee, K. and Lauffer, M.A. (1966) Biochemistry 5, 1957-1964.

17. Fraenkel-Conrat, H. and Williams, R.C. (1955) Proc. Natl. Acad. Sci., U.S. 41, 690-698.

18. Lippincott, J.A. and Commoner, B. (1956) Biochim. Biophys. Acta, 19, 198-194.

19. Srinivasan, S. and Lauffer, M.A. (1970) Biochemistry 9, 2173-2180.

20. Srinivasan, S. and Lauffer, M.A. (1973) Arch. Biochem. Biophys. 158, 53-66.

21. Schramm, G. (1947) Z. Naturforsch, 2b, 249-257.

22. Franklin, R.E. (1955) Biochim. Biophys. Acta 18, 313-314.

23. Caspar, D.L.D. (1963) Advan. Protein Chem. 18, 37-131.

24. Markham, R., Frey, S. and Hills, G.J. (1963) Virology 20, 88-102.

25. Finch, J.T., Leberman, R., Yu-Shang, C. and Klug, A. (1966) Nature (London) 212, 349-350.

26. Champness, J.N., Bloomer, A.C., Bricogne, G., Butler, P.L.G. and Klug, A. (1976) Nature (London) 259, 20-24.

27. Lauffer, M.A., Shalaby, R.A.F. and Khalil, M.T.M. (1967) Chimia (AARAU) 21, 460-462.

28. Durham, A.C.H. and Klug, A. (1972) J. Mol. Biol. 67, 315-322.

29. Durham, A.C.H. and Klug, A. (1971) Nature New Biol. 229, 42-46.

30. Lauffer, M.A. (1962) In: The Molecular Basis of Neoplasia (A collection of papers presented at the Fifteenth Ann. Symp. on Fundamental Cancer Research, 1961, at the University of Texas M.D. Anderson Hospital and Tumor Institute at Houston, Texas, pp. 180-206, University of Texas Press, Austin.

31. Ansevin, A.T., Stevens, C.L. and Lauffer, M.A. (1964) Biochemistry 3, 1512-1518.

32. Khalil, M.T.M. and Lauffer, M.A. (1967) Biochemistry 6, 2474-2480.

33. Lauffer, M.A. and Shalaby, R.A.F. (1977) Arch. Biochem. Biophys. 178, 425-434.

34. Scheele, R.B. and Lauffer, M.A. (1967) Biochemistry 6, 3076-3081.

35. Eiskamp, J.G. (1969) A Study of the Titrimertic, Sedimentation and Optical Behavior of the Structural Protein of Tobacco Mosaic Virus in Various States of Aggregation. Ph.D. Dissertation. Eugene, Oregon.

36. Paulsen, G. (1972) Z. Naturforsch. 27b, 427-444.

37. Butler, P.J.G., Durham, A.C.H. and Klug, A. (1972) J. Mol. Biol. 72, 1-18.

38. Vogel, D. and Jaenicke, R. (1974) Eur. J. Biochem. 41, 607-615.

39. Scheele, R.B. and Schuster, T.M. (1975) J. Mol. Biol. 94, 519-525.

40. Durham, A.C.H. and Hendry, D.A. (1977) Virology 77, 510-519.

41. Shalaby, R.A.F. and Lauffer, M.A. (1977) J. Mol. Biol. 116, 709-725.

42. Durham, A.C.H., Vogel, D. and DeMarcillac, D.G. (1977) Eur. J. Biochem. 79, 151-159.

43. Stevens, C.L. and Loga, S. (1979) in press)

44. Lauffer, M.A. (1944) J. A. C. S. 66, 1195-1201.

45. Shalaby, R.A.F., Banerjee, K. and Lauffer, M.A. (1968) Biochemistry 7, 955-960.

46. Banerjee, K. and Lauffer, M. A. (1971) Biochemistry 10, 1100-1102.

47. Gallagher, W. and Lauffer, M.A. (unpublished results).

48. McMichael, J.C. and Lauffer, M.A. (1975) Arch. Biochem. Biophys. 169, 209-216.

49. Lauffer, M.A. (1971) Tobacco Mosaic Virus and Its Protein. In Subunits in Biological Systems (S.N. Timasheff, G.E. Fasman, Eds.), Part A, Chap. 4, pp. 149-199. New York: Dekker.

50. Gorin, M.H. (1942) in: Electrophoresis of Proteins and the Chemistry of Cell Surfaces by H.A. Abramson, L.S. Moyer and M.H. Gorin, p. 126. New York: Rheinhold.

51. Hill, T.L. (1955) Arch. Biochem. Biophys. 57, 229-239.

52. Kramer, E. and Wittmann, H.G. (1958) Z. Naturforsch 13b, 30-33.

53. Smith, C.E. and Lauffer, M.A. (1967) Biochemistry 6, 2457-2465.

54. Shalaby, R.A.F. and Lauffer, M.A. (1967) Biochemistry 6, 2465-2473.

55. Lauffer, M.A. and Shalaby, R.A.F. (in preparation).

56. Green, A. A. (1931) J. Biol. Chem. 93, 495-516.

57. Stauffer, H., Srinivasan, S. and Lauffer, M.A. (1972) Biochemistry 9, 193-200.

58. Brandts, J.F. (1964) J. Am. Chem. Soc. 86, 4302.

59. Melchers, G. (1940) Biol. Zbl. 60, 527-531.

60. Wittmann, H.G., Wittmann-Leibold, B. and Jaurequi-Adell, J. (1965) Z. Naturforsch 20b, 1224-1234.

61. Shalaby, R.A.F. and Lauffer, M.A. (1979) Manuscript in preparation.

62. Stevens, C.L. (1975) FEBS Letters 56, 12-15.

63. Wittmann, H.G. (1964) Z. Vererbungsl 95, 333-344.

64. Erickson, J.W., Bancroft, J.B. and Horne, R.W. (1976) Virology 72, 514-517.

65. Driedonks, R.A., Krigsman, P.C.J. and Mellema, J.E. (1977) J. Mol. Biol. 13, 123-140.

66. Tremaine, J.H., Ronald, W.P. and Valcic, A. (1976) Phytopathology 66, 34-39.

67. Tremaine, J.H. and Ronald, W.P. (1976) J. Gen. Virol. 30, 299-308.

68. Ide, G.E., Barksdale, A.D. and Rosenberg, A. (1976) J. Am. Chem. Soc. 98, 1595-1596.

69. Nakamura, P., Sugita, Y., Hashimoto, K., Yoneyama, Y. and Pisciotta, A.V. (1976) Biochem. Biophys. Res. Comm. 70, 567-572.

70. Valdes, R., Jr. and Ackers, G.K. (1977) J. Biol. Chem. 252, 74-81.

71. Valdes, R., Jr. and Ackers, G.K. (1977) J. Biol. Chem. 252, 88-91.

72. Ip, S.H.C. and Ackers, G.K. (1977) J. Biol. Chem. 252, 82-87.

73. Osborne, J.C., Jr., Palumbo, G., Brewer, H.B., Jr., and Edelhoch, H. (1976) Biochemistry 15, 317-320.

74. Gwynne, J., Palumbo, G., Osborne, J.C.,Jr., Brewer, H.B.,Jr. and Edelhoch, H. (1975) Arch. Biochem. Biophys. 170, 204-212.

75. Schultz, R.M., Konovessi-Panayotatos, A. and Peters, J.R. (1977) Biochemistry 16, 2194-2202.

76. Aune, K.C., Goldsmith, L.C. and Timasheff, S.N. (1971) Biochemistry 10, 1617-1622.

77. Reisler, E. and Eisenberg, H. (1971) Biochemistry 10, 2659-2663.

78. Gauper, S.P., Markau, K. and Sund, H. (1974) Eur. J. Biochem. 49, 555-563.

79. Dosako, S., Kaminogawa, S. and Yamauchi, K. (1975) Agric. Biol. Chem. 39, 2347-2351.

80. Teresi, J.B. and Luck, J.M. (1952) J. Biol. Chem. 194, 823-834.

81. Boyer, P.D., Funston, G.L., Ballou, G.A., Luck, J.M. and Rice, R.G. (1946) J. Biol. Chem. 162, 181-198.

82. Teresi, J.B. (1950) J. Am. Chem. Soc. 72, 3972-3978.

83. Rodrigues de Miranda, J.F., Eikelbloom, T.D. and van Os, G.A.J. (1976) Molecular Pharmacology 12, 454-462.

84. Hofstee, E.H.J. (1949) J. Gen. Physiol. 32, 339-349.

85. Finchman, J.R.S. (1957) Biochem. J. 65, 721-728.

86. Shukuya, R. and Schwerdt, G.W. (1960) J. Biol. Chem. 235, 1658-1661.

87. Pullman, M.E., Penefsky, H.S., Datta, A. and Racker, E. (1960) J. Biol. Chem. 235, 3322-3329.

88. Penefsky, H.S. and Warner, R.C. (1965) J. Biol. Chem. 240, 4694-4702.

89. Raijman, L. and Gisola, S. (1964) J. Biol. Chem. 239, 1272-1276.

90. Kirkman, H.N. and Hendrickson, E.G. (1962) J. Biol. Chem. 237, 2371-2376.

91. Shuster, C.W. and Duodoroff, M. (1962) J. Biol. Chem. 237, 603-607.

92. Graves, D.J., Seabock, R.W. and Wang, J.H. (1965) Biochemistry 4, 290-296.

93. Dua, R.D. and Burris, R.H. (1963) Proc. Nat. Acad. Sci. 50, 169-175.

94. Numa, S. and Ringelmann, E. (1965) Biochem. Z. 343, 258-268.

95. Havir, E.A., Tamir, H., Ratner, S. and Warner, R.C. (1965) J. Biol. Chem. 240, 3079-3088.

96. Jarabak, J., Seeds, A.E.,Jr., and Talalay, P. (1966) Biochemistry 5, 1269-1279.

97. Irias, J.J., Olmsted, M.R. and Utter, M.F. (1969) Biochemistry 8, 5136-5148.

98. Hardie, G. and van Regenmortel, M.H.V. (1975) Immunochemistry 12, 903-908.

99. Szewczuk, M.R. and Makkur, T.K.S. (1977) Immunology 32, 111-119.

100. Henry, J.P. and Monny, C. (1977) Biochemistry 16, 2517-2525.

101. Lee, J.C. and Timasheff, S.N. (1976) Biochemistry 15, 5183-5187.

102. Lee, J.C. and Timasheff, S.N. (1977) Biochemistry 16, 1754-1764.

103. Lee, J.C., Harrison, D. and Timasheff, S.N. (1975) J. Biol. Chem. 250, 9276-9282.

104. Highsmith, S. (1977) Arch. Biochem. Biophys. 180, 404-408.

Catsimpoolas, ed. Physical Aspects of Protein Interactions

APOLIPOPROTEIN A-I: SELF-ASSOCIATION AND LIPID BINDING

ANGELO M. SCANU
Departments of Medicine and Biochemistry, The University of Chicago Pritzker School of Medicine and The Franklin McLean Memorial Research Institute, Chicago, Illinois 60637, U.S.A.

ABSTRACT

Apolipoprotein A-I (apo A-I), the major protein constituent of plasma high density lipoproteins, has been shown to play an important role in their structure. In man, apo A-I consists of a single linear 243-residue polypeptide chain. Predictive analyses of secondary structure from primary sequence data have indicated that apo A-I is an amphiphilic molecule having repeating α helixes which are interconnected by proline-containing β turns. This structural feature may account for the high tendency of apo A-I to occupy the air-water interface and to fold and unfold readily both in solution and at the interface. In aqueous media, apo A-I undergoes a concentration-dependent, reversible oligomerization which best fits the model monomer $\rightleftharpoons$ dimer $\rightleftharpoons$ tetramer $\rightleftharpoons$ octamer. Oligomerization of apo A-I decreases its capacity to bind lipids, which is maximal for apo A-I monomers. The protein-concentration dependence of both apo A-I self-association and lipid binding, as well as the common hydrophobic basis of these two processes, make studies on apoprotein-lipid interactions difficult to interpret unless the experimental conditions are carefully chosen.

BACKGROUND

Apolipoprotein A-I (apo A-I) represents about 70% of the total protein mass of the plasma high density lipoproteins (HDL) of d 1.063-1.21 g/ml[1-3]. When prepared in its pure, lipid-free state, this apoprotein is soluble in aqueous media; this has stimulated investigations on its properties in solution and its mode of interaction with lipids. In man, apo A-I is a single polypeptide chain which, according to the most recent sequence analysis, contains 243 amino acid residues and has a molecular weight of 28,000[4]. It has no cysteine or cystine, a low content of methionine and tryptophan, and no carbohydrates (Table 1).

At least two polymorphic forms of apo A-I have been recognized[1,2]; the chemical basis for these forms, however, has not been established. The amino acid sequences reported thus far differ somewhat[4,5], but these differences are not substantial. Analyses of the primary-sequence

TABLE 1

PROPERTIES OF APOLIPOPROTEIN A-I[a]

No. of amino acid residues	243
NH_2 terminus	Aspartic
$COOH_2$ terminus	Glutamine
Carbohydrates	0
Molecular weight	28,000
Polymorphic forms	2
Functions	LCAT activation, lipid binding, others?

[a]According to Brewer et al.[4]

data have shown that apo A-I does not have long regions of hydrophobic or hydrophilic amino acid residues, but rather cycles of 11 repeating units, particularly between residues 94 and 239[6,7]. An example of such a segment is that between residues 124 and 134: -Pro-Leu-Arg-Ala-Glu-Leu-Gln-Glu-Gly-Ala-Arg. According to Fitch[6], each of these repeating units is coded by a single ancestral gene which, by undergoing duplication, generates 22-residue units having amphiphilic properties. The 22-residue regions are interrupted by tetrapeptides that contain one proline residue. Based on spectroscopic criteria, apo A-I has a high α-helical content (about 55%), which increases about 20% in the presence of lipids, and α-helix-promoting solvents such as 2-choroethanol[1,2]. In accordance with the spectroscopic data, Segrest et al.[8], using published sequence data, predicted amphiphilic α-helical regions characterized by positively and negatively charged amino acids along the edges of the polar surface of each helix. More recent analyses based on the predictive method of Chou and Fasman[9] as well as on the solvation free energy of the amino acid side chains[10] have indicated that apo A-I is totally amphiphilic at the surface, and that it has β turns between α helixes, thus permitting a spatial orientation of the helical segments (Figure 1).

Recent studies based on calorimetric data and on the course of denaturation in urea or guanidine hydrochloride[11,12] have pointed to the very low free energy of stabilization of apo A-I in solution (2.4 kcal/mole) and to its great structural flexibility. Subsequent work[13] has shown, however, that the course of apo A-I denaturation is sensitive to the initial apoprotein concentration, and most probably to self-association. Thus, it is evident that the characteristics of this process must be known when the behavior of apo A-I in solution is examined.

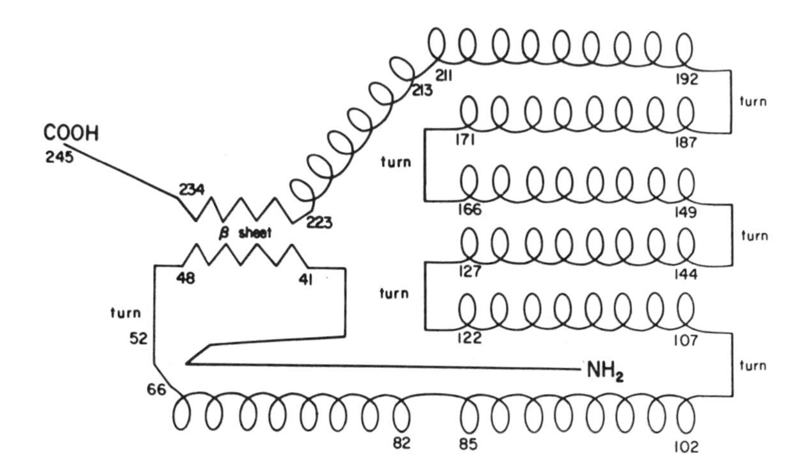

Fig. 1. Secondary structure of human apo A-I, as predicted from sequence analyses[10]. A proline residue is present in each turn. α-helix, ~~~~ ; random, ———; turn,].

SELF-ASSOCIATION OF APO A-I

In early investigations[14], apo A-I was reported to be monomeric at concentrations of up to 1 mg/ml. Later studies, however, revealed the tendency of apo A-I to self-associate and stimulated analyses of this process[15-18] (Table 2). Most of these analyses were carried out by the method of sedimentation equilibrium; however, experimental conditions such as rotor speed, protein concentration, and buffer varied from study to study. Stone and Reynolds[16] found that their results conformed most closely to a monomer-dimer-trimer or a monomer-dimer-trimer-tetramer model. On the other hand, Vitello and Scanu[17] obtained results consistent with a monomer-dimer-tetramer-octamer model in rapid equilibrium (Table 3).

In more recent studies[19], we have examined apo A-I self-association by the technique of sedimentation velocity. At concentrations ranging between 0.05 mg/ml and 0.6 mg/ml, apo A-I in 0.02 M EDTA, pH 8.6, had a rapid increase in its apparent sedimentation coefficient: extrapolation to zero protein concentration gave an $s^{o}_{20,w}$ value of 2.0 S for the apo A-I monomer. At apo A-I concentrations above 1.5 mg/ml, two well-separated, apparently symmetrical Schlieren peaks were observed, with sedimentation coefficients of 4S and 6S, respectively. These findings were taken to support the view that apo A-I self-associates at very low protein concentrations, and that the 2S and 4S components are in rapid

TABLE 2

STATE OF ASSOCIATION OF HUMAN APO A-I IN AQUEOUS SOLUTIONS, ASSESSED BY EQUILIBRIUM SEDIMENTATION

Conditions used	Results and test-fit models	References
10 mM Tris, 10 mM acetate, pH 7.36 Other conditions not specified	Monomeric up to 20 μM	Gwynne et al.[14]
10 mM Phosphate buffer, 10 mM NaN_3, pH 8.2 Other conditions not specified	Monomeric up to 1 mg/ml (35 μM)	Reynolds and Simon[15]
10 mM Tris-HCl, 1 mM EDTA, 5 mM NaN_3 (μ=0.045), pH 8.0 20°C; protein concentration = 0.5 and 3.2 mg/ml Rotor speed: 10,000 to 14,000 rpm	Monomer-dimer-tetramer- or Monomer-dimer-trimer-tetramer	Stone and Reynolds[16]
20 mM EDTA, 1 mM NaN_3 (μ=0.12), pH 8.6 20°C; protein concentration = 0.2 to 0.9 mg/ml Rotor speed: 6,000 to 13,000 rpm	Monomer-dimer-tetramer-octamer	Vitello and Scanu[17]
10 mM Tris, 1 mM NaN_3, pH 7.4, KCl ranging from 0 to 0.5 M KCl (μ 0.01-0.51), 21°C; protein concentration = 0.1 to 1.4 mg/ml Rotor speed: 10,000 to 22,000 rpm	Monomer-dimer-tetramer-octamer	Formisano et al.[18]

equilibrium, unlike the slow interconversion between 4S and 6S components. The results of this study, together with the values of molecular weight, partial specific volume, and fluorescence measurement provided estimates of the shape of the apo A-I monomer in solution. For a hydration value of 0.3 gm H_2O/gm protein, apo A-I was regarded as a prolate ellipsoid having an axial ratio of about 6:1 (Table 4).

Formisano et al.[18] recently reinvestigated the problem of apo A-I self-association by combining sedimentation equilibrium ultracentrifugation and gel-permeation chromatography. These authors noted changes

TABLE 3

EQUILIBRIUM CONSTANTS OF HUMAN APO A-I IN SOLUTION ACCORDING TO THE MONOMER ⇌ DIMER ⇌ TETRAMER ⇌ OCTAMER MODEL (0.02 M EDTA; pH 8.6, 20°C)[a]

K_2	3.3×10^4 M^{-1}
K_4	3.2×10^{13} M^{-1}
K_8	2.0×10^{33} M^{-1}

[a]From Vitello and Scanu[17].

TABLE 4

SHAPE OF HUMAN APO A-I (MONOMER) IN SOLUTION[a]

Ultracentrifugal studies	$s^o_{20,w}$	Frictional ratio f/f_o		Axial ratio a/b
	2.0	1.43; 1.40[b]		5.4
Fluorescence studies	**Fluorescence lifetime (τ) (ns)**	**Rotational relaxation time ρ_h (ns)**	**Rotational relaxation ratio ρ_h/ρ_o**	
	10.6	66.8	2.10	6.5

[a]From Barbeau et al.[19]
[b]From Reynolds et al.[28]

in apparent-weight average molecular weight as a function of ionic strength and decreasing rotor speed. The results were interpreted as being consistent with a monomer-dimer-tetramer-octamer association scheme, provided that pressure-dependent effects based on changes of the partial specific volume were assumed. These changes, however, were not directly documented. In the absence of these measurements, and in view of the fact that Reynolds et al.[15,16] and Scanu et al.[17,19] failed to detect pressure effects, acceptance of this interpretation by Formisano et al. must await further experimental work.

Swaney and O'Brien[20] have examined the problem of apo A-I self-association by cross-linking techniques, in which they used the bifunctional reagent dimethylsuberimidate, followed by separation of the cross-linked products by means of sodium dodecyl sulphate polyacrylamide gel electrophoresis. The authors considered their findings as supporting polymerization up to tetramers and pentamers. The discrepancy between these results and the ultracentrifugal data remains unexplained; however, recent studies in our laboratory[19] on the process of apo A-I elongation, in which we used a combination of ultracentrifugal, viscometric, and fluorescence data appear inconsistent with the presence of odd-number oligomers.

Regardless of the mechanism of oligomerization, self-association of apo A-I in aqueous solution is an established process which acquires significance in the study of the functions of this apoprotein, including ligand binding. The association of apo A-I with lipids will be examined below.

APO A-I SELF-ASSOCIATION AND LIPID BINDING

Initially, studies on binding of apo A-I with phospholipid vesicles did not take into account the effect of protein concentration. Under conditions which are now recognized to favor oligomerization, Assman and Brewer[21], Verdery and Nichols[22], and Middlehoff and Brown[23] found that apo A-I did not associate with lipids to any significant extent. On the other hand, apo A-I interacted well with lipids when experiments were carried out in the presence of urea[24], lysolecithin[22], or apo A-II[25], which are expected to influence the process of apo A-I oligomerization[1,2]. That the initial apo A-I concentrations in solution influence lipid binding was recently shown in our laboratory. In reconstitution studies on human[26] and rhesus monkey[27] apolipoprotein, apo A-I at various concentrations was equilibrated with the whole HDL lipids (i.e., phospholipids, cholesterol, cholesteryl esters, and triglycerides). The highest recovery of reconstituted HDL was observed at low apo A-I concentrations (0.01 to 0.2 μM); it was significantly less at concentrations of apo A-I between 1 and 6.5 μM. The results, examined in the context of the ultracentrifugal data on the self-association of apo A-I, indicated that the apoprotein monomer is preferred in lipid binding.

The requirement of apo A-I monomer for lipid binding is in keeping with the results of recent studies on the solution properties of this apoprotein. These studies suggest that apo A-I is asymmetric in solution[19,28], and that binding with the alkyl chains of phospholipids occurs preferentially along the long axis of apo A-I monomer, leading to the formation of less asymmetric protein-lipid complexes[28]. Studies on the interaction between apo A-I and L-α-dodecanoyl phosphatidyl choline have provided evidence that the total hydrophobic volume of the lipids bound to apo A-I is constant, thus indicating the existence of "pockets" in the polypeptide chain. Since it has been suggested that the dimerization of apo A-I also occurs along the long axis of the apo A-I monomer and involves the hydrophobic surface[19], it follows that lipids interfere with the process of dimer formation, and that dimers have a more limited surface available for binding.

It should be clear, however, that the degree of self-association of apo A-I is not the only determinant in lipid binding, and that additional factors such as protein-lipid stoichiometry and geometry of the lipid vesicles are recognized[1,2,21,29]. Nonetheless, knowledge of the extent of self-association can explain why, in early studies, apo A-I failed to interact with lipids; it is likely that, under the experimental conditions which were used, the dissociation of oligomers into monomer was not favored. We believe that, since self-association of apo A-I is reversible, all of the reacting protein and lipid should be incorporated into a complex if sufficient time is provided for interactions to take place[26].

If optimal interactions between apoprotein and lipids are to occur, the experimental conditions must be carefully chosen so that, by a kinetically driven process, a lipid-protein complex having properties distinct from those of the individual reactants is formed. The stabilizing effect of lipids on apoprotein conformation was documented long ago[3] and is supported by recent studies, all of which point to the hydrophobic nature of this process[30,31]. The general mode of interactions with lipids in vitro exhibited by apo A-I appears to apply to other plasma apolipoproteins as well, although some differences have been encountered, for example, between apo A-I and apo A-II. The known selectivity of the apoproteins for given lipoprotein classes is likely to depend on the surface properties of these apoproteins and on their capacity to occupy the curved amphiphilic lipoprotein interface.

RELATIONSHIP OF THE ABOVE OBSERVATIONS TO HDL STRUCTURE

It is now apparent that mature HDL is the result of structural remodeling which occurs in the circulation, starting from nascent precursors generated in the liver or intestine[22,32]. It is possible that many of the factors governing protein-lipid interactions in vitro are also operative in vivo. It has been suggested that, in the rat, apo A-I, by associating with a precursor bilayer disc, changes the interfacial tension of the latter, thus rendering it sensitive to attack by the enzyme lecithin-cholesterol acyl transferase[32]. It has also been shown by in vitro experiments that the apo A-I monomer in solution is in equilibrium with the HDL surface[10], and that apo A-I has an essential role in HDL structure[1,26,33,34]. Thus, factors regulating lipid-protein interactions in vivo appear to have physiological importance. The initial stoichiometry of the reactants may, for example, determine the size of the HDL particles by modulating their surface properties. At present, we do not know whether lipid-free apo A-I monomers actually exist in the circulation. If the observations in vitro apply in vivo as well, we may surmise that the apo A-I monomer is the functionally active unit, either at the lipoprotein surface or in the aqueous environment. Thus, it is apparent that HDL formation, interconversion, and degradation recognize regulatory processes one of which is the initial interaction between apo A-I and lipid. This process may be of importance in determining the future metabolic fate of the resulting lipid-protein complex.

ACKNOWLEDGEMENTS

The work in our laboratory cited in this paper was conducted by L. Vitello, Ta-Lee Teng, D. Barbeau, and C. Edelstein. The latter two also provided constructive criticism during the preparation of the manuscript. The author acknowledges support by USPHS grants HL-08727 and HL-06481. The Franklin McLean Memorial Research Institute is operated by The University of Chicago for the U.S. Department of Energy under Contract No. EY-76-C-02-0069.

REFERENCES

1. Scanu, A.M., Edelstein, C. and Keim, P. (1974) in The Plasma Proteins, Volume 1, 2nd edition, Putnam, F. ed., Academic Press, New York, pp. 317-391.
2. Osborne, J.C., Jr. and Brewer, H.B., Jr. (1977) Adv. Prot. Chem., 31, 253-337.
3. Smith, L.C., Pownall, H.P. and Gotto, A.M., Jr. (1978) Annu. Rev. Biochem., 47, 751-777.
4. Brewer, H.B., Jr., Fairwell, T., LaRue, A., Ronan, R., Houser, A. and Bronzert, T.J. (1978) Biochem. Biophys. Res. Commun., 80, 623-630.
5. Baker, H.N., Gotto, A.M., Jr. and Jackson, A.M. (1975) J. Biol. Chem., 250, 2725-2738.
6. Fitch, W.M. (1977) Genetics, 86, 623-644.
7. McLachlan, A.D. (1977) Nature, 267, 465-466.
8. Segrest, J.P., Jackson, R.L., Morrisett, J.D. and Gotto, A.M., Jr. (1974) FEBS Lett., 38, 247-253.

9. Chou, P.Y. and Fasman, G.D. (1978) Annu. Rev. Biochem., 47, 251-276.

10. Edelstein, C., Kézdy, F., Scanu, A.M. and Shen, B.W. (1979) J. Lipid Res., in press.

11. Tall, A.R., Shipley, G.C. and Small, D.M. (1976) J. Biol. Chem., 251, 3749-3755.

12. Reynolds, J.A. (1976) J. Biol. Chem., 251, 6013-6015.

13. Edelstein, C. and Sakoda, N. (1978) Fed. Proc., 37, 1481.

14. Gwynne, J., Brewer, H.B., Jr. and Edelhoch, H. (1974) J. Biol. Chem., 249, 2411-2416.

15. Reynolds, J.A. and Simon, R.H. (1974) J. Biol. Chem., 249, 2937-2940.

16. Stone, W.L. and Reynolds, J.A. (1975) J. Biol. Chem., 250, 8045-8048.

17. Vitello, L. and Scanu, A.M. (1976) J. Biol. Chem., 251, 1131-1136.

18. Formisano, S., Brewer, H.B., Jr. and Osborne, J.C. (1978) J. Biol. Chem.,253, 354-360.

19. Barbeau, D.L., Jonas, A., Teng, T.L., and Scanu, A.M. Submitted for publication.

20. Swaney, J.B. and O'Brien, K.J. (1976) Biochem. Biophys. Res. Commun., 71, 636-643.

21. Assman, G. and Brewer, H.B., Jr. (1974) Proc. Natl. Acad. Sci. USA, 71, 989-993.

22. Verdery, R.B. and Nichols, A.V. (1974) Biochem. Biophys. Res. Commun., 57, 1271-1278.

23. Middlehoff, G. and Brown, W.V. (1974) Circulation, 50, Suppl. III, 112.

24. Stoffel, W., and Därr, W. (1976) Hoppe-Seyler's Z. Physiol. Chem., 357, 127-137.

25. Ritter, M.C., Kruski, A.W. and Scanu, A.M. (1974) Fed. Proc., 33, 1585.

26. Ritter, M.C. and Scanu, A.M. (1977) J. Biol. Chem., 252, 1208-1216.

27. Barbeau, D. and Scanu, A.M. (1974) Circulation, 50, Suppl. III,260.

28. Reynolds, J.A., Tanford, D. and Stone, W.L. (1977) Proc. Natl. Acad. Sci. USA, 74, 3796-3799.

29. Morrisett, J.D., Jackson, R.L. and Gotto, A.M., Jr. (1977) Biochim. Biophys. Acta, 472, 93-133.

30. Gennis, R.B. and Jonas, A. (1977) Annu. Rev. Biophys. Bioeng., 6, 193-238.

31. Gwynne, J., Brewer, H.B., Jr. and Edelhoch, H. (1975) J. Biol. Chem., 250, 2269-2274.

32. Hamilton, R.L. (1978) in Disturbances in Lipid and Lipoprotein Metabolism, Dietschy, J.M., Gotto, A.M., Jr. and Ontko, J.A., eds., Waverly Press, Inc., Baltimore, pp. 155-171.

33. Scanu, A.M. (1972) Biochim. Biophys. Acta, 265, 471 508.

34. Scanu, A.M. (1978) Trends in Biochem. Sci., 3, 202-205.

Catsimpoolas, ed. Physical Aspects of Protein Interactions

INTERACTIONS OF CLOSELY RELATED PROTEINS IN THE SOLID STATE

JACINTO STEINHARDT AND MARYANN MCD. JONES
Chemistry Department, Georgetown University, 37th & O Streets, N.W., Washington, D.C.

ABSTRACT

When crystals are formed from aqueous solutions of two mixed very similar globular proteins A and B, and the major component A is the less soluble, its solubility relative to that with B absent may either increase or decrease. A decrease may be ascribed to either interactions in the dispersed phase, as the result of e.g. excluded volume effects on the thermodynamic activity of A, or the formation of solid solutions in the condensed phase. An increase may show that lattice imperfections have been produced, or that insufficiently specific methods of assay may include the solubility of B in the estimate of A. All four cases are known in the crystalline protein literature. More recently it has been postulated that the enhancing effect of non-gelling forms (B) of human hemoglobin on the gelling or crystallization of deoxy sickle-cell hemoglobin (A)[#], represents simply a salting-out effect, due primarily to an excluded volume effect of B. We have suggested that an enhancing effect on solubility, at higher amounts of B, is due to the production of lattice defects incident to cocrystallization. Application of two variants of a new turbidity method of determining solubility obviates the difficult necessity of measuring pellet volume and composition in distinguishing between these alternatives. The results to date indicate that admixture of Hb°A (normal) or COHbS to Hb°S, under non-hybridizing conditions, have typical salting-out effects on the solubility (mgc) of Hb°S. Thus substantial excluded volume effects exist. However with relatively high concentrations of the isomorphous Hb°A (but not with the non-isomorphous COHbS) they are offset by an increase in lattice imperfections, an example of a dispersed phase interaction counteracted by a solid phase interaction; thus the solubility of Hb°S goes through a minimum as the weight fraction of Hb°A is increased.

[#]The abbreviations used in this paper are as follows: Hb°S, deoxygenated sickle hemoglobin; Hb°A, deoxygenated normal adult hemoglobin A; COHbS, carbonmonoxy-sickle hemoglobin; mgc, "minimum gelling concentration"; IHP, inositol hexaphosphate; DPG, diphosphoglycerate; HbF, fetal hemoglobin; Hb^+, methemoglobin; O_2HbS, oxygenated sickle hemoglobin.

Suggestions are made as to how part of Hb°S may be converted to a solubilizing isomorphous form within the intact erythrocyte.

INTRODUCTION

Early solubility determinations on the hemoglobins of a number of species, while very crude, tended to show that the solubilities were additive, and that thus the solubilities were independent of the presence of one another; thus there was no need to postulate a complex solid phase.

Between about 1920 and 1940 there was considerable work on the solubility of crystalline preparations of proteins, not only as a means of aiding in the separation of mixtures of proteins by fractional crystallization, but also as a means of testing and demonstrating the purity of the resulting crystals[=]. Even after exhaustively fractionating (by salting in and out) it had been obvious to many, but notably to Sörensen and his colleagues at the Carlsberg Laboratory, that with both crystalline and non-crystalline protein preparations the "solubility" was never wholly independent of the ratio of protein to solvent. The dependence on this ratio was very marked with some of the serum proteins. There was reluctance to invoke impurity as an explanation after the many apparently clear cut discontinuities in salting-out or in mixed solvent extraction. A number of schemes for accounting for this variability were offered, the most noteworthy by Sörensen[1a,1b], and the most difficult to refute by Bonot[2a,2b]. The latter depended on the idea that closely similar proteins - in a sense, isomers, - formed solid solutions in one another by interactions in the solid phase.

Interactions in the dispersed phase were well-known to all physical chemists - in fact the concept of thermodynamic activity furnished a scale with which to measure it. Application of this scale to substances much simpler and more homogeneous than proteins had been facilitated by the assignment of a constant activity to the solid with which the solutions are equilibrated: the condensed phase. Additions of third substances to suspensions in solvent of slightly soluble substances changed the amounts dissolved; these changes were properly interpreted as being inversely proportional to the ratio of the activities in the absence and in the presence of the additives.

[=]The importance of this work may be judged by the fact that so skilled a preparative biochemist as Willstätter has reported purified materials with enzymatic activity which were free of protein!

On this basis the effects of different valence types of salts, or of neutral molecules on the activity of others were commonly assessed. The method was one of those by which the Debye-Hückel theory was experimentally verified during the mid-twenties.

With the success of this theory, and following the well-known tendency to oversimplify in applying physical chemistry to biochemistry, the same method was soon applied to sparingly soluble proteins, in spite of the fact that there was no constant solubility base to stand on. This daring is easier to understand in view of the attempts by Northrop and his collaborators to demonstrate the purity of the crystalline enzymes they isolated - starting with pepsin - by means of "phase-rule" solubility studies[3].

These phase rule studies were based on the notion, illustrated by Fig. 1A, that the amount of a pure substance dissolved would be equal to the total amount introduced into a solvent until the "solubility" was attained, beyond which point there would be no further increase in the amount dissolved. In the presence of a single impurity there would be a stepped line, (Fig. 1B) and if many impurities were present the result might appear to be a continuous curve.

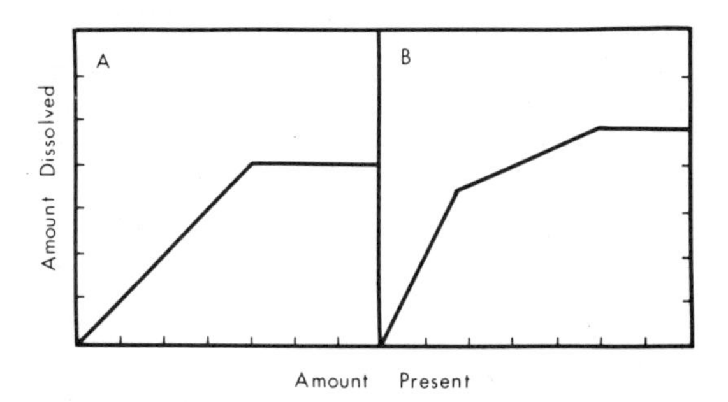

Fig. 1. A- Solubility as a function of concentration of a single pure component. B- Solubility as a function of concentration for a mixture of two pure components with independent solubilities.

This test came to be applied routinely, and sometimes quite blindly. Thus diagrams like 1A were sometimes imposed on scattered points that would have better been fitted with diagrams like 1B. It escaped the notice of at least one very successful investigator (and his Editor) that a diagram like 1A had no significance as a test of purity if the angle at the origin (with equal scales on both axes) were not 45°.

Such studies carried conviction largely because of the desire to believe them. The crystalline enzyme preparations tested showed constant activities and solubilities ±20 percent or better, under fixed conditions which usually included a large excess of solid. However a very exhaustive investigation by Steinhardt[4,5] showed that beautifully crystalline pepsins prepared from two different commercial sources of crude enzyme differed in solubility in identical solvents by 500 percent! At about the same time, Hewitt[6] extracted from crystals of serum albumin, by repeated recrystallization, a very highly soluble tryptophan-rich glycoprotein which would not crystallize, leaving a much less soluble crystalline material containing no sugar and only one tryptophan per mole.

Steinhardt's work clearly showed that the supernatants or filtrates contained protein that differed systematically in composition and specific activity from one successive equilibration to the next, as the solubility diminished. He was successful in analyzing his data on the basis of a model which postulated that the crystals were solid solutions of a more soluble component in a less soluble material. Simple algebraic transformation of the resulting equations permitted application of the model to experiments in which the ratio of crystals to solvent volume was directly varied. This formulation was also successful in describing many of the experiments at the Carlsberg Laboratory. The latter success is important since the fact that Steinhardt's experiments were conducted with a protease, and some hydrolysis demonstrably occurred, would always raise the possibility that his results were due to decomposition during the successive extraction, even though all the compositions and solubilities were expressed in terms of active enzyme and of protein. However, work by Steinhardt (unpublished) with horse CO hemoglobin, prepared by numerous successive crystallizations, showed that this familiar well-characterized protein exhibited the same anomalous solubility behavior as pepsin when examined by the same technique.

It has since been recognized that all the protein preparations of those days, whether crystalline or amorphous, were very impure.

Crystalline pepsin itself contains a number of different active components - Perlmann[7] has shown that the best way to get the pure enzyme is to purify pepsinogen, converting it to pepsin only at the last moment when pepsin itself is to be studied. It is noteworthy that all the crystallized proteins for which high purity was so avidly claimed in the thirties, have since been fractionated by chromatographic means, into several components. It is remarkable that with the availability of such more highly purified material, the interest in solubility seems to have largely disappeared since that time. Perhaps a younger "inexperienced" generation will soon rediscover its usefulness.

Our interest in the old experiments with impure crystals of other proteins was revived by our recognition that the gelling or crystallization of Hb°S resulted from a difference in solubility of Hb°S and other forms of hemoglobin. It has long been known that the solubility of Hb°S, the only form of HbS to gel or crystallize at concentrations below about 30 percent, is diminished on the addition of small amounts of either unliganded HbA or liganded forms of either HbA or HbS. This observation is usually described as reducing the "minimum gelling concentration". This change is contrary to the expectation if solid solutions of a soluble protein and a less soluble protein are formed, or if other forms of contamination (cocrystallization) occur. Both of these effects, as every organic chemist knows, lower freezing points and increase solubility.

This is not primarily a paper on sickle-cell hemoglobin. The rest of this presentation discusses experiments on the latter in the general context of protein interactions in the condensed phase. It is reasonable to start by asking why one should want to study the solubility of protein crystals grown from a mother-liquor which contains two or more proteins? The experiments are not easy, especially if anaerobic, and if one wants to know the composition of the crystals that form. The earlier work on protein solubility had as its purpose to determine whether the crystals were pure - hopefully one would find that no mixtures were involved. However the crystals were actually impure and interesting solubility relationships were developed. In the present case, with Hb°S, it is desired to understand why the presence of "extenders" (non-Hb°S additives) reduces the minimum gelling concentration, not only for scientific reasons, but because Hb°S *in vivo* is always found in the erythrocte with oxyhemoglobin and with other hemoglobins such as fetal hemoglobin (HbF), and, in the case of heterozygotes, with normal hemoglobin (both oxy-

and deoxy-) as well. The action of these "extenders" as well as pH, ionic strength and of the allosteric effector diphosphoglycerate (DPG), has potential practical chemical significance. In addition, as Benesch _et al_[8] and others have shown, it should be possible with data on mixtures with a number of mutants to draw inferences as to the residues which furnish contact points between constituent tetramers in the gels or crystals, which would shed light on why this particular mutant protein has a much lower solubility than Hb°A.

EXPERIMENTAL

When a solution of Hb°S is too dilute to crystallize (or gel) the light scattered at a given wavelength (630 n.m. in our work) has a characteristic value which depends on the concentration and scattering-coefficient of the molecules. The _turbidity_, which is indistinguishable from the optical density, therefore also depends on the path length. When the concentration is high enough for crystals to form, their much higher scattering-coefficient causes the measured turbidity to rise. This increase, which is easily measured by comparing the turbidity of the crystalline suspension with the turbidity of the same solution when the crystals have been dissolved, depends on the amount of crystals formed: Fig. 2 shows that ΔO.D. at 0.1 mm path length rises by 0.10 units per gram of crystals per deciliter formed. On measuring several ΔO.D. values with various initially supersaturated concentrations of the protein, a very accurate linear least-square extrapolation intercepts the abscissa (concentration scale) at the solubility.* This method of extrapolation has numerous advantages. The two most important are that no corrections need be made for the trapping of solution in the crystals, which would require difficult estimates of both volume and composition; and, equally important, that no correction need be made for the change in concentration of the solution due to removal of solute into the crystals. The slope, 0.10, includes the second correction, and makes the first one operationally meaningless.

Use of this technique has previously permitted us to measure the solubilities and changes in free-energies, enthalpies, and entropies on crystallization, as a function of temperature between 15° and 30°,

---

*A linear relationship between the turbidity (ΔOD_{630}) and the mass of crystals per unit volume is not only established empirically[9] but is theoretically demonstrable for large asymmetric particles[10,11] over the range of turbidity increments ($<$ 0.6) in our experiments.

and has permitted determination of these quantities both in the absence of inositol hexaphosphate (IHP) and in its presence[9]. IHP halves solubility, but has almost no effect on the thermodynamic parameters just listed.

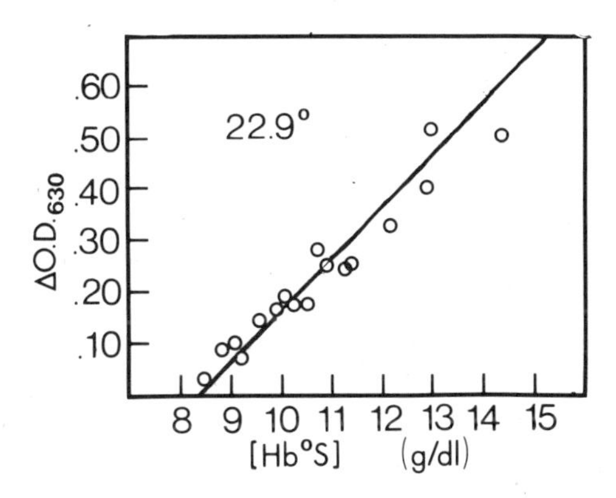

Fig. 2. Excess turbidity due to crystals of Hb°S in the presence of IHP at 22.9°C. The least squares determined slope is 0.1008 with a correlation coefficient of 0.974. The extrapolated solubility is 8.42 g/dl.

In order to analyze the solubilities as affected by mixing, it is necessary to specify all of the components of the mixtures. In many forms of human and other hemoglobins, tetramers (which normally predominate) and dimers are in labile equilibrium especially under some conditions such as low concentration or pH, presence of urea and other amides, or of guanidine salts. As a result, hybridized tetramers may form when hemoglobins are mixed, and may enter the aggregates or structures to be studied: their effect on aggregation and solubility must then be taken into account. Fortunately human deoxyhemoglobin has a very low tendency to hybridize. Consequently if the deoxyhemoglobin to be used should be formed before the other hemoglobins are added hybridization will not occur. It is enough to convert to deoxyhemoglobin before mixing with e.g. COHb. The effectiveness of the procedure may be tested by comparing the results obtained with and without prior deoxygenation of Hb (or reduction of methemoglobin [Hb^+] with dithionite). One may also choose to use a markedly unlabile hemoglobin such as the cyanmet derivative as one of the components. None of the recently reported results appears to be affected by hybridization[12,13,14].

The composition of mixtures of unliganded Hb with liganded forms, before and after crystallization, have been determined by spectrophotometric observations at three wavelengths and calculation of three simultaneous equations in three unknowns. The extinction coefficients used were those measured for chromatographically purified Hb under our conditions. Independent determination of such coefficients was necessary because of the large effect of IHP, present in our solutions, on the spectrum of methemoglobin. With mixtures of Hb°S and Hb°A the differential analysis was conducted by application of electrophoresis in acrylamide gels. All operations - mixing, dilutions, and transfer - were carried out strictly anaerobically. Matched cells of 0.10 mm path length were used.

All solutions contained phosphate buffer (pH 7.0, 0.1 ionic strength), 0.1 mM EDTA, and 5 mM IHP. Use of the latter permitted crystallization at relatively low concentrations and conserved material. The solubilities are higher when phosphate is not used.

Application of the turbidity method to the determination of solubilities in solutions of mixed hemoglobins was made by two distinct kinds of experiments. In the first, one component was kept constant and the second varied. While in the second, solubility was measured as a function of the addition of successive amounts of a fixed ratio of Hb°S to additive mixture. This latter procedure is directly analogous to the pure Hb°S solubility experiments, just described, but cannot be analyzed, or directly compared with, the plethora of mixed solubility data in the literature obtained in other ways.

RESULTS

We first describe the data obtained when successive increments in the concentration of Hb°A are added to a fixed concentration of Hb°S at constant volume. The results of this procedure are easier to analyze than when A and S are added together at a series of fixed ratios. In these experiments the total amount of protein is not constant.

The results are shown in Figs. 3 and 4. Both have a feature in common: as Hb°A is added the crystal yield first increases, then falls again at still higher concentrations of Hb°A. In Fig. 3 no crystals appear until the solution contains almost 30 percent as much A as S. The composition of the crystals formed remains to be determined since both A and S are present, and the initial concentration of Hb°S was too low to induce crystallization by itself. However, when crystals are present from the start and more crystals are induced to form by additions of Hb°A (Fig. 4) one cannot necessarily conclude that Hb°A is included in the crystals.

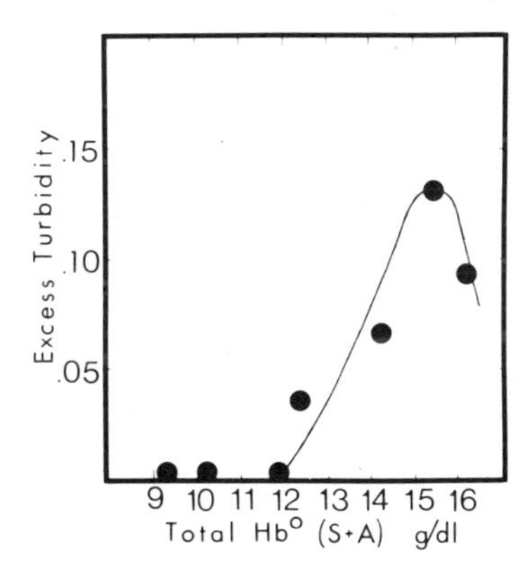

Fig. 3. Excess turbidity as a function of total hemoglobin (Hb°S + Hb°A) concentration. The concentration of Hb°S at each total concentration was 9.3 g/dl.

The data of Fig. 4 permit further observations and conclusions. The initial decrease in solubility when Hb°A is added can be understood in terms of the non-ideality of the very concentrated solutions used[15]. Concentrations may be converted to activities by using the virial coefficients calculated by Ross and Minton[15] derived from data in the literature. These investigators, as well as Williams[16], have pointed out the calculated virial coefficients express principally excluded-volume effects. The effect of added non-crystallizing protein is therefore essentially a salting-out effect.

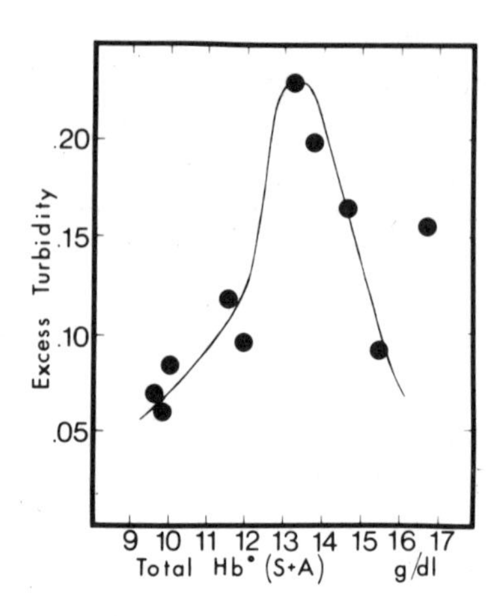

Fig. 4. Excess turbidity as a function of total hemoglobin (Hb°S + Hb°A) concentration. The concentration of Hb°S at each total concentration was 9.76 g/dl.

If adding Hb°A were to contribute only an excluded-volume effect the solubility of the crystals should continue to diminish as more foreign protein is added. Clearly it does not do so. Other kinds of interactions must be involved; at least one of which must be an interaction in the condensed phase. Concepts and names for such condensed phase interaction include (a) regular cocrystallization; (b) formation of solid solutions (these would be expected to reduce rather than increase the solubility still further); (c) induction of lattice imperfections which weaken the integrity of the crystal. One way of thinking about the condensed phase interaction is that the pure component crystals, salted out by initial aliquots of Hb°A, are gradually transformed, with more Hb°A, to a more soluble complex. With enough Hb°A the driving force of the salting-out process is overcome.

With different initial fixed amounts of Hb°S, the results found do not differ greatly.

If we drop all the virial coefficients except the second from the non-ideality model, we may deduce that for a given solubility, Hb°S concentration may be altered without changing the excess turbidity (Δ O.D.$_{630}$) if a compensatory change is made in Hb°A concentration so that the sum of [Hb°S] plus [Hb°A] is kept constant. This approximation becomes more valid, the smaller the total protein present.

It follows that for the amount of Hb°S remaining in solution, the solubility will be invariant for all compositions K_i in which

$$[Hb°S] + [Hb°A] = [K_i] \qquad \text{(Eq. 1)}$$

Thus varied mixtures of Hb°S and Hb°A which add up to various totals $[K_i]$ will all be along a series of horizontal lines, the levels of which correspond to the $[K_i]$. Such adjustments would improve the smoothness of the data shown in Fig. 4.

Determining solubilities by making mixtures of Hb°S and additives in fixed ratios is more directly comparable to single component solubility determinations[9]. Such experiments facilitate straightforward speculation about both solution phase and solid phase interactions between different hemoglobins. A series of dilutions from a concentrated stock solution of Hb°S plus Hb°A is made to a constant final volume. The solutions in the series then contain varying amounts of Hb°S at total Hb concentrations such that the ratio of Hb°S to Hb°A is fixed. If there were no interactions between Hb°S and Hb°A either in the solution or in the crystals (or gels), the solubilities of each would be independent of one another. One would then expect a linear extrapolation of the relationship between turbidity and total Hb concentration to a "solubility" equal to (Hb°A + Hb°S / Hb°S) multiplied by the solubility of Hb°S alone. The slope of that linear relationship would be lower than that for pure Hb°S by the inverse of this factor, (Hb°S / Hb°S + Hb°A) (Fig. 5-A1). If turbidity were plotted against [Hb°S], rather than [total Hb°], the slope and intercept would be unaltered relative to that for pure Hb°S (Fig. 5-A2).

If instead, a single kind of interaction, the excluded volume effect of the additive on the dispersed phase, were to occur, a more complicated result should be expected. This expectation is a direct result of the fact that aliquots of added Hb above the solubility of the mixture consist partly of material which crystallizes (Hb°S) and therefore makes no contribution to the dispersed phase, and partly of material (Hb°A) which accumulates and causes an increasing amount of the crystallizable material to come out of solution.

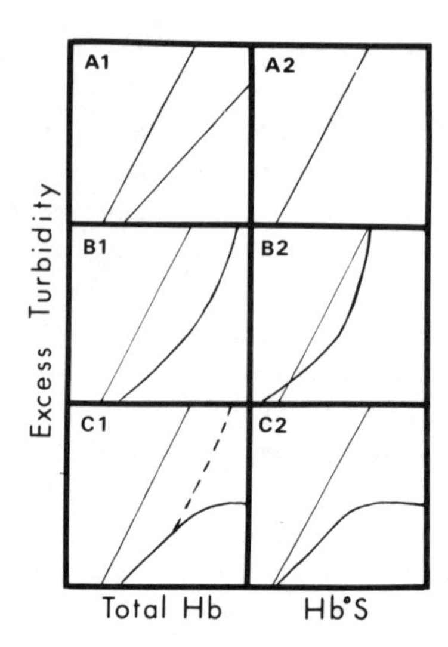

Fig. 5. The expected relationships between excess turbidity and either total hemoglobin concentration (1) or Hb°S concentration (2) under the following circumstances: A- no interactions between Hb species occur in either the solution or condensed phase; B- excluded volume effects of Hb°A on the dispersed phase occur; C- both excluded volume effects on the dispersed phase and interactions between species in the condensed phase occur.

The result of such joint effects on the graphical depiction of the relationship between turbidity and total Hb appear in Fig. 5-B1. All Hb°S added (above the solubility concentration) would increase the crystal yield and therefore the turbidity of the solutions, at a rate with respect to concentration equal to Hb°S / Hb°A + Hb°S times the slope for pure Hb°S. However, the progressively increasing concentration of Hb°A left in solution would cause the salting-out of the less soluble Hb°S at a constantly increasing slope, i.e. a curve turning sharply upward. If turbidity were plotted against Hb°S concentration alone, the curved effect would still be obvious but the resultant x-axis intersection would be at a solubility concentration below that for solutions of pure Hb°S (Fig. 5-B2).

However, if in addition to excluded volume effects, the competing influence of interactions in the condensed phase (cocrystallization or complex formation) were taken into account, the curved portions of the turbidity versus Hb concentration relationship should be progressively affected in the opposite sense (Fig. 5-C1). Depending on the magnitude of the condensed phase interactions, the curvature could be reduced, or eliminated, i.e. the rate of increase in turbidity with respect to Hb concentration would be reduced causing a flattening out of the curve, or the curve could, in fact, turn downward. Such general contour changes would probably also occur in plots of turbidity against Hb°S concentration as well. However the most important effect would be in the extrapolated solubility concentration of Hb°S (Fig. 5-C2). With extensive cocrystallization, a higher solubility threshold than for Hb°S alone could result.

The possible relationships depicted in Fig. 5 correspond fairly well with actual results found for three different fixed ratio mixtures of Hb°S and Hb°A (Fig. 6). The relationship between turbidity

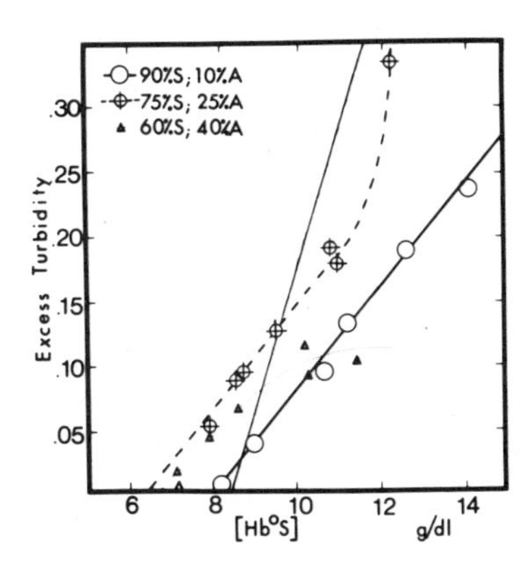

Fig. 6. Excess turbidity as a function of Hb°S concentration for three fixed ratio mixtures of Hb°S and Hb°A: ○ - 90% Hb°S/ 10% Hb°A; ⊕ - 75% Hb°S/ 25% Hb°A; △ - 60% Hb°S/ 40% Hb°A. The solid light line represents the turbidity vs concentration relationship for pure Hb°S.

and total protein is linear up to about 15% protein for both the 90% Hb°S/ 10% Hb°A and the 75% Hb°S/ 25% Hb°A mixtures, before turning slightly upward, while the 60% Hb°S/ 40% Hb°A mixture shows strong signs of curve flattening or reversal. However the slopes of the linear portions are all considerably lower than the previous discussion would have predicted if no interactions were occurring (0.04 for 90% Hb°S; 0.04 for 75% Hb°S, and, initially, about 0.025 for 60% Hb°S mixture) when plotted against Hb°S concentration (Fig. 6). Excluded volume effects should have produced a slope discrepancy in the opposite direction. Such anomalously low slopes correspond to the special case of Fig. 5C in which the slope and gradual curvatures predicted are reduced by condensed-phase interactions just enough to produce a good approximation to a straight line over a wide range of total Hb concentration.

Determination of the relative proportions of Hb°A and Hb°S in mixtures and supernatants by the electrophoretic separation of these components allows an indirect estimation of the composition of the crystals in such fixed ratio mixtures (Table I), and some indication that the kind of condensed phase interactions postulated do indeed occur. In each of the four mixtures analysed, the proportion of Hb°S to Hb°A decreased in the supernatant relative to that in the mixture. However, when these percent compositions were applied to the total concentration of mixture and supernatant and the composition of crystal calculated, it was found that some Hb°A had to exist in each crystal crop. The change in total concentration between mixture and supernatant ranged between 3.33 g/dl and 4.92 g/dl while the excess turbidity levels found were less than 0.100. This suggests that the mixed crystals formed are either smaller or less asymmetrical than crystals of pure Hb°S (the crystals of which produce a change in turbidity of 0.100 units per g/dl) and consequently scatter light differently.

As anticipated for additives known as "extenders", the presence of Hb°A in these fixed ratio mixture experiments produces a decrease in the solubility of Hb°S relative to that in its absence (Fig. 6). The solubility of Hb°S does increase to 6.75 g/dl for the 60% Hb°S/ 40% Hb°A mixtures relative to the lowest solubility observed - 6.52 g/dl for the 75% Hb°S/ 25% Hb°A mixtures, but an increase beyond that for pure Hb°S is not observed. It should be noted that the effect of Hb°A on the extrapolated solubility of Hb°S was small, compared with the effect on slope, for all levels of Hb°A used, implying that changes in slope convey more information about these interactions

TABLE I

ELECTROPHORETIC ANALYSIS OF THE COMPOSITION OF Hb°S/Hb°A MIXTURES AND SUPERNATANTS

Mixture Composition			Supernatant Composition		Crystal Composition[a]	Excess Turbidity
Relative Proportion		g/dl	Relative Proportion	g/dl	g/dl	(ΔO.D.$_{630}$)
S	69.10%	12.84	63.64%	8.70	4.14	.059
A	30.90%	5.74	36.36%	4.97	.77	
S	69.34%	11.84	62.12%	7.73	4.11	.076
A	30.66%	5.34	37.88%	4.71	.63	
S	55.45%	9.74	49.55%	7.05	2.69	.018
A	44.55%	7.82	50.45%	7.18	.64	
S	55.18%	10.53	49.23%	8.24	2.29	.040
A	44.82%	9.69	50.77%	8.50	1.19	

[a]Inferred by subtracting the composition of supernatant from the composition of mixture. No correction has been made for change in volume accompanying crystallization.

than do the solubilities themselves.

Since none of the abscissa in our figures represents equilibrium concentrations, but simply the amount of Hb present when the turbidity was determined, no purpose would be served by applying appropriate activity coefficients in order to analyze the data.

Experiments with a liganded form

When Hb°S is mixed with a liganded form, such as COHbS, instead of Hb°A, and spectrophotometric analysis is used in place of electrophoresis, data is obtained more easily, and the analysis can be carried further than in any of the foregoing.

Fig. 7 shows the results of adding increasing amounts of COHbS to Hb°S at three different concentrations of Hb°S, when excess turbidity is plotted against total hemoglobin concentration.

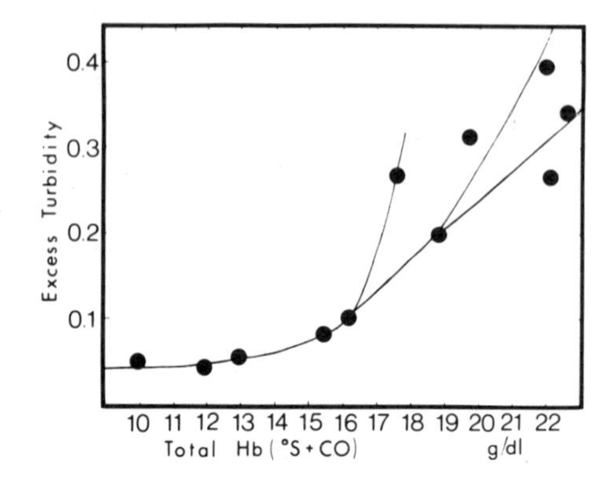

Fig. 7. Excess turbidity as a function of total hemoglobin (Hb°S + COHbS) concentration. Hb°S concentration was 12.91, 11.94, or 9.98 g/dl. The concave upward segments correspond to the Hb°S concentrations in that order.

The most striking feature of Hb°S/COHbS mixtures is that the data fall on monotonic curves which are slightly concave upward. There is no sign of a minimum solubility such as characterized Hb°S/Hb°A mixtures. There is no need therefore to invoke a force other than salting-out due to excluded volume terms. It appears therefore that COHbS mixtures do not result in appreciable cocrystallization. This conclusion, which agrees with that of Ferrone _et al_[17], from experiments carried out by a different procedure, is also supported by the differences in relative concentration of Hb°S and COHbS in the supernatant brought about by crystallization (See Table II). The concentration of the former in the supernatant always falls relative to that in the mixture itself as the result of crystallization, and the concentration of COHbS rises.[¢]

[¢] An exact interpretation of those concentrations requires correction for the volume changes brought about by the separation of proteins (as crystal) from the solution phase. The uncorrected data serve to establish that Hb°S is strongly _preferentially_ crystallized relative to COHbS.

TABLE II

COMPOSITION OF MIXTURES AND SUPERNATANTS IN THE 63% Hb°S/ 37% COHbS FIXED RATIO EXPERIMENT, DETERMINED BY SPECTROPHOTOMETRIC ANALYSIS

Mixture Composition g/dl		Supernatant Composition g/dl	Crystal Composition g/dl	Excess Turbidity
Hb°S	8.80	3.84	4.96	.232
COHbS	7.94	8.48		
Hb°S	10.35	4.96	5.39	.278
COHbS	6.52	7.11		
Hb°S	8.24	3.90	4.34	.190
COHbS	5.38	5.57		
Hb°S	6.58	3.67	2.91	.054
COHbS	3.46	6.27		
Hb°S	11.22	4.93	6.29	.350
COHbS	6.01	6.72		

It appears therefore that there is less cooperation between COHbS and Hb°S in the condensed phase than there is between Hb°A and Hb°S in their crystals.[&]

Experiments with mixtures of Hb°S and COHbS in fixed ratios

Fixed ratio experiments have been performed with Hb°S/COHbS mixtures taking advantage of the differential spectrophotometric assay (see Experimental) as a replacement for electrophoresis. Two nearly fixed ratios were set up by successively diluting two different stock mixtures. There was some variation in the ratios achieved. The members of one set with an average ratio of 73% Hb°S/ 27% COHbS deviated only 1 to 3 percent from this value. Members of the other set, which had an average lower ratio of 63% Hb°S/ 37% COHbS varied more widely, although the values clustered close to the means.[+] Data obtained with the latter set are given in Fig. 8. The

[&]The quality of the assay data just used is not high enough to investigate the relation between turbidity and mass of crystals.

[+]Hb^+ not included in ratio usually very small - under 5% of total.

turbidity data, especially when plotted against total Hb°S, falls on straight lines. The slope against total protein is 0.0381 and against Hb°S it is 0.0580. As explained earlier, the slope to be expected with no interaction in either phase is 0.063. A reference line of slope 0.10 is included, obtained with pure S, and indicating a solubility of the latter in the pure state of 8.42 g/dl. The solubility of Hb°S in this mixture extrapolates to 5.19 g/dl; the figure is lower than for the pure substance, due to salting-out by COHbS.

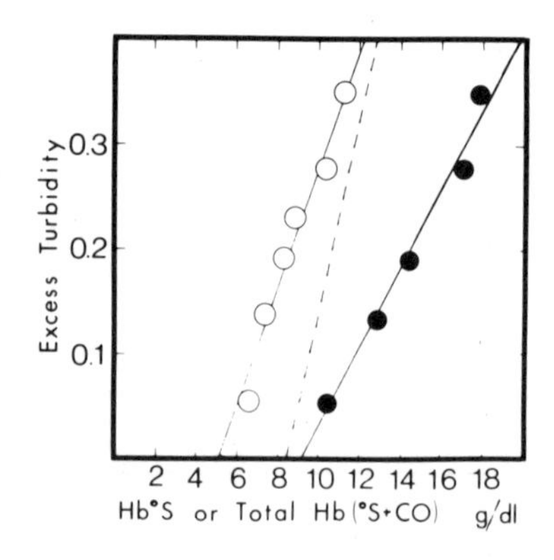

Fig. 8. Excess turbidity as a function of total Hb (●) or Hb°S concentration (○) from mixtures of 63% Hb°S and 37% COHbS. The slope of the line when plotted against Hb°S is 0.0580 with a correlation coefficient of .982 and intercepts the x-axis at 5.19 g/dl. The slope against total Hb is 0.0381 with a correlation coefficient of .992 and intercepts the x-axis at 9.30 g/dl. The dotted line represents the same relationship for pure Hb°S.

The data just described provide an excellent opportunity to test the validity of the assumption made in much of this work: that turbidity is proportional to mass of crystal, with a proportionality constant of 0.10 between turbidity and mass, when the latter is measured in grams per deciliter. Fig. 9 shows a test of this assumption - a linear relation clearly prevails, and the slope is .087. This close agreement not only supports the validity of the turbidity technique, but also the conclusion that little or no COHbS cocrystallizes with Hb°S.

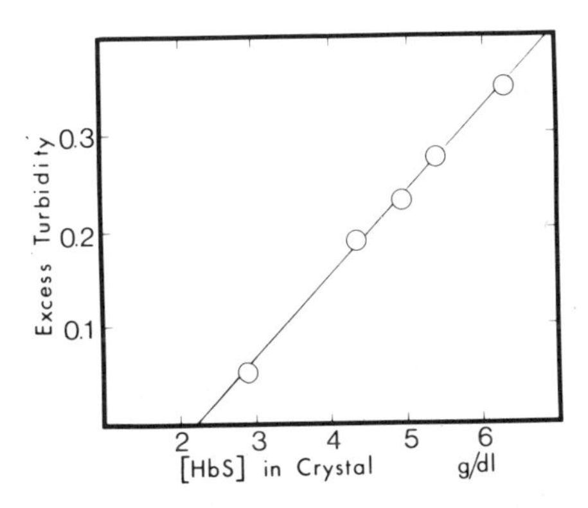

Fig. 9. Excess turbidity as a function of Hb°S concentration in the crystals formed from mixtures of Hb°S and COHbS. The slope of the line is 0.0874 with a correlation coefficient of 0.998.

With the 73% Hb°S/ 27% COHbS ratio the line relating turbidity to total protein has a slope (least squares) of 0.041 somewhat steeper than any which might be fitted to the 63% Hb°S/ 37% COHbS data (Fig. 10). This line can be extrapolated to 8.05 g/dl, just below the solubility of pure Hb°S under these conditions. The slope of the line for turbidity against [Hb°S] is somewhat higher 0.0516 and extrapolates to a lower solubility, 5.54 g/dl, because of the salting-out effect of the constant ratio of added COHbS.

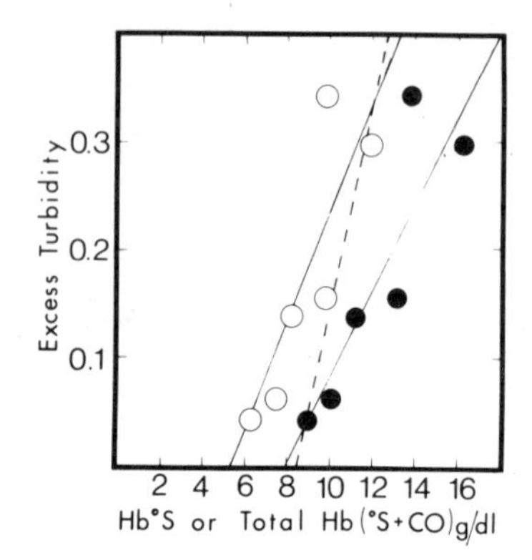

Fig. 10. Excess turbidity as a function of total Hb (●) or Hb°S concentration (○) from mixtures of Hb°S and COHbS in 73%/27% ratio. The slope of the line when plotted against Hb°S is 0.0516 with a correlation coefficient of .837 and intercepts the x-axis at 5.54 g/dl Hb°S. The slope when plotted against total Hb is 0.0413 with a correlation coefficient of .886 and intercepts the x-axis at 8.05 g/dl Hb. The dotted line represents the relationship for pure Hb°S.

DISCUSSION

It has long attracted notice that adding certain other kinds of hemoglobin to Hb°S reduces the concentration of the latter at which it will gel. Understanding this observation has been properly viewed as an important step in understanding the molecular aspects of the process of gelling. A number of important investigations have recently been concerned with it[12,18,19,23,24] and have met with partial success in explaining it[18,20,25]. The concept of non-ideality applied to Hb°S solubility[9,15,18] is an essential element in any analysis of the mixed hemoglobin gelling process.

We have undertaken the present studies of the corresponding phenomenon with crystals, partly because they illuminate the great variety of solubility effects produced by "impurities" in general, but also because they promise to simplify our thinking about the causes and control of Hb°S aggregation, whether crystallization or gelling. The simplification lies in regarding both of these phenomena as a manifestation of changes in the low solubility of Hb°S relative to other hemoglobins. We have formed crystals rather than gels not only because it is more convenient to work with crystals but also for the assurance we gain that we are dealing with manifestations of true equilibria rather than of kinetic freezes[9,21].

Thus the problem of understanding the pathological effects of Hb°S becomes the problem of why this mutant molecule becomes so much less soluble than Hb°A in the deoxy (T) state. Other related problems are: (a) why is the solubility lowered still further in the presence of DPG or IHP, both which are preferentially bound by the T conformer? (b) Are all T conformers less soluble than corresponding R conformers? If so they are nevertheless soluble enough not to gel under physiological conditions. The case of methemoglobin is pertinent, since it can be induced to gel when its concentration is high enough but only in the presence of IHP. Spectral changes induced by IHP shows that unless IHP is added, some of the methemoglobin is not in the T state.

Our mixture experiments could be considerably improved. There is evidence that our materials, although chromatographically pure, still show slight deviations from the Gibbs phase rule (plotting the turbidities against [Hb°S] in the supernatants gives a very slight deviation from a vertical line). In addition, more data are needed, in order to improve the statistical reliability of the relations expressed in our graphs. More differential analyses of crystal and solution composition are especially needed for mixtures of Hb°S and Hb°A. It would be particularly helpful if crystal composition could be measured directly instead of by differential analysis of original solution and of supernatant after crystallization [as, for example, by centrifugation into an underlying immerscible layer[22]].

Some of our hypotheses remain to be proved:

1. The solubilizing force of the isomorphous additives may reside in the opportunities they permit for any lattice location for a single isolated Hb°A molecule to have other Hb°A molecules adjacent to it, or entirely around it. Since Hb°A is much more soluble the crystal is filled with weakened bonds, sufficient to increase the fugacity of the whole structure.

2. More work is needed to establish the partial interchangeability of Hb°S, previously mentioned, with either Hb°A or COHbS in salting-out. Determining the extent of its validity of this approximation will permit evaluation of the contribution of the third and higher virial coefficients to Hb°S solubility, and thus to the <u>specific</u> aspects of the salting-out effectiveness of each "extender".

3. The anomalously low slopes of turbidity against concentration found in some of our mixture experiments require further explanation.

We are now discussing second-order tests. Our principal finding is that interactions between similar (possibly isomorphous) proteins do occur in the solid state and affect the solubility. In the case studied, the result is an increase in solubility of the less soluble component of the crystals or of the crystals as a whole, which overcomes the salting-out effect of smaller additions of other proteins, whether isomorphous or not. The latter are mediated by more familiar interactions in the dispersed phase. In other cases, such as when solid-solutions are formed, a decrease is produced in the solubility of the more soluble component. Thus, the solubility behavior of Hb°S in mixtures depends on the particular hemoglobin with which it is mixed. In our work, the isomorphous Hb°A (T form) was incorporated into crystals of Hb°S, also a T conformer. COHb, an R conformer, was not. Further work is needed to determine whether HbF (presumed by some not to promote gelation of Hb°S) will behave similarly, distinguishing clearly between the salting-out and solubilizing effects. Both methemoglobin A and S may prove to cocrystallize with Hb°S since they are both at least partly in the T state. The effect of cyanide on them (well-known) and of IHP should also bear on this point. Since methemoglobin can be easily formed *in vivo* at the expense of O_2HbS and Hb°S the possibility arises of dissolving gels or preventing their occurence, by limited manipulation in homozygotes of methemoglobin levels. We shall put this idea to experimental test with intact erythrocytes. There are numerous related questions which would make it desirable to investigate the solubility behavior of Hb°S in other mixtures, and to investigate the cases presented here much more thoroughly.

ACKNOWLEDGEMENTS

The work was supported by the National Heart and Lung Institute (grant HL12856).

REFERENCES

1. (a) Sörensen, S.P.L. (1930) Compt.-rend. trav. Lab. Carlsberg, 18, No. 5, 1.

   (b) Sörensen, S.P.L. and Sörensen, M., (1933) Compt.-rend. trav. Lab. Carlsberg, 19, No. 11, 1.

2. (a) Bonot, A. (1934) J. Chim. Phys., 31, 258, 301, 383.

   (b) Bonot, A. (1937) Ann. de Chim., 8, 425.

3. Northrop, J.H. (1930) J. Gen. Physiol., 13, 739, 767.

4. Steinhardt, Jacinto (1938) Cold Spring Harbor Symposia on Quantitative Biology, Volume VI, 301-317.

5. Steinhardt, Jacinto (1939) J. Biol. Chem. 129, No. 1, 135-170.

6. Hewitt, L.F. (1936) Biochem. J., 30, 2229.

7. Perlmann, G.E. and Blumenfeld, O.O. (1959) J. Gen. Phys. 42, 533.

8. Benesch, R.E., Yung, S., Benesch, R., Mack, J., and Schneider, R.G., (1976) Proceedings of the Symposium on Molecular and Cellular Aspects of Sickle-Cell Disease (Edited by J.I. Hercules, G.L. Cottam, M.R. Waterman, and A.N. Schecter) 113-132.

9. Jones, M. McD. and Steinhardt, J. (1978) submitted to J. Mol. Biol. for publication.

10. van de Hulst, H.C. (1957) Light Scattering by Small Particles, New York: John Wiley and Sons, Inc.

11. Kerker, M. (1969) The Scattering of Light and Other Electromagnetic Radiations, New York: Academic Press.

12. Goldberg, M.A., Husson, M.A., and Bunn, H.F. (1977) J. Biol. Chem. 252, 3414-3421.

13. Bookchin, R.M., Nagel, R.L., and Balazs, T. (1975) Nature 256, 667-668.

14. Jones, M. McD. and Steinhardt, J., unpublished results.

15. Ross, P.D. and Minton, A.P. (1977) J. Mol. Biol. 112, 437-452.

16. Williams, R.C., Jr. (1973) Proc. Nat. Acad. Sci. USA 70, 1506.

17. Ferrone, F.A., Hofrichter, J., and Eaton, W.A. (1978) Biophysical J. 21, 50a.

18. Ross, P.D., Hofrichter, J., and Eaton, W.A. (1977) J. Mol. Biol. 115, 111-134.

19. Benesch, R.E., Benesch, R., Edalji, R., and Kwong, S. (1978) Biochem. Biophys. Res. Comm. 81, 1307-1312.

20. Behe, M.J. and Englander, S.W. (1978) Biophys. J. 23, 129-145.

21. Pumphrey, J.G. and Steinhardt, J. (1977) J. Mol. Biol. 112, 359-375.

22. Bendet, I.J., Smith, C.E., and Lauffer, M.A. (1960) Arch. Biochem. Biophys. 88, 280-286.

23. Bertles, J.F., Rabinowitz, R., Döbler, J. (1970) Science 169, 375.

24. Noguchi, C.T. and Schechter, A.N. (1977) Biochem. Biophys. Res. Comm. 74, 637-642.

25. Sunshine, H.R., Hofrichter, J., and Eaton, W.A. (1978) Fed. Proc. 37, 1581.

Catsimpoolas, ed. Physical Aspects of Protein Interactions

LIMITS OF PERFECTABILITY: FUNCTIONAL OPTIMIZATION IN MULTIFUNCTIONAL COMPONENTS

DONALD B. WETLAUFER
Department of Chemistry, University of Delaware, Newark, Delaware 19711

ABSTRACT

It is impossible to simultaneously perfect competing or conflicting functions in biological macromolecules.

In the fields of operations research, graph theory, and game theory, a system is recognized as a collection of related elements. While it is often possible to optimize for some system (or global) goal by manipulation of the elements, it is recognized that optimization of a particular element is likely to take place at the expense of the system goal and other elements. For this reason it makes no sense to optimize subsystems before optimizing the whole system.

Biological macromolecules are multifunctional.* This multifunctionality inevitably leads to compromise in the effectiveness of the functions.* It has become a fairly common viewpoint among some enzymologists to evaluate the catalytic function of an enzyme as if it operated in isolation from the other functions. We believe this to be a mistake, in that it can lead to unrealistic expectations of the behavior of biological systems and their components.

Let us briefly examine a few brief biographies of individual proteins to demonstrate that multiplicity of physiological function is indeed the rule. The proteolytic enzyme trypsin is synthesized in certain cells of the mammalian pancreas as a zymogen. It may, like insulin, even be synthesized as a pre-zymogen. Self-assembly of the nascent polypeptide chain--the process of protein folding--is one of the first physiological functions of trypsinogen. Trypsin is not immediately secreted, but is stored intracellularly for secretion in response to

---

*The question of how a function is properly defined could generate a lengthy discussion. We will defer that. Instead, we will here use function to mean a physiologically relevant function, acknowledging that validation of the assignment of such a function depends only on plausible evidence and argument. Thus, for the present discussion, we do not subscribe to a set of functions as extensive as those of Zuckerkandl[1,2] who includes parameters such a pI, charge density, and pK's.

an appropriate signal. Following secretion, catalytically inactive trypsinogen is activated by limited proteolysis to catalytically active trypsin. Its secretion into the duodenum is coordinated with the arrival of substrates in an unorderly mixture of masticated, pepsin-processed lunch, plus bile. It is the poorly controlled physical and chemical nature of this melange that puts greater stability demands on trypsin than on an intracellular enzyme. Trypsin, working in concert with other digestive hydrolases, works to reduce a large fraction of the bolus to assimilable units. But the story is not yet over. In a relationship symmetrical to the activation of trypsinogen by trypsin, trypsin autolyzes in bimolecular encounters with trypsin and other proteases. Without autolysis, the high rate of turnover of the surface cells lining the gut would surely be higher still.

Now let us examine the general case of an inducible intracellular allosteric enzyme such as aspartic transcarbamylase. Its minimal set of functions will include rapid self-assembly to a functional form, catalytic activity, allosteric modulation, and intracellular inactivation and degradation. The latter two functions can be viewed as the other side of the coin (=function) labeled "stability."

Finally, let us look at hemoglobin, probably the most intensively studied multi-subunit protein. Its early life history includes folding and heme insertion in the α and β chains and association to yield the functional $\alpha_2\beta_2$ tetramer. Studies of these processes in vitro show that the products are highly pathway-dependent[3]. The tetramer exhibits a well-known set of linked binding equilibria seen in O_2-binding cooperativity, diphosphoglycerate modulation of O_2-binding, and the Bohr effect. Closer scrutiny reveals that hemoglobin also facilitates CO_2 transport. Further, after adventitiously losing an electron to become methemoglobin, the macromolecule can effectively interface with glutathione/TPNH methemoglobin reductase to accept an electron and return its O_2-binding function. The relatively high stability of hemoglobin (or better, the functional stability of the erythrocyte and its contents) is also commonly viewed as a physiological function. This property is most clearly seen as a "function" in the perspective of a hemoglobin variant whose decreased stability leads to pathological consequences. A similar situation exists for the function that is seen to be deficient in a homozygous hemoglobin S (HbS). Unless we were familiar with something of the mechanism of HbS malfunction--the formation of HbS precipitates which decrease the plasticity of the erythrocyte--it might seem far-fetched to call HbA's solubility properties an important "function."

Without attempting to be exhaustive, we believe that the foregoing examples illustrate our view that multiplicity of function is the rule rather than the exception for globular proteins. Similar arguments might be constructed for fibrous proteins. Further, it is clear that, for tRNA molecules, multiplicity is a central theme in their functional analysis. The idea of more than one physiological function for an enzyme was dramatically established by Monod and Jacob in their introduction of allostery[4]. Although some investigators** have gone no further than to accept catalysis and control as functions, it seems likely that folding and degradation are also general functions for intracellular proteins.

It seems to be necessary to reiterate[6] that it is probably a mistake to view maximum stability as a desirable feature of a protein. At least one recent commentator on protein structure believes otherwise. In speculating on the development of protein folding pathways, Richardson says[7],

> "During its evolution, once a protein finds a reasonably stable, kinetically accessible, minimum energy conformation, it is then subject to natural selection for stability. Selection will dig that particular local energy minimum as deep as possible by adjusting the amino acid sequence to approach optimal fit for the native conformation, and at the same time probably will raise all the other local energy minima whose stability is not being selected for. The result is very likely to be a genuine global energy minimum, in spite of the influence of kinetic requirements."

However, turnover remains a fact of life[8,9], and a way of successfully responding to many challenges. Selection for maximum stability is in all likelihood selection against facile turnover.

The crucial question now is, "Is it possible to simultaneously optimize multiple functions in a macromolecule?" We believe that the general answer is no. While we can imagine that certain pairs of functions are independent, it seems likely that in the general case the multiple functions of a protein will exhibit interdependence. In such a case, attempts optimize overall design inevitably result in

---

**A recent example is found in the very interesting paper of Albery and Knowles[5] titled "Evolution of Enzyme Function and the Development of Catalytic Efficiency." We quote therefrom, "Obviously, any system that falls short of perfection must either be susceptible of further evolutionary refinement, or have sacrificed simple catalytic efficiency to the possibly higher good of control." We would prefer the latter half of this sentence to read, "or to have compromised simple catalytic efficiency in the development of other physiological functions."

compromises. This result is also familiar to architects and designers of macroscropic tools. Design elements favorable for functions i and j are antagonistic to functions p an q. Attempts to optimize all functions of multiple-function molecule end in the compromise of at least some functions.

However, we should recognize that extensive linkage of two functions can occur without conflict. For example, the secondary structure of mRNA is likely to play a role in recognition and control processes involving these molecules. It is in principle possible to enumerate the set of all messages, M, consistent with the observed amino acid sequence for some protein, but in practice this set is computationally too large to be of use, being on the order of Avogadro's number for even a small protein (hen egg lysozyme). However, it is possible to select two messages from M without explicit enumeration. These are:

$\overline{m}$ - the potential message with maximal secondary structure, and

$\underline{m}$ - the potential message with minimal secondary structure.

It turns out that $\overline{m}$ is entirely tied up in hairpin loops while $\underline{m}$ is virtually devoid of loop structure. Thus it appears that the set M is sufficiently rich that evolutionary changes in the protein need not occur at the expense of structural constraints on the message[10]. It seems unlikely that multiple functions within a globular protein will have linkage as loose as that of between mRNA secondary structure and amino acid sequence.

This work was supported by USPHS Grant No. 7RO1 GM23713-03.

REFERENCES

1. Zuckerkandl, E. (1976a) J. Mol. Evol. 7, 167-183.

2. Zuckerkandl, E. (1976b) J. Mol. Evol. 7, 269-311.

3. Yip, Y. K, Waks, M., and Beychok, S. (1977) Proc. Nat. Acad. Sci. (Wash.) 74, 64-68.

4. Monod, J., Changeux, J.-P., and Jacob, F. (1963) J. Mol. Biol. 6, 306-329.

5. Albery, W. J. and Knowles, J. R. (1976) Biochemistry 15, 5631-5640.

6. Wetlaufer, D. B. (1973) J. Food Science 38, 740-743.

7. Richardson, J. S. (1977) Nature 268, 495-500.

8. Schoenheimer, R. (1942) Dynamic State of Body Constituents, Harvard University Press, Cambridge, Mass.

9. Schimke, R. T. (1969) in Current Topics in Cellular Regulation, Vol. 1, Horecker, B. L. and Stadtman, eds., Academic Press, New York, pp. 77-124.

10. Rose, G. D. (1976) Ph.D. Dissertation, Oregon State University.

Catsimpoolas, ed. Physical Aspects of Protein Interactions

HYDROPHOBIC AND ELECTROSTATIC PARAMETERS IN AFFINITY CHROMATOGRAPHY

M. WILCHEK, G. TOMLINSON, W. SCHELLENBERG and T. VISWANATHA
Department of Biophysics, The Weizmann Institute of Science, Rehovot (Israel) and Department of Chemistry, University of Waterloo, Waterloo, Ontario N2L 3G1 (Canada)

INTRODUCTION

Since its introduction in 1968[1], the technique of affinity chromatography has become an extremely powerful tool in the isolation and purification of biologically active molecules[2-4]. The method is generally based on the original observation that compounds containing free amino groups can be covalently attached to water-insoluble polysaccharides that have been activated by prior treatment with cyanogen bromide[5]. By appropriate choice of ligands advantage can be taken of the unique biological specificities of a variety of cellular components in their isolation from complex mixtures. The resourcefulness of this technique in the facile isolation of a variety of biologically active substances has been well documented during the past decade. Despite numerous successes, the physical and chemical characteristics of affinity gels are just beginning to be understood. It has become evident that the complexity of the affinity process extends beyond that of a simple association-dissociation phenomenon. In addition to the biospecificity of the attached ligand, possible roles for the "spacer arm" through which the ligand is coupled to the gel, for the type of chemical linkage at the gel-ligand interface and for the gel itself have been documented. Both hydrophobic[6,7] and ion exchange[7,8] effects may be operative and these may either diminish or abolish the biospecificity of the ligand[9]. The influence of each of these effects is discussed below.

Hydrophobic effect

From the many studies on affinity chromatography, the necessity for the introduction of an arm or a spacer separating the affinity ligand from the insoluble support has become evident[1]. Since the spacer is usually a hydrocarbon chain, interference due to excess hydrophobicity and charge arising from its attachment to the matrix has to be minimized. In the early studies on chymotrypsin and nuclease, we prepared ligands with long hydrocarbon chain prior to their attachment to agarose. This approach was adopted in our laboratory for the preparation of affinity adsorbents[10]. A distinct advantage of such a procedure is the facility for the assessment of the binding capacity of the ligand-spacer unit with the macromolecule in question. For example, we have found that ε-aminocaproyl-D-tryptophan methyl ester (the ligand-spacer unit) is a stronger inhibitor (K_i = 1.0 μM) of chymotrypsin than the free ligand, D-tryptophan methyl ester, which is characterised by a K_i value of approximately 100 μM. Similar enhancement in the affinity upon attachment of spacer to the ligand has also been noted by Singer[11] who showed that decanoyl choline was more effective than d-tubucurarine in protecting acetyl choline receptors from P-(trimethylammonium) benzene diazonium fluoroborate induced inactivation. It is unlikely that the strong binding of either ε-aminocaproyl-D-tryptophan methyl ester to chymotrypsin or of decanoylcholine to the receptor is due to the specific affinity of the proteins to the hydrocarbon chain. Hence the above findings suggest that there is a significant contribution to nonspecific, hydrophobic binding by the spacer to a surface close or adjacent to the active site of the protein. Such hydrophobic groups with nonspecific binding increments are usually incorporated into the ligands to enhance the specific purification of proteins. By initially coupling the spacer to the ligand, the amount of hydrophobic side chains can be readily controlled since the number of spacers introduced is stoichiometric with the ligands incorporated. Thus, it will be possible to control the hydrophobic interactions arising from the spacer either by lowering the amount of ligand incorporated or by appropriate dilution of the affinity gel with unmodified agarose. Either of the above mentioned procedures can be successfully adopted for the purification of the protein primarily on the basis of its bio-specificity. However, such manipulations are difficult to achieve when the preparations of the affinity gels involves the coupling of the spacer (to the support) prior to the introduction of biospecific ligand. Under these conditions, the hydrocarbon chains of the spacer will always be in large excess. The resultant hydrophobic effect will be detrimental to the protein being purified. Indeed, the unsubstituted spacers may function as insoluble detergents and effect denaturation of the protein. Our studies on the purification of leucine aminopeptidase on tyramine-agarose gels serve to confirm the above view (Table 1).

TABLE 1

RECOVERY OF LEUCINE AMINOPEPTIDASE FROM DIFFERENT SOURCES AFTER CHROMATOGRAPHY ON TYRAMINE-AGAROSE

The enzyme units represents μmoles of product formed from 2 mM leucine *p*-nitroanilide at 25°C per min. The activity of the purified enzyme was determined after the elimination of salt by dialysis against 0.05 M Tris buffer (pH 8.5)-5mM $MgCl_2$

Source of enzyme	Protein applied	Protein recovered from the active fractions
Swine kidney: 80% $(NH_4)_2SO_4$	11.25 mg with 0.696 units/mg = 7.84 units	2.60 mg with 0.555 units/mg = 1.44 units
Swine kidney commercial preparation	1.0 mg with 56.6 units/mg = 56.6 units	0.375 mg with 35.3 units/mg = 13.3 units
Bovine eye-lens crystalline preparation	0.300 mg with 71.9 units/mg = 21.57 units	0.280 mg with 62.5 units/mg = 17.5 units

Such denaturation may occur even in gel preparations containing smaller numbers of hydrocarbon chains, not so much due to adsorptions but by the elution of the protein by "polarity reducing solvents" such as 20-50% ethylene glycol, organic solvents or neutral salts[7].

The hydrophobic columns clearly are active by virtue of their aversion to interact with water and hence bind to the nonpolar regions of the protein molecule. If the residues comprising such apolar regions were to be in the interior of the protein molecule, alteration in its conformation can be expected so as to facilitate its binding to the gel. Such changes in protein conformation may lead to its eventual irreversible denaturation. The lower polarity of the environment resulting from the interaction between hydrocarbon chain and the protein may cause an increase in the stabilization of hydrogen bonds, a situation which can lead to partial or total denaturation of biologically active proteins during their recovery from gels with organic solvents. The interference due to hydrophobic effect can be overcome either by directly coupling ligands containing spacers to agarose or by using hydrophilic spacers, e.g. hydrazide or polyglycine. The effectiveness of hydrazide spacer in overcoming the hydrophobic effects is clearly demonstrated by the data shown in Fig. 1.

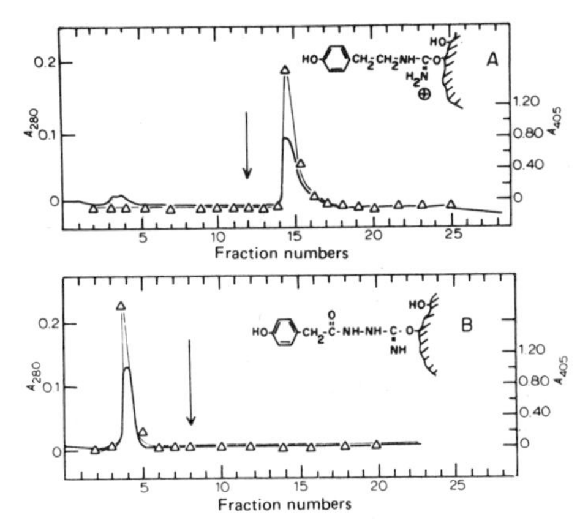

Fig. 1. Chromatography of bovine lens leucine aminopeptidase on tyramine-agarose (A) and 4-hydroxyphenylacetyl hydrazide-agarose (B). 0.5 mg of enzyme was applied to a Pasteur pipette filled with about 0.6 g of moist agarose Derivative A or B. The columns were eluted with 0.05 M Tris-HCl buffer (pH 8.5)-0.1 M NaCl. Fractions of 0.5 ml were collected. At the position indicated by the arrow, 1 M NaCl in the same buffer was applied. Δ, relative activity of 100 μl of the fractions with 1mM leucine *p*-nitroanilide.

The ion-exchange effect

Recently we have shown that N-substituted isourea is the major product of the reaction between amines and polysaccharide activated with cyanogen-bromide. Treatment of α-N-acetyl ornithine coupled to Sepharose with dilute ammonium hydroxide has been found to yield α-N-acetyl arginine in about 80% yield (Scheme 1).

SCHEME 1: CONVERSION OF ORNITHINE TO ARGININE

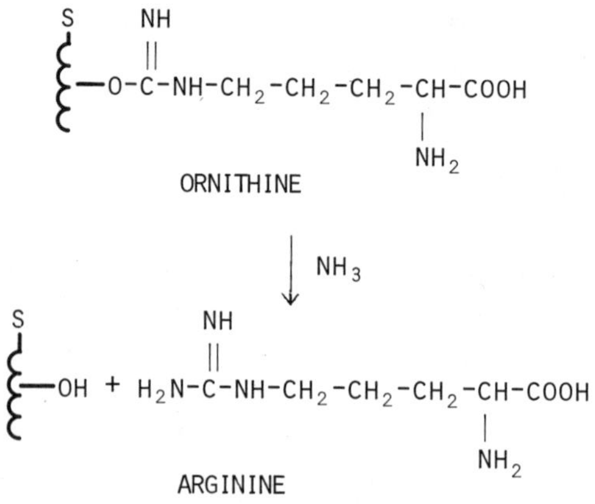

$$\text{S}\!-\!O\!-\!\underset{\|}{\overset{NH}{C}}\!-\!NH\!-\!CH_2\!-\!CH_2\!-\!CH_2\!-\!\underset{\underset{NH_2}{|}}{CH}\!-\!COOH$$

ORNITHINE

$$\downarrow NH_3$$

$$\text{S}\!-\!OH + H_2N\!-\!\overset{NH}{\overset{\|}{C}}\!-\!NH\!-\!CH_2\!-\!CH_2\!-\!CH_2\!-\!\underset{\underset{NH_2}{|}}{CH}\!-\!COOH$$

ARGININE

These studies show that the charge is maintained upon coupling of amines to agarose. While this may be advantageous in cases where the charge is essential for the function of the ligand, it also endows the affinity gel with ion-exchange properties. Consequently, significant non-specific adsorption of the protein could occur and thus invalidate the prospects for chromatography based on biospecificity. The presence of charge itself is generally not destructive and its effect can be countered by increasing the salt concentration or by coupling the ligands through other linkage (whenever possible) or by acetylation of the isourea moieties present in the affinity gel (Scheme 2).

SCHEME 2: ACETYLATION OF ISOUREA-SEPHAROSE

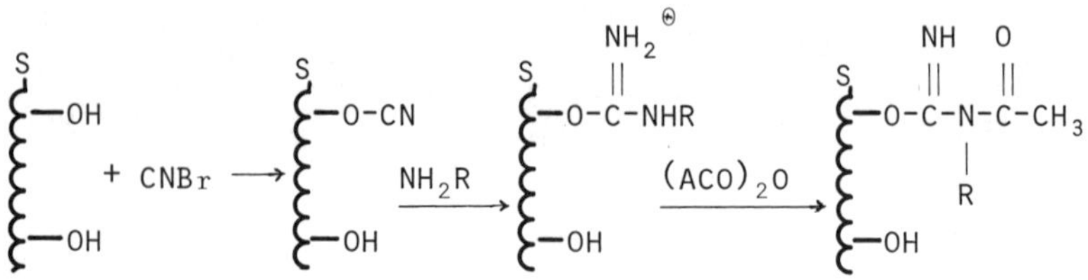

$$\text{S}\begin{cases}-OH\\-OH\end{cases} + CNBr \longrightarrow \text{S}\begin{cases}-O-CN\\-OH\end{cases} \xrightarrow{NH_2R} \text{S}\begin{cases}-O-\overset{NH_2^{\oplus}}{\overset{\|}{C}}-NHR\\-OH\end{cases} \xrightarrow{(ACO)_2O} \text{S}\begin{cases}-O-\overset{NH}{\overset{\|}{C}}-\underset{\underset{R}{|}}{N}-\overset{O}{\overset{\|}{C}}-CH_3\\-OH\end{cases}$$

The effectiveness of acetylation as a means of minimizing non-specific interactions of the affinity-gel is demonstrated by the studies on the purification of chymotrypsin (Table 2).

TABLE 2

AFFINITY CHROMATOGRAPHY OF α-CHYMOTRYPSIN

One gram of each adsorbent was filled into a Pasteur pipette. The column was equilibrated with 0.05 M Tris-HCl buffer pH 8.0. 1 mg of α-chymotrypsin was applied in the same buffer. One-milliliter fractions were collected and assayed. α-Chymotrypsin was eluted with 0.1 M acetic acid.

Sepharose derivative	% Adsorbed	% Eluted	Specific activity (units/mg protein)	Purification factor
Agarose	0	-	33	1
Agarose-ε-aminocaproyl D-tryptophan methyl ester	90	100	46	1.4
Acetyl-agarose-ε-caproyl D-tryptophan methyl ester	70	100	63	1.9

The coupling of hydrazides is analogous to that of alkyl amines and provides derivatives which are also uncharged at neutral pH[12]. These affinity gels (containing hydrazide spacers) fail to adsorb proteins by hydrophobic interaction due to the elimination of the charge.

In this communication we describe experiments designed to explore the relative contributions of both specific and non-specific interactions in the binding of α-chymotrypsin and other proteins to affinity gels. The results provide further insight into the binding properties of both the derivatized gels and of proteins themselves, and serve to emphasize the need for careful investigation of these properties prior to using a given gel preparation in the separation of α-chymotrypsin from inactive species.

RESULTS AND DISCUSSION

Several affinity gels with either phenylbutylamine (PBA) or ε-aminocaproyl-D-tryptophan methyl ester as ligands were prepared according to published procedures[13]. In the preparation of these gels, attempts were made to vary the extent of substitution either by controlling the amount of CNBr used for the activation of Sepharose or by varying the concentration of the ligand during the final coupling step. The extent of substitution in the gels was assessed by the determination of guanido compound released upon treatment of the gel with aqueous ammonia[8,14]. The ability of these gels to retain chymotrypsinogen, chymotrypsin and its inactive derivatives (TPCK and DFP inhibited chymotrypsins) was determined. These studies revealed that the property of gel preparation was, to a large extent, dependent on the amount of the affinity ligand bound. In general, gel preparations with less than 15 μmoles of ligand bound per ml of gel were found to be specific in that they retained chymotrypsin and exhibited little or no affinity for either chymotrypsinogen or the inactive derivatives of the enzyme. Furthermore, none of the cationic proteins (trypsin, subtilisin and ribonuclease) were retained by these gels. These results are shown in Figures 2a and 2b. Since the adsorbed enzyme can be readily recovered

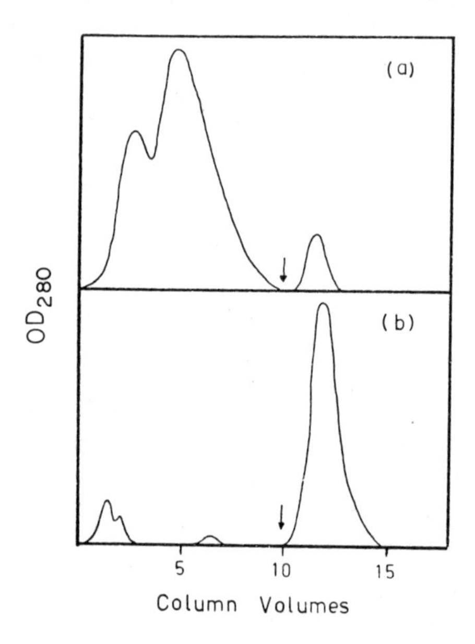

Fig. 2. Binding of chymotrypsinogen and α-chymotrypsin to Sepharose 4B containing less than 15 μmoles PBA. (a) Chymotrypsinogen A; (b) α-chymotrypsin. Arrow indicates change of buffer from 0.05 M Tris, pH 8.0 to 0.1 M acetic acid.

by elution of the gel with 10 mM solution of the free ligand (phenylbutylamine), the adsorption of the protein appears to be governed by the phenomenon of biological specificity (Fig. 3a). The fact that free ligand can successfully compete with that bound to the gel for the enzyme provides direct evidence for the contribution from biospecificity to the binding process. However, bioaffinity does not seem to be the sole contributing factor, since the adsorbed enzyme can also be eluted with a solution of sodium chloride (Fig. 3b). These observations suggest that ion-exchange properties also play a role in the binding of the protein. When these ionic interactions are suppressed (by inclusion of salt in the buffer), bioaffinity by itself is insufficient to retard the protein. Conversely, the ion-exchange properties alone

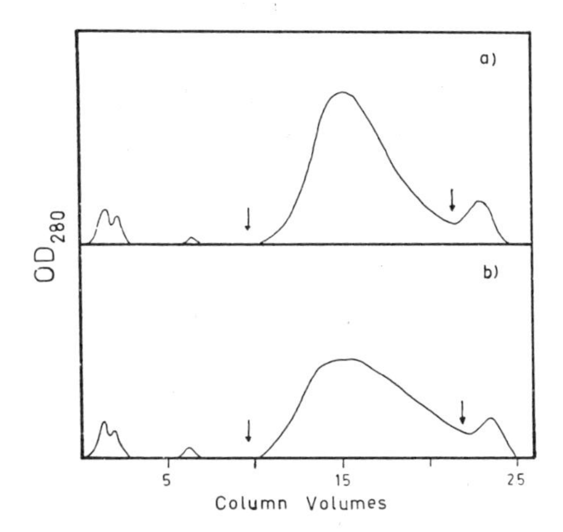

Fig. 3. Effect of 0.01 M PBA and 0.10 M NaCl on α-chymotrypsin binding to PBA gel (Fig. 2). (a) Elution buffer 0.05 M Tris-0.01 M PBA, pH 8.0; (b) elution buffer 0.05 M Tris-0.10 M NaCl, pH 8.0. Applied buffer was 0.05 M Tris, pH 8.0, in both cases.

are also insufficient as demonstrated by the ability of the free ligand to displace the protein from the gel. Thus, both the biospecificity of the ligand and the ion-exchange property of the affinity gel play a major role in the adsorption of the protein. Contributions from hydrophobic interactions between the gel and the protein are virtually nonexistent since the adsorbed enzyme can be quantitatively recovered without the use of organic solvents.

On the other hand, affinity adsorbents bearing greater than 15 μmoles of ligand per ml of gel were less selective in that they were capable of adsorbing even the zymogen and the inactive derivatives of the enzyme. Neither the buffer containing only the ligand nor the one having sodium chloride alone was effective in releasing the adsorbed protein from the gel (Fig. 4a and 4b). The protein could only be recovered by elution with acid, a condition which may not be desirable in view of the possible denaturation of the proteins.

In order to further investigate the hydrophobic nature of the gel, the adsorbed protein was eluted with polarity reducing solvent, 50% ethylene glycol. The results obtained with the two types of gels are illustrated in Fig. 5a, b and c. While the material on the affinity column containing less than 15 μmoles of ligand/ml gel could be readily recovered by the solvent, such was not the case with extensively substituted gel preparations. The elution profile from these latter gel preparations exhibited

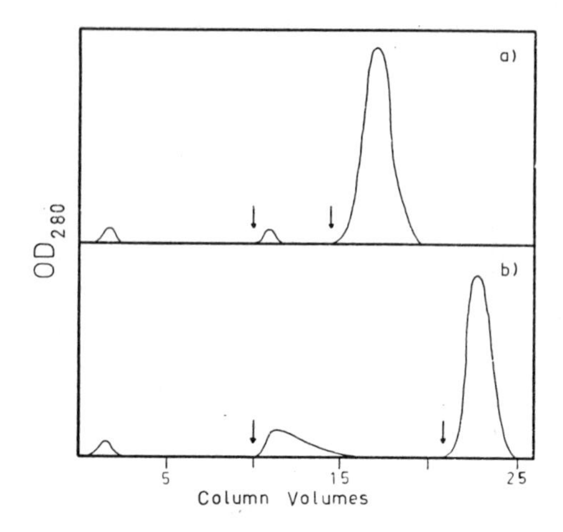

Fig. 4. Effect of PBA and NaCl on binding of α-chymotrypsin and its derivatives to gels containing more than 15 μmoles PBS. (a) First elution buffer 0.05 M Tris-0.01 M PBA, pH 8.0; (b) first elution buffer 0.05 M Tris-1.0 M NaCl, pH 8.0. Applying buffer was 0.05 M Tris, pH 8.0; second elution buffer was 0.1 M acetic acid, in both cases.

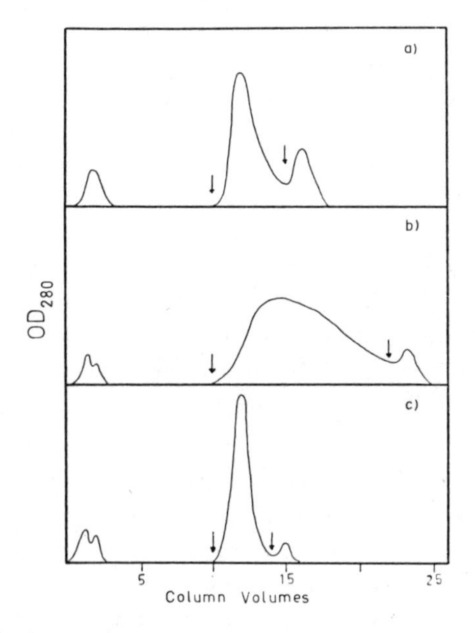

Fig. 5. Effect of mixed solvents on α-chymotrypsin binding to PBA gels. Applying buffer was 0.05 M Tris, pH 8.0; (a) Gel from Fig. 2, elution buffer 0.05 M Tris pH 8.0 -ethylene glycol (1:1); (b) gel from Fig. 4 elution buffer as in (a); (c) gels from Fig. 2 and Fig. 3 elution buffer 0.05 M Tris-1.0 M NaCl, pH 8.0-ethylene glycol (1:1). Second elution buffer was 0.1 M acetic acid in all cases.

considerable tailing in contrast to the sharp profile noted with gels containing lesser amount of the ligand. Inclusion of sodium chloride in the solvent was essential for the complete and facile elution of the protein. The strong adsorption of proteins by these gels does not appear to be only due to hydrophobic interactions. Were this to be the case, the presence of salt in the medium would have resulted in maximising these hydrophobic interactions [6] and facilitated binding of the proteins. Indeed, the binding is further reduced indicating that forces derived from interactions other than those of hydrophobic origin contribute significantly to the binding of the proteins to the gels. An alternative explanation for this phenomenon of nonspecific binding could be that the hydrophobic side chains markedly enhance ion pair formation between carboxyls on the protein and the isourea moieties on the gel. Such enhancement of ion-pair formation by hydrophobic chain is a known phenomenon: e.g. the equilibrium constant for ion pair formation in water for benzene sulfonic acid and N-trimethyl amino decane is characterized by K_{ass} value of 7.1 ℓ/g ion, compared with the K_{ass} value of 42,000 ℓ/g ion for anthracene sulfonic acid and the same cation [15]. These observations serve to reemphasize the view that hydrophobic interactions are not the only contributory factors to the adsorption of the protein by affinity gel or hydrophobic gels. The phenomenon is a consequence of a variety of interactions which, besides hydrophobicity, include biospecific and ion-exchange properties of the affinity adsorbent.

In summary, affinity chromatography procedure was developed with the sole idea of exploiting the unique specificities displayed by biologically active substances. The specific interaction between the biologically active molecule and its ligand encompasses a variety of structural features - hydrophobic, electrostatic and stereochemical in nature. The affinity is simply an expression of how well these forces augment each other in promoting complimentarity between two interacting molecules. Prior to the introduction of affinity chromatography procedure, the isolation of a biologically active substance depended largely on its physical and/or chemical properties. These procedures were often time consuming and resulted in low yields primarily due to denaturation of the substance under conditions of isolation. The affinity procedure afforded a simple, rapid, often a single step procedure for the isolation of biologically active compounds. In view of the multiplicity in the factors contributing to biospecificity, it is hardly necessary to reemphasize the advantage of exploiting the hydrophobic as well as the ion-exchange properties of the affinity adsorbent for maximizing its interaction with the substance to be isolated. Methods of isolation based solely on either ion-exchange property or hydrophobicity of the affinity adsorbents, while proving occasionally successful, do not in general share the advantages afforded by methods based on biological specificity.

REFERENCES

1. Cuatrecasas, P., Wilchek, M. and Anfinsen, C.B. (1968) Proc. Natl. Acad. Sci. USA, 61, 636-643.

2. Jakoby, W.B. and Wilchek, M. eds. (1974) Methods in Enzymology, Vol. XXXIV, Academic Press, New York.

3. Lowe, C.R. and Dean, P.D.G. (1974) Affinity Chromatography, John Wiley, London.

4. Dunlap, R.B. ed. (1974) Adv. Exp. Med. Biol. 42, Plenum Press, New York.

5. Axén, R., Porath, J. and Ernback, S. (1967) Nature 214, 1302-1304.

6. Wilchek, M. (1974) Adv. Exp. Med. Biol. 42, 15-30.

7. Hofstee, B.H.J. (1973) Analyt. Biochem. 52, 430-448.

8. Wilchek, M., Oka, T. and Topper, Y.J. (1975) Proc. Natl. Acad. Sci. USA, 72, 1055-1058.

9. O'Carra, P., Barry, S. and Griffin, T. (1974) FEBS Lett. 43, 169-175.

10. Wilchek, M. and Rotman, M. (1970) Israel J. Chem. 8, 172P.

11. Singer, S.J. (1970) in Molecular Properties of Drug Receptors, Porter, R. and O'Conner, M. eds., Churchill, London, 229- .

12. Jost, R., Miron, T. and Wilchek, M. (1974) Biochim. Biophys. Acta 362, 75-82.

13. Tomlinson, G., Shaw, M.C. and Viswanatha, T. (1974) Methods in Enzymology, 34, 415-420.

14. Tomlinson, G. and Viswanatha, T. (1974) Analyt. Biochem. 60, 15-24.

15. Packter, A. and Denbrow, M. (1962) Proc. Chem. Soc. (London), 220-221.

Catsimpoolas, ed. Physical Aspects of Protein Interactions

THERMODYNAMIC EXAMINATION OF THE SELF-ASSOCIATION OF BRAIN TUBULIN TO MICROTUBULES AND OTHER STRUCTURES

SERGE N. TIMASHEFF
Graduate Department of Biochemistry, Brandeis University, Waltham, Massachusetts, 02154, U.S.A.

ABSTRACT

A thermodynamic study has been carried out of three modes of association of purified calf brain tubulin: to microtubules, to 42 S closed ring structures and to vinblastine-induced indefinite polymers. The techniques used were quantitative turbidimetry, and velocity sedimentation with analyses of the results in terms of the most rigorous current theories. The deduced stoichiometries were further examined in terms of the thermodynamic state variables, with particular emphasis on the effects of ligands, both small and macromolecular, and of solvent components on the course of the reactions. The three modes of tubulin self-association can be used as good examples of ways in which proteins can associate and of the methods available for studying these associations.

INTRODUCTION

Microtubules

Cytoplasmic microtubules are widely distributed organelles which, as their name implies, have the general structure of long hollow cyclinders. The outside and inside diameters of these cylinders are 25 and 15 nm. A schematic representation based on structural analysis (1,2,3,4) is shown in Figure 1. Microtubules are found to consist of protofilaments of the subunit protein, tubulin, aligned longitudinally along the axis of the cylinder, 13 protofilaments being arranged around the circumference. The protofilaments are ordered axially with respect to each other in such a manner that, viewed from the side, neighboring subunits are displaced from horizontal juxtaposition by an angle of 10°. This results in the appearance of a helical structure, each helical strand making a complete turn on the fourth level, so that a microtubule may be regarded as a 3-start helix with 13 subunits to a turn. The subunit protein, consists of two almost identical molecules, α- and β-tubulin, with a molecular weight of close to 55,000 (5). The stable form of isolated tubulin is a dimer of 110,000 molecular weight and a sedimentation coefficient, $s^{\circ}_{20,w}$, of 5.8S (6). It is generally believed that the predominant dimer is α-β (7).

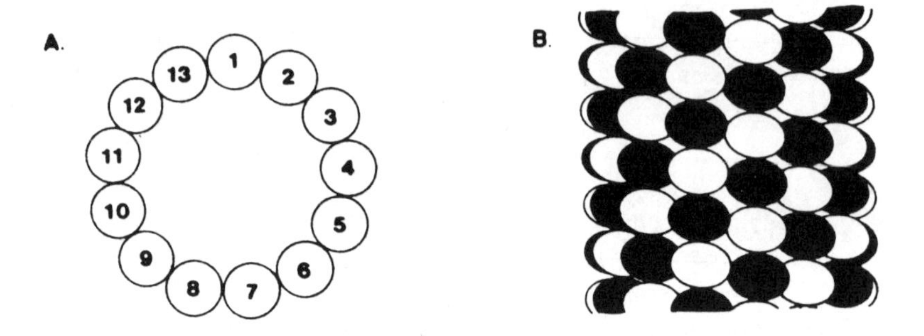

Fig. 1. Schematic representation of the structure of a cytoplasmic microtubule. A. Cross-section. B. Side view. Black and white ellipsoids represent α- and β-subunits. (Adapted from Snyder and McIntosh, ref. 4).

Modes of Tubulin Self-Assembly

Ever since the discovery of cytoplasmic microtubules, and especially following the isolation of their constituent subunit protein, tubulin, (8), a question that has focused the attention has been that of the manner in which these organelles are assembled from the tubulin subunits. The long elusive breakthrough came in the summer of 1972, when Weisenberg (9) first discovered conditions at which microtubules would assemble in a test tube from a partially purified brain extract. The key to Weisenberg's success was his finding that calcium ions inhibit this process and that, therefore, they had to be sequestered in order to permit assembly. As could be expected, this discovery was followed by an eruption of activity on this general problem, resulting in the rapid accumulation of a very vast literature with, not surprisingly, much confusion and many apparent contradictions. The main thrusts of the efforts were the elucidation of the pathway of microtubule assembly from tubulin, and the determination of the nature of the factors which control this assembly, in particular of the possible role of biological effectors in this process and of the manner in which certain drugs interfere with microtubule assembly. Our studies have concentrated on a quantitative analysis of three modes in which highly purified tubulin can self-associate. These are: the formation of microtubules, that of closed cyclic oligomers with a molecular weight of close to three million and $s^{\circ}_{20,w}$ of 42S, and that of assemblies mediated by the anti-cancer drug vinblastine. Since until recently our studies have been the only ones directed at pure tubulin, this chapter will be restricted almost solely to work from our laboratory on tubulin prepared from calf brain. Physico-chemically, these three self-association reactions represent three distinct ways in which protein polymerization can take place and their examination requires the application of different types of approaches. As such, therefore, the work described in this chapter can be regarded not only as a description of the tubulin-microtubule system, but also as a presentation of different possible modes of protein self-association and of methods which can be used to analyze them.

Tubulin Preparations

The vast majority of solution studies on tubulin described in the literature have been performed on brain protein. In present practice, there are two principal methods of isolation of brain tubulin. The first, used in our laboratory, is a classical biochemical technique, first developed by Weisenberg and co-workers (8) and further extended in our laboratory (5,6,10). It involves ammonium sulfate precipitation and DEAE chromatography. It yields a tubulin which is 99% pure by the criterion of SDS gel electrophoresis (5). The second method, which was adopted almost universally after its description in 1973, is an elegant extension by Shelanski et al. (11) of Weisenberg's conditions (5) for *in vitro* microtubule reconstitution. This method consists of a set of cycles of one dimensional recrystallizations of tubulin, namely of assemblies and disassemblies of microtubules, keeping, in turn, the pellet and the supernatant. The cycle procedure yields tubulin which contains 10-25% of other proteins on SDS gel electrophoresis, distributed over 15-30 bands of different molecular weights (12,13,14,15, 16,17). These proteins, called MAPs (microtubule associated proteins), vary in quantity and molecular weight distribution depending on the exact details of the preparation procedure. The most important ones are a family with molecular weights close to 350,000 (called HMW) (16,18), and one with molecular weights close to 60,000 (known as tau factor) (13,19). Although originally not much significance was ascribed to the differences between the methods of tubulin preparation, these turned out to be of major importance in the direction of research in this field.

CAN PURIFIED TUBULIN SELF-ASSEMBLE INTO MICROTUBULES?

Background

Our interest in the problem of the *in vitro* microtubule reconstitution stems from 1974, when we were completing the equilibrium thermodynamic study of the assembly of purified calf brain tubulin in the presence of magnesium ions to form the 42S oligomers (6,20). Namely, we were struck by the similarity, if not identity, of the electron micrographs of two structures: 1. The double rings formed as the end product of the rapidly reversible self-association of 5.8S tubulin at pH 7.0 in phosphate buffer at room temperature in the presence of Mg^{++} ions (6,21); and 2. Double rings identified as the end product of the depolymerization of reconstituted microtubules by exposing them to 4°C (12). The latter, however, were described as not being dissociable into 5.8S tubulin (12,22). In fact, at the time, there was a general consensus that pure tubulin is totally devoid of the ability to form microtubules (22,23,24), and that preexisting ring-structured nucleation centers (25) and other high molecular weight components were required for *in vitro* microtubule assembly. (12,14).

It seemed logical to us, however, that a pathway to the formation of microtubules from pure tubulin should exist if both pure tubulin and microtubules could give rise to essentially identical ring structures, even though the rings need not be intermediates in the microtubule assembly process.

Purified Tubulin Can Self-Associate Into Microtubules

To test our hypothesis, we decided that the best chances of success lay in a combination of conditions known to be most favorable for ring formation together with factors known to favor microtubule formation. Therefore, the buffer in which the ring formation was strongest, namely pH 7.0, 0.01M Kphosphate, 10^{-4}M GTP and 1.6 x 10^{-2}M $MgCl_2$, was modified by adding a calcium sequestering agent, EGTA, (10^{-3}M) and a high concentration of glycerol, which is known to stabilize microtubules (5). Purified calf brain tubulin was then dissolved in this medium, and upon heating to 37°C, it assembled successfully into microtubules (26,27)

Quantitative Method For Following Self-Assembly Into Microtubules

The self-assembly of tubulin into microtubules was monitored by turbidity measurements at 350nm and by electron microscopy. The turbidimetric methods, which, at present, seems to be the method of choice for following this process, was introduced first by Gaskin, Cantor and Shelanski in 1974 in an important paper (28) which contains a theoretical Appendix by Berne (29) and lays a solid theoretical foundation for the use of this technique as a method of measuring quantitatively the mass of tubulin polymerized into microtubules. In his theoretical analysis of the light scattering by rod shaped structures, Berne (29) showed that turbidity is a direct measure of the mass of material polymerized if the assembled structure has the geometry of a long thin rod (with respect to the wavelength of the radiation). This theoretical conclusion is valid for this specific geometry, within the realm of the applicability of the Rayleigh-Gans approximation, since within these limits the light scattering law reduces to a dependence of the turbidity on the inverse third power of the wavelength instead of the more familiar dependence on the inverse fourth power of the wavelength. This simplification fails for short rods, and need not hold for other geometric models. Furthermore, the rods are assumed to be monodisperse and conditions must be set so as to exclude external interference (30) due to strong repulsive forces and the decrease in the freedom of motion of particles in solution due to long-range ordering effects (31). This theory was tested for *in vitro* microtubule formation by Gaskin et al. (28) who showed that this process conformed to the criteria specified by Berne (29).

SELF-ASSEMBLY OF PURIFIED TUBULIN INTO MICROTUBULES

The Self-Assembly Phenomenon

Purified tubulin was dissolved in a pH 7.0, 0.1M phosphate buffer, which contained 1.6×10^{-2}M $MgCl_2$, 10^{-4}M GTP, 3.4 M glycerol and 10^{-3}M EGTA (PMG assembly buffer). Heating of this solution to 37° was followed within a few minutes by the development of turbidity, which could be reversed by cooling to 20° (27), as shown in Figure 2. Addition of 10^{-3}M $CaCl_2$ or of colchicine inhibited the generation of turbidity, as did the absence of GTP from the medium (27), namely the system exhibited behavior characteristic of microtubules.

The structure of the aggregates formed was identified by electron microscopy. Typical results, shown in Figure 3A, indicate the presence of filaments. On higher magnification (Figures 3B and 3C) these appear to have a regular structure. The diameter of the observed filaments was 29 ± 3nm, similar to the dimensions of intact native microtubules (32). The presence of protofilaments, such as are seen in normal microtubules, is unmistakable as well as the pictoral appearance of diagonal striations consistent with helices. In fact, these pictures are strikingly similar to those of reconstituted microtubules described in the literature (12,14). No filaments could be observed after cooling to 20° or below. The results of these turbidity and electron microscopy

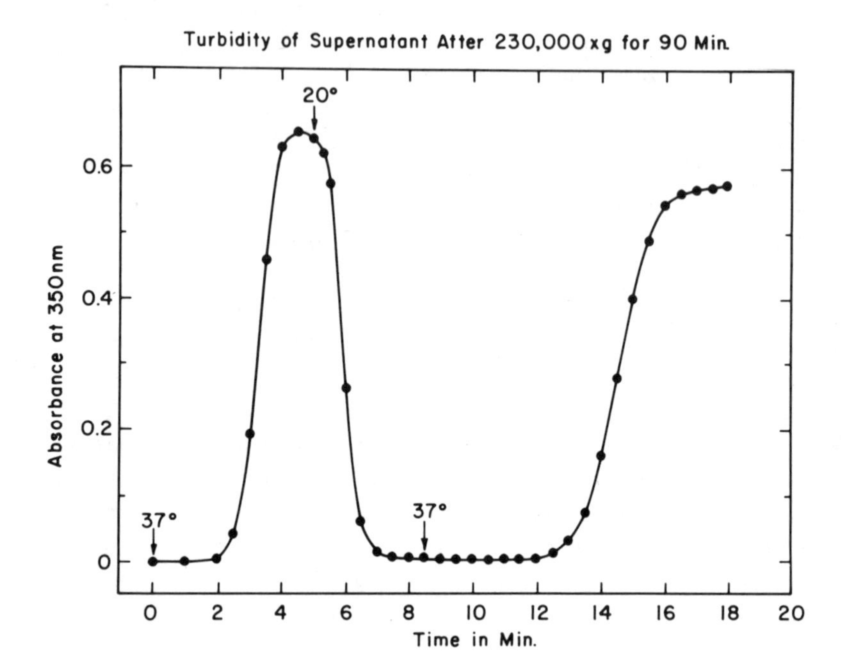

Fig. 2. Effect of temperature on the turbidity of calf brain tubulin prepared by the modified Weisenberg procedure and sedimented in a partition cell of the analytical ultracentrifuge until only 5.8S material remained. The reconstitution buffer was PMG, defined in the text.

experiments clearly demonstrated for the first time that, notwithstanding literature claims to the contrary, highly purified 5.8S tubulin is fully competent to assemble into microtubules. The reconstituted structures are morphologically similar to intact native microtubules and they possess properties that are characteristic of microtubules, namely disruption of structure by low temperature or the addition of calcium ions, as well as inhibition of filament formation by the presence of low concentrations of Ca^{2+} or of colchicine.

The requirement for high molecular weight components, such as pre-existing nucleation centers, tubulin oligomers or non-tubulin proteins, which have been reported as necessary for assembly into microtubules, was tested by high speed centrifugation of the tubulin preparation in a separation cell of a Beckman Model E analytical ultracentrifuge, keeping only the 5.8S component. The resulting protein, which showed no

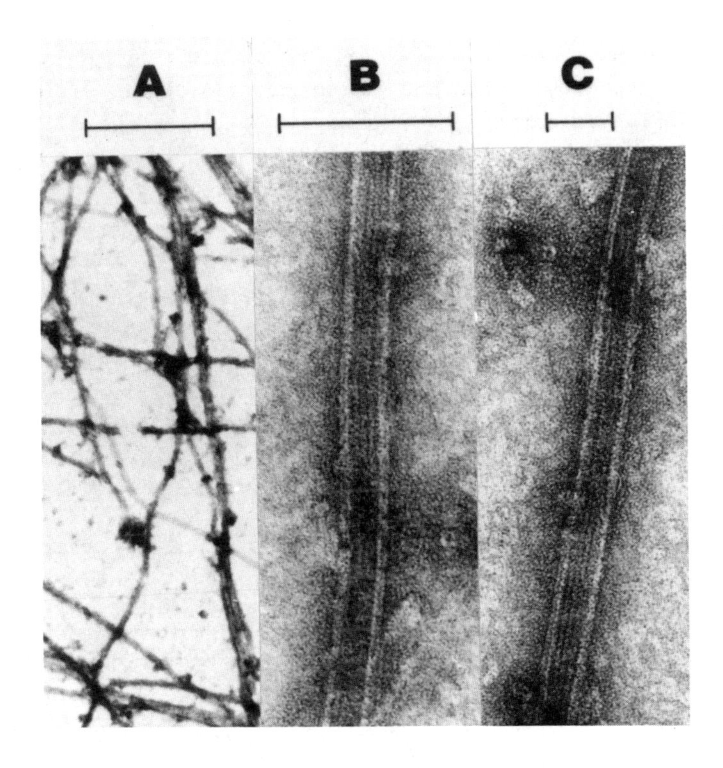

Fig. 3. Electron micrographs of assembled structures of tubulin in PMG assembly buffer at 37°C. A and B in the presence of 0.5% by weight poly-L-lysine at two different magnifications. C. Pure tubulin after high-speed sedimentation. The bar in A corresponds to 500nm; in B, to 100nm; in C to 40nm.

trace of high molecular weight components on SDS gel electrophoresis at a loading concentration of 350μg, could form microtubules as easily as the non-centrifuged material (27), establishing that tubulin must contain all the information necessary for the formation of microtubules. Since then, this finding has been confirmed in a number of laboratories (17,33,34,63).

Our discovery that microtubules can be assembled *in vitro* from purified tubulin opened the way to a systematic solution physico-chemical analysis of the process of the *in vitro* self-assembly of tubulin into microtubules and to the examination of questions, such as what are the minimal requirements of this process? What are the effects of solution variables, such as temperature and solvent components? What effectors or inhibitors control this process? The method chosen for the quantitative studies was that of turbidity, as first introduced by Gaskin et al. (28), with an analysis of the results in terms of the theory developed by Oosawa and co-workers (35,36,37,38) for the helical polymerization of actin.

Thermodynamic Methodology

The Oosawa Theory of Helical Polymerization. The applicability of the Oosawa et al. theory (35) to the equilibrium thermodynamic analysis of microtubule reconstitution was a consequence of the observation made by Gaskin et al. (28), that, for this process, a plot of turbidity as a function of total tubulin concentration, C_o, extrapolated not to the origin but to a finite protein concentration which, by analogy with micelle formation, can be equated with a critical concentration, C_r. Such behavior is characteristic of a highly cooperative process. A typical set of data, drawn from our studies, is shown in Figure 4.

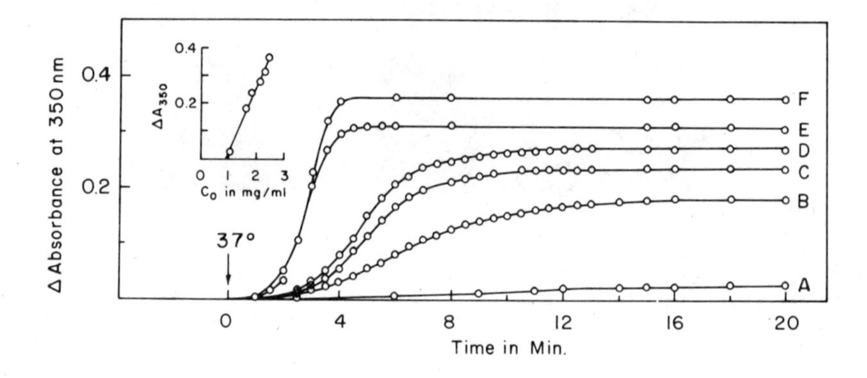

Fig. 4. Determination of the critical concentration of tubulin self-assembly into microtubules at 37°C in PMG assembly buffer (Lee, J.C. and Timasheff, S.N., previously unpublished data).

According to the theory of Oosawa and co-workers (35), helical polymerization can be described in terms of a two step process, depicted schematically in Figure 5*. The first step is a nucleation step. It consists of a linear polymerization with the formation of a single contact between subunits, the addition of each subunit being characterized by an association constant, K_n. This leads to the formation

*There are many possible variations in this scheme, and there is no intention here to imply that the scheme of Figure 5 describes accurately the self-association of microtubules. In fact, that process must be more complicated because of the complex nature of the microtubule helix.

of a nucleus of degree of polymerization, n. Once the nucleus is formed, addition of one further subunit closes the first turn of the helix, two contacts being formed between the $(n+1)^{th}$ subunit and the growing structure. All further subunits are then added on in similar manner, the assembly growing in helical array. The association

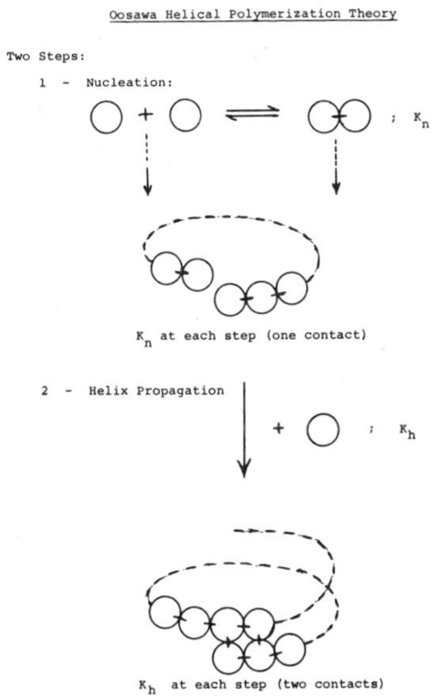

Fig. 5. Schematic representation of the Oosawa nucleated helical polymerization mechanism.

constant for the helix growth process, K_g, for the addition of each subunit must be greater than K_n because of the free energy contribution of the second intersubunit contact. The requirement that $K_g > K_n$ renders the process cooperative* and has three consequences. First, there must exist a concentration, called critical concentration, C_r, below which the protein is found essentially only in the states of monomer and of linear polymers with a degree of polymerization $\leq n$, large polymers being present in vanishingly small amounts. Second, all protein in excess of C_r is incorporated into high polymers, the concentration of protein not so incorporated remaining constant at C_r. Third, the

*Conversely, the existence of a critical concentration points necessarily to cooperativity, which could not exist if $K_g < K_n$.

critical concentration is related to the growth constant, K_g, by the simple relationship

$$K_g = C_r^{-1} \quad (1)$$

where C_r is expressed in moles of tubulin dimer per liter. The standard free energy of growth is, then, given by

$$\Delta G_g^\circ = -RT\ln K_g = RT\ln C_r \quad (2)$$

Dependence on State Variables. The thermodynamic analysis of the microtubule growth process was carried out using the turbidimetric method and assuming the validity of equation 2. The dependence on state variables of the apparent standard free energy for the microtubule growth reaction is given by the standard thermodynamic relationship

$$d\Delta G_g^\circ = \Delta V_g dP - \Delta S_g^\circ dT + \sum_{i=1}^{k} \left(\frac{\partial \Delta G^\circ}{\partial m_i}\right)_{T,P,m_{j\neq i}} dm_i \quad (3)$$

where ΔV_g is the change in volume per mole of dimeric tubulin incorporated into a microtubule, ΔS_g° is the standard entropy change, P is pressure, T is thermodynamic temperature, and m_i is the molal concentration of component i. The changes in volume and entropy and hence enthalpy, ΔH_g°, and heat capacity, $\Delta C_{p,g}$, are easily measurable by following the reaction as a function of pressure and temperature at constant composition, since

$$\Delta V_g = -RT\left(\frac{\partial \ln K_g}{\partial P}\right)_{T,a_i} \quad (4)$$

$$\Delta H_g^\circ = -R\left(\frac{\partial \ln K_g}{\partial (1/T)}\right)_{P,a_i} \tag{5}$$

$$\Delta S_g^\circ = \frac{\Delta H_g^\circ - \Delta G_g^\circ}{T} \tag{6}$$

$$\Delta C_{p,g} = \left(\frac{\partial \Delta H_g^\circ}{\partial T}\right)_{P,a_i} = \left(\frac{\partial \Delta S_g^\circ}{\partial \ln T}\right)_{P,a_i} \tag{7}$$

where a_i is the activity of component i.

The Wyman Linked Functions. The effect of ligands can be examined rigorously in terms of the Wyman linked function relations (39) that, at constant T and P, and at a standard state of unliganded reactant,

$$-\left(\frac{\partial \Delta G_g^\circ}{\partial \mu_x}\right)_{a_{j\neq x}} = \left(\frac{\partial \ln K_g}{\partial \ln a_x}\right)_{a_{j\neq x}} = \Delta\bar{\nu} = \left(\frac{\partial m_x}{\partial m_t}\right)^{(Mt)}_{\mu_x, m_{j\neq t}} - \left(\frac{\partial m_x}{\partial m_t}\right)^{(Tb)}_{\mu_x, m_{j\neq t}} \tag{8}$$

where subscript x denotes ligand, subscript t denotes tubulin, a is activity on the molal scale, m is molal concentration; the superscripts (Mt) and (Tb) indicate tubulin incorporated into microtubules and in the free 5.8S state, respectively; μ_i is the chemical potential,

$$\mu_i = RT\ln m_i + RT\ln \gamma_i + \mu_i^\circ(T,p) \qquad (9)$$

where γ_i is the activity coefficient and μ_i° is the standard chemical potential. According to equation 8, a double logarithmic plot of the equilibrium constant vs the activity of the ligand yields the parameter $\Delta\bar{\nu}$, i.e. the difference between the numbers of ligand molecules bound to a tubulin dimer incorporated into a microtubule and one in the free state. In fact, it is the change in the preferential interaction between the protein and solvent components which is measured by this treatment (40), namely the change in the balance between ligand and water molecules "bound" to the tubulin (41), since, assuming that the binding of the various ligands is independent of each other, namely that the binding of ligand x does not affect the binding of ligand y, in other words that there is no linkage between the two reactions (42),

$$\bar{n} \equiv \left(\frac{\partial m_x}{\partial m_t}\right)_{\mu_x} = \bar{\nu}_x - \left(\frac{m_x}{m_{H_2O}}\right)\bar{\nu}_{H_2O} = -\frac{(\partial\mu_x/\partial m_t)_{m_x}}{(\partial\mu_x/\partial m_x)_{m_t}} \qquad (10)$$

and

$$\Delta\left(\frac{\partial\mu_3}{\partial m_2}\right)_{T,P,m_3} = \Delta\left(\frac{\partial\mu_2}{\partial m_3}\right)_{T,P,m_2} = -RT\Delta\bar{\nu}\left(\frac{1}{m_3} + \frac{\partial\ln\gamma_3}{\partial m_3}\right) \qquad (11)$$

The preferential interaction (or "binding") parameter, $\bar{n}$, is a true thermodynamic quantity since it measures the effect of changing protein concentration on the macroscopic chemical potential of the ligand and vice-versa. "Binding" here has its most general meaning, namely that of the perturbation of the activity coefficient of one solution component by another and it encompasses the full spectrum of interactions, from strong complexing at specific sites to momentary perturbations of the freedom of rotation or translation of a molecule of one component by another.

Effects of Solution Variables on Microtubule Self-Assembly.

Effect of Temperature. The van't Hoff plot of the microtubule growth reaction (43), shown in Figure 6, is characterized by a pronounced curvature. This indicates that this reaction is accompanied by a change in heat capacity. Analysis of the data yielded a value of ΔC_p for microtubule growth of -1500 cal/deg-mole of tubulin dimer added. Recently this value has been fully confirmed by direct microcalorimetric measurements (44). As shown in Figure 7, both the

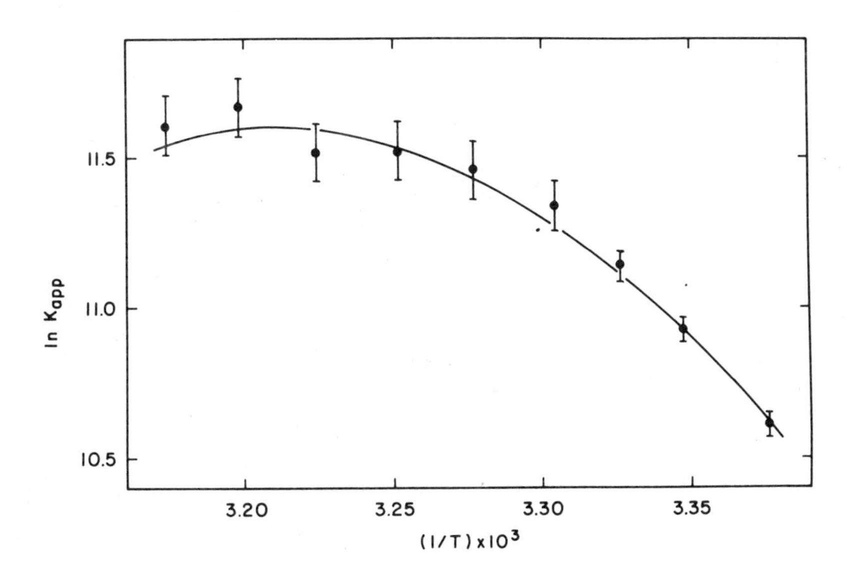

Fig. 6. Van't Hoff plot of the microtubule reconstitution reaction in PMG assembly buffer (Reproduced from Ref. 43).

enthalpy and entropy changes are temperature dependent, conforming to equation 7, and the reaction can be regarded as entropy-driven (43). The results are summarized in Table I. The thermodynamic parameters found, namely positive $\Delta S°$ and negative ΔC_p, are diagnostic of an association mechanism which involves the departure of more strongly ordered water molecules from the site of interaction to less ordered domains in bulk solvent. This conclusion, which, as will be shown below, is supported by the effect of concentrated glycerol on this reaction, is consistent with the involvement of hydrophobic, or/and electrostatic interactions (45). Furthermore, these thermodynamic results are qualitatively consistent with parameters reported for the *in vivo* formation of microtubules (46,47,48).

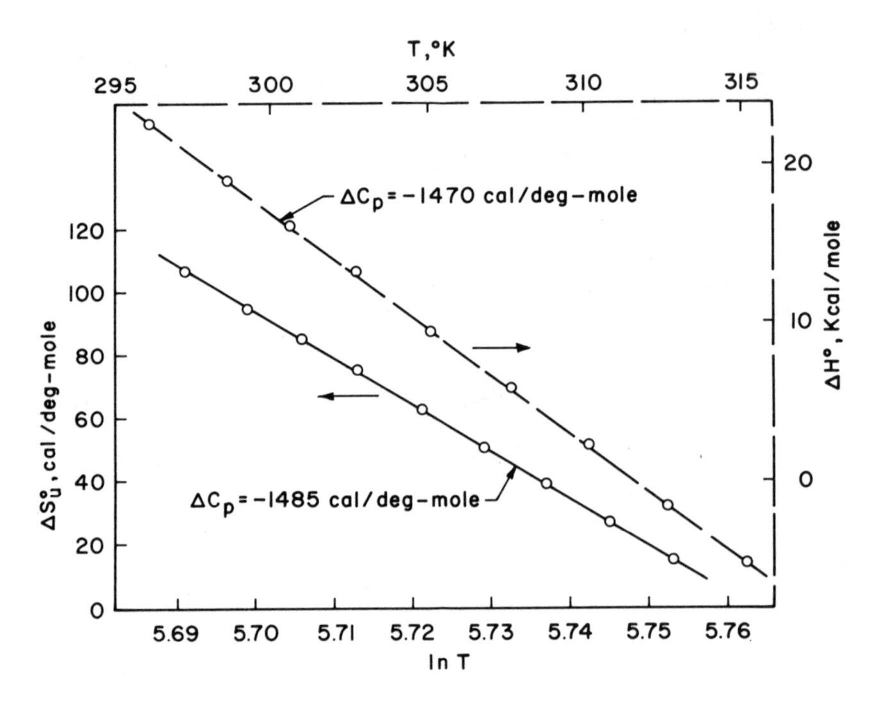

Fig. 7. Temperature dependence of the standard enthalpy and standard unitary entropy changes in the microtubule reconstitution reaction in PMG assembly buffer.

TABLE 1

THERMODYNAMIC PARAMETERS OF MICROTUBULE ASSEMBLY

(PMG assembly buffer at 37°C)

ΔG° = -7.1 kcal/mole	ΔH° = 2.2 kcal/mole
ΔS° = 32 e.u.	ΔC_p (van't Hoff) = -1500 cal/mole deg
ΔS°_u = 40 e.u.	ΔC_p (microcal.) = -1600 cal/mole deg

The values of the determined thermodynamic parameters are all based on the Oosawa-Kasai relation, $K_p = C_r^{-1}$. Returning to considerations of the polymerization theory of Oosawa and Kasai, the critical concentration, C_r, can be expressed as (35),

$$C_r = K_g^{-1} \left(\frac{K_g - K_n}{K_g} \right)^{-2} \left[1 + s \left(\frac{K_n}{K_g} \right)^{n-1} \left(1 - \frac{K_n}{K_g} \right)^2 \right] \quad (12)$$

(RTlns = - ΔG^* is the unfavorable free energy of distorting the linear polymer into a one-turn helix in the nucleation reaction.) Since $K_g >> K_n$, and n is a large number, the value of $\Delta H°$ deduced from the van't Hoff analysis, $\Delta H°_{app}$, is defined as (43)

$$\Delta H^{\circ}_{app} = - \Delta H^{\circ}_g + 2R \frac{d \ln (K_g - K_n)}{d(1/T)} \quad (13)$$

This reduces within a very close approximation to $\Delta H^{\circ}_{app} = \Delta H^{\circ}_g$, i.e. the enthalpy measured by this method is essentially the enthalpy change of the helix growth reaction.

Effect of Magnesium. Experiments as a function of ligand concentration were analyzed (43) according to equations 8 and 10. In the case of $MgCl_2$, a plot of $\ln K_g$ as a function of $\ln a_{\pm}$ ($MgCl_2$) resulted in a straight line with a slope of 0.78. What is the meaning of this number? In a self-association reaction, if ligand binding is an integral part of the reaction with only liganded monomer capable of entering into the polymerization reaction,the ligand binding equilibrium constant must be considered when the method of determining the association constant involves measurements of mass distribution of the species (20,43,29). This condition is true of the turbidimetric method for determining $K_g = C_r^{-1}$, since C_r is the total concentration of all protein species both liganded and unliganded with a degree of polymerization below that of the nucleus. Therefore, K_g, operationally defined in the manner, is only an apparent equilibrium constant, K_{app}. Let us consider the overall reaction, with an additional assumption that ligand binding must precede the addition of subunits to the growing helical polymer:

$$\begin{aligned} T + X &\rightleftharpoons TX;\ K_1 \\ TX + X &\rightleftharpoons TX_2;\ K_1 \\ &\vdots \\ TX_{n-1} + X &\rightleftharpoons TX_n;\ K_1 \\ TX_n + M &\rightleftharpoons MTX_n;\ K_2 \end{aligned} \tag{14}$$

where T is tubulin dimer, X is ligand, K_1 is the ligand binding constant. K_2 is the propagation constant for adding a liganded subunit (TX_n) to a growing microtubule M, and MTX_n is a microtubule elongated by one liganded tubulin subunit (in the next step MTX_n becomes M, since the number of microtubules does not increase, but only their size). Then

$$K_{app} = \frac{[MTX_n]}{[M]\sum_{i=0}^{n}[TX_i]} = \frac{K_1^{\,n}K_2\,[X]^n}{\sum_{i=0}^{n}(K_1[X])^i} \tag{15}$$

Combining the definition of K_{app} given in equation 15 with the requirement of the Wyman linked function theory (39) that the standard state be taken as unliganded monomer defines the slope of the $\ln K_{app}$ vs $\ln a_x$ plot not simply as $\Delta\bar{\nu}$, but as (43)

$$\frac{d \ln K_{app}}{d \ln a_x} = \Delta\bar{\nu} - \frac{d \ln \sum_{i=0}^{n}(K_1 a_x)^i}{d \ln a_x} \tag{16}$$

Introduction of the measured value of $K_1 = 1.06 \times 10^2$ liters per mole (6) into equation 15, showed that the best value of n which gives a constant value of K_2 is 1.0 (43). Substitution into equation 16 then showed that the experimental value of the slope - 0.78 corresponds to $\Delta\nu$= 1.02, i.e. the slope of the correct Wyman plot in a standard state of unliganded reactant corresponds to the incorporation of one magnesium ion per step in the growth reaction. It should be emphasized that these results in no way establish direct participation of magnesium ions in the pathway of the reaction, nor is the order of events known, and the binding of ions to the negatively charged macromolecule may only reflect a change in the surface electrostatic free energy of tubulin when its state of polymerization is altered.

A more rigorous analysis of the system should be carried out in terms of the complete reaction scheme (43):

$$\begin{array}{ccc} M + T + X & \overset{K_1}{\rightleftharpoons} & M + TX \\ K_4 \updownarrow & & \updownarrow K_2 \\ MT + X & \underset{}{\overset{K_3}{\rightleftharpoons}} & MTX \end{array} \tag{17}$$

The apparent propagation constant is now given by:

$$K_{app} = \frac{1}{C_r} = \frac{K_1K_2a_x}{(1 + K_1a_x)}\left(1 + \frac{1}{K_3a_x}\right) = \tag{18}$$

$$= \frac{(K_1K_2a_x + K_4)}{(1 + K_1a_x)} = \frac{K_1K_2}{K_3}\,\frac{(1 + K_3a_x)}{(1 + K_1a_x)}$$

Equations 17 and 18 describe the linkage between the association constants of liganded and unliganded tubulin and the binding constants of ligand to tubulin in the two states, namely monomeric and polymerized. Since Mg^{++} ions enhance the growth reaction, $K_2 > K_4$. As a consequence, $K_3 > K_1$, and the strength of binding of the Mg^{++} ion linked to microtubule formation is greatly increased by incorporation into microtubules.

Other Ligands. The same situation can be expected to exist for other ligands which enhance microtubule formation. These effects are simply direct consequences of the linkage between ligand binding and self-association. In descriptive terms, these relations can be regarded as examples of the utilization of the free energy of polymerization for the enhancement of the tightness of ligand binding in the ligand-mediated polymerization, and of the reciprocal situation in the ligand-facilitated reaction.

In similar fashion, it was shown that one proton becomes incorporated per tubulin dimer added to the assembled structure (43), although a detailed analysis could not be carried out due to the limited zone of stability of tubulin, between pH 6.3 and 7.1. The results on the effect of ligands are summarized in Table 2. The effects of Ca^{++} and GTP could not be analyzed quantitatively since their binding constants to tubulin are not known, nor is the stoichiometry in the case of Ca^{++}. That these interactions are strong can be assessed from the fact that calcium is inhibitory at concentrations close to equimolar with tubulin and that GTP is effective at a similar level.

TABLE 2

EFFECT OF LIGANDS ON MICROTUBULE ASSEMBLY

Ligand	$\Delta\bar{\nu}$	Ligand	Effect
Mg^{++}	1.02	GTP	Required
H^{+}	0.86	Ca^{++}	Inhibits
Glycerol	0.96	Polycations	Enhance

The GTP requirement for microtubule growth from pure tubulin is evident: (1) its decrease below equimolar with tubulin lowers K_g and (2) it is consumed during the reconstitution process (43). In fact, the hydrolysis of at least one GTP molecule per tubulin dimer incorporated into microtubules has been recently demonstrated by David-Pfeuty et al. (50,51).

When ligands inhibit a reaction, such as calcium ions and colchicine in the case of microtubule formation, the linked function relation (equation 8) shows that ligand must become dissociated from the protein during the course of polymerization. In other words, as shown by equations 14,17 and 18 inhibiting ligands must be bound more strongly to monomer than to polymer, $K_3 > K_1$, $K_2 > K_4$ in the scheme of equation 17, and the predominant species are TX and MT. Hence, if binding measurements are performed at concentrations not higher than those at which the inhibitors, such as calcium and colchicine, are known to bind to dimeric tubulin, their liganding to the assembled system might go undetected. It is clear, then, that in such cases it is not necessary to invoke conformational changes, steric hindrance or removal of liganded or unliganded binding sites from contact with solvent in order to explain the apparent lack of binding or unbinding of ligand in one of the end states of a self-association reaction, or indeed of many other reactions. It is true that such various kinetic effects may exist, and may indeed be the mechanistic causes of the change in strength of ligand binding, but no information can be obtained on them from strictly thermodynamic data such as a difference between the strength of ligand binding in the two end states of the reaction or an apparent lack of ligand action in one of the end states. In fact, linkage (positive or negative) or, stated differently, the use of the free energy of one reaction for enhancing another reaction may be brought about without any conformational, steric, or burying effects, for example by a non-specific general rearrangement of the charge distribution on the surface of the protein molecule due to the difference in the protein electrostatic free energies of the two end states (52).

Measured and Intrinsic Thermodynamic Parameters. Similar considerations apply to the analysis of the enthalpy, and hence the entropy, and heat capacity changes calculated for an assembly reaction from equilibrium data when the association constant for the addition of each subunit is based on measurements of mass distribution of species. In such a case, the enthalpy and heat capacity changes also become complex functions in which ion binding must be taken into account specifically (20,43). For the case of a polymerization reaction which is linked to the binding of n molecules of ligand per chain growth step, the standard enthalpy of addition of a subunit to the growing structure, ΔH_2°, is shown in equation 19, where ΔH_1 is the enthalpy of ligand binding, for example of magnesium binding to tubulin.

The heat capacity change, $\Delta C_p = d\Delta H/dT$, is also a complex function. For the case of microtubule growth, the binding of one magnesium ion is linked to each step in the reaction, and $\Delta C_{p,2}$, the intrinsic heat

$$\Delta H_2^\circ = \Delta H_{app}^\circ - \Delta H_1^\circ \frac{\sum_{i=0}^{n} \left[(n-i)(K_1 a_x)^i\right]}{\sum_{i=0}^{n} (K_1 a_x)^i} \tag{19}$$

capacity change of microtubule growth is related to the apparent experimental heat capacity change, $\Delta C_{p,app}$, by (43)

$$\Delta C_{p,2} = \Delta C_{p,app} - \frac{\Delta C_{p,1}}{(1 + K_1\, a_{Mg})} + \left(\frac{\Delta H_1^{\,2}}{RT^2}\right)\left(\frac{K_1 a_{Mg}}{(1 + K_1\, a_{Mg})^2}\right) \tag{20}$$

where $\Delta C_{p,1}$ is the heat capacity of magnesium binding. While at present the values of ΔH_1 and $\Delta C_{p,1}$ for magnesium binding to tubulin are not known, an estimate of their contribution to the observed quantities can be made from literature values of these parameters for other protein systems (53,54). A preliminary microcalorimetric measurement of the heat of binding of magnesium to tubulin indicates that ΔH_1 is of the order of 1.0 kcal/mole (H.-J. Hinz and S.N. Timasheff, unpublished). The results of such estimates indicate that the intrinsic thermodynamic parameters of the growth reaction will remain little changed from those given in Table 1 and Figures 6 and 7.

Effect of Glycerol A ligand of particular interest is glycerol. Its effect on microtubule growth is summarized in Table 3. Glycerol is known not to be required for microtubule reconstitution to proceed as has been shown both in our laboratory (43) and in the studies of Herzog and Weber (34) who worked with phosphocellulose purified tubulin in a different assembly buffer. Assembly proceeds in its absence with a critical concentration of 8 mg/ml in our assembly buffer (43). Use of equation 8 leads to the formal conclusion that glycerol binding to tubulin increases by one molecule per tubulin dimer incorporated into

TABLE 3

EFFECT OF GLYCEROL ON MICROTUBULE GROWTH

% Glycerol (v/v)	$\ell n a_{gly}$	C_{crit} mg/ml	ΔG_{app} kcal/mole
15	0.9407	2.0	-6.6
20	1.3079	1.45	-6.8
25	1.6233	1.10	-7.0
30	1.8942	0.75	-7.1
0		8.0	-5.7

microtubules, since $\Delta\bar{\nu}$ = 0.96. Yet, this effect can hardly be one of specific binding. Let us examine the reasons for this conclusion. Specific binding implies strong binding. In order to exert its activity, glycerol must be present at high concentration, 1-4M. A corresponding binding constant would be $\sim 1M^{-1}$, with a standard free energy of binding of $\sim$0 kcal/mole. The explanation for the glycerol effect must, therefore, be sought elesewhere. The positive value of $\Delta\bar{\nu}$ is the difference between the preferential interactions of glycerol with the product and the reactant. Examination of equation 8 shows that this value can be obtained in two ways: (1) either both terms on the right hand side of the equation are positive, the first one being greater than the second, or (2) both terms are negative, with the absolute value of the second being greater. For the tubulin-aqueous glycerol system, it is known that $\bar{n}^{(Tb)}$ is negative (55), as it is for a variety of proteins (56), meaning that the concentration of glycerol in the immediate domain of tubulin is lower than that in the bulk solvent, i.e. glycerol is preferentially excluded from contact with tubulin. To obtain a positive value of $\Delta\bar{\nu}$, it is sufficient that preferential exclusion of glycerol be reduced per tubulin dimer when these become incorporated into microtubules. Why should such a reduction in glycerol exclusion (or non-contact with the protein if we wish) favor the polymerization reaction? Examination of equations 10 and 11 shown that a negative value of $\bar{n}^{(Tb)}$ corresponds to a positive change in the chemical potential of the ligand induced by introduction of tubulin into the solvent. This is a thermodynamically unfavorable situation. In order to relieve this, the system may undergo a variety of changes, for example the protein may undergo a conformational change, or the extent of glycerol exclusion from the protein may be decreased simply by reducing the surface of contact per tubulin subunit between the protein and glycerol. One way of accomplishing this is by inducing the protein to self-associate. Indeed, if this co-solvent exclusion is sufficiently strong, phase separation may result as is true in the crystallization of ribonuclease A by 2-methyl-2,4-pentanediol (57). This effect can be made lucid by the very simple model of Figure 8.

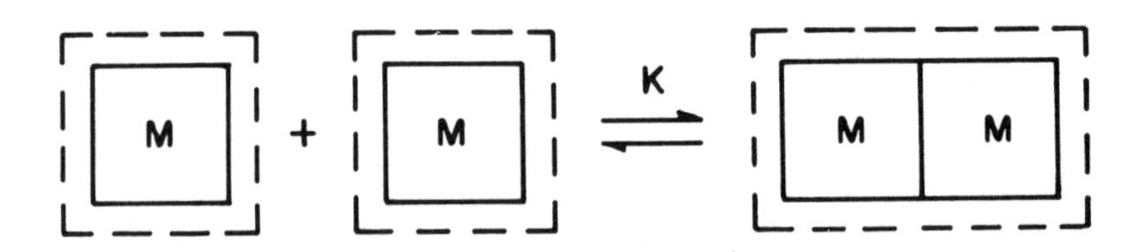

Fig. 8. Schematic representation of the non-specific enhancement of self-association by a solvent component which is preferentially excluded from contact with the macromolecule.

Consider two protein monomers depicted as squares, each with a concentric zone of solvent from which one of the solvent components is excluded. Assembly of these into a dimer must reduce the zone of exclusion, and, as a consequence, make less negative the value of the preferential interaction with protein of the excluded solvent component. As shown in Table 3, this effect is weak, and glycerol is able only to enhance an already existing association reaction. It acts in this scheme in a totally non-specific manner, as a thermodynamic booster which is not itself involved in the reaction, and which, in fact, need not come into contact with the self-associating protein.

The same effect could be achieved by squeezing out a patch of water molecules from contact with protein on association, without affecting the number of glycerol molecules in the domain of the protein. In either case, as is evident from equations 10 and 11, addition to the system of the intert solvent component will cause the self-association to proceed more strongly without that solvent component itself participating in the reaction. This is why we call it a thermodynamic booster. In the present case, by making the free energy of the reaction more negative by approximately 1 kcal/mole, 3.4M glycerol boosts the system to where it is easily manageable in the laboratory (C_r = 1 mg/ml); in the absence of glycerol, C_r = 8 mg/ml, which is a difficult concentration to handle in the case of tubulin. Furthermore, the value of $\Delta\bar{\nu}$ - 0.96 indicates that a minimum of 14 water molecules must leave the immediate domain of the dimeric tubulin when it becomes incorporated into microtubules (43). This conclusion is fully consistent with the observed changes in ΔS°_g, ΔH°_g and $\Delta C_{p,g}$ (43) and ΔV (58).

Effect of Macroligands. Another microtubule assembly-enhancing type of ligand which has received much attention recently consists of a class of proteins which copurify with tubulin (13,14,18,19,59,60), when the latter is prepared by the Shelanski et al. (11) cycle procedure. The observation that these proteins cocrystallize with tubulin at a constant composition through a series of assembly-disassembly cycles has led to the conclusion (13,15,16,18,19,61,62) that their interaction with tubulin is specific. Fractionation of cycle tubulin preparations on ion exchange columns separates tubulin, which elutes as an anionic protein, from the other proteins which behave as polycations on the column (18,63). Cycle tubulin reconstitutes readily into microtubules. It is the failure to achieve reconstitution in certain assembly buffers, once MAPs are separated from tubulin, which has led to the erroneous conclusion of the absolute requirement for assembly of tau factor or HMW proteins, with pure tubulin being unable to assemble into microtubules. A more likely explanation (64) is that, in the buffers used, assembly is weak and the tubulin concentrations employed were below the critical concentration required to make microtubules. The effective action of MAPs (HMW or tau factor) is to enhance the reaction to such a level that it can be readily measured in the laboratory. In fact, in order to achieve a decrease in C_r by a factor of 10 it takes only a small boost in the standard free energy of assembly, -1.4 kcal/mole.

The assembly-boosting action of MAPs has been clearly demonstrated and results from our studies (64) are summarized in Table 4. It is seen that, when working in our assembly buffer, removal by phosphocellulose ion exchange chromatography of MAPs from the cycle-prepared tubulin (C-tubulin) raises the critical concentration to the same value as that of our tubulin purified by the modified Weisenberg procedure (W-tubulin). Addition of MAPs back to the phosphocellulose chromatography purified tubulin (PC-tubulin) or to W-tubulin lowers the critical concentration of both to the value of C-tubulin. It is clear, therefore, that the addition of MAPs to purified tubulin acts only to enhance the assembly, but is not required to induce it.

TABLE 4

EFFECT OF MAPs ON TUBULIN ASSEMBLY IN PMG BUFFER

Protein	C_r, mg/ml
W-Tubulin	0.9
W-Tubulin + 4% by weight of MAP fraction	0.2
C-Tubulin	0.2
PC-Tubulin	0.8
PC-Tubulin + MAP fraction	0.3
W-Tubulin + 0.5% by weight poly-L-lysine	0.1

How do these proteins boost the self-assembly reaction? While at present the mechanism is unclear, other than they complex with the assembled structures, a simple possibility could be electrostatic interactions with constellations of charged groups on the surface of formed tubulin aggregates, reducing the unfavorable electrostatic free energy of the assembled structure (52) through the formation of additional intertubulin linkages in a macroligand-facilitated pathway. The complexation with the assembled structure should be cooperative, the interaction free energy per tubulin dimer being small. Such complexation would stabilize the structure against dissociation, for example by colchicine (65). The interaction could be either specific or non-specific, depending on whether specifically matching constellations of charges on the two macromolecules are required or not. This mechanism becomes quite clear if we write out a totally general model reaction which makes no assumptions either about specificity of interaction or about the nature of the forces involved in the complexation (64). Suppose that the solution contains tubulin, T, which can associate into microtubules with an association constant K_g, for the addition of each subunit to the growing microtubule and a second protein, A, which can complex with assembled structures by some interaction which may be either specific or non-specific and the nature of which need not be specified, one molecule of A adding for each n tubulin molecules incorporated into microtubules, with a binding constant, K_a. If M represents a growing microtubule at a given stage of elongation and $T^{(i)}$ are tubulin dimers consecutively adding to it, this model can be expressed through the set of reactions

$$\begin{array}{llll}
M + T^{(1)} & \rightleftharpoons & MT^{(1)} & ;\quad K_g \\
MT^{(1)} + T^{(2)} & \rightleftharpoons & MT^{(1)}T^{(2)} & ;\quad K_g \\
\vdots & & \vdots & \\
MT^{(1)}\ldots\ldots T^{(n-1)} + T^{(n)} & \rightleftharpoons & MT^{(1)}\ldots\ldots T^{(n)} = MT_n & ;\quad K_g \\
MT_n + A & \rightleftharpoons & MT_nA & ;\quad K_a
\end{array} \qquad (21)$$

For this scheme, the experimentally determined growth constant, K_g^{app}, will be

$$K_g^{app} = \frac{1}{C_r} = K_g \ (1 + K_a[A])^{1/n} \qquad (22)$$

Therefore, as a consequence of the binding of A to microtubules, the observed critical concentration will be lower than that corresponding to the intrinsic growth constant, K_g. This results not only in an enhancement of the microtubule growth process, but also will lead to the observed constancy of composition of assembled structures and the appearance of a definite stoichiometry. This can be easily demonstrated with a simple calculation, using our data on the effect of poly-L-lysine on microtubule assembly given in Table 4 (64). As will be shown below, poly-L-lysine acts as a strong booster of microtubule reconstitution. Saturation with poly-L-lysine (PLL) occurs at a molar ratio of PLL to tubulin of 1 to 100. This sets n = 100 in equations 21 and 22. The values of K_g and K_g^{app}, derived from the critical concentrations of Table 4, are 1.2×10^5 and 1.1×10^6 liters/mole of tubulin dimer, respectively, giving $K_a \simeq 10^{100}$ liters/mole of PLL. This corresponds to a ΔG° of binding of PLL to a microtubule of ~ -140 kcal/mole of poly-L-lysine or ~ -1.5 kcal/mole of tubulin dimer incorporated in microtubules. The interaction with each dimer is weak, the binding constants of PLL to 5.8S tubulin being 10-100 liters/mole, and no complexation will be observed when the system is dissociated. Yet, whenever microtubules are assembled, all of the PLL will complex with them and will be found in constant composition in a series of cycles of assembly and disassembly. Therefore, the interaction will appear to be stoichiometric when the initial composition of the system is varied. The strong stabilizing effect results from the cooperativity of the interaction since the positively charged lysine residues are covalently linked in PLL. Monomeric L-lysine does not enhance the assembly reaction (43).

Similar calculations can be made for the case when the non-tubulin component is the MAP fraction. Using the experimental data of Table 4 and assuming n = 5 in equation 22, the amount of non-tubulin protein, or MAP, remaining free in solution during each cycle of polymerization can be estimated as ~ 0.1 μg/ml, i.e. the MAPs will co-purify with tubulin in constant ratio through a number of assembly cycles. From these simple calculations in which no assumptions were made concerning specificity, it is evident that, while constancy of composition is consistent with specificity of interaciton, it is not a sufficient condition, and the conclusions reached on the basis of this argument need not be valid. Therefore, the question whether any of the MAPs are specific or not must be regarded as still open, a conclusion also reached by Williams, Jr. (66) on the basis of two-dimensional electrophoresis experiments.

The specificity of participation of the MAPs in the microtubule system has been questioned also by observations of Calissano and coworkers (67), Erickson and Voter (63, 68), and others (64,69,70), that a variety of polycationic macromolecules, such as histones, ribonuclease, DEAE dextran and poly-L-lysine, can mimic the effect of MAPs in microtubule reconstitution. Results of typical experiments, drawn from our studies (64), are shown in Figure 9. It is evident that a

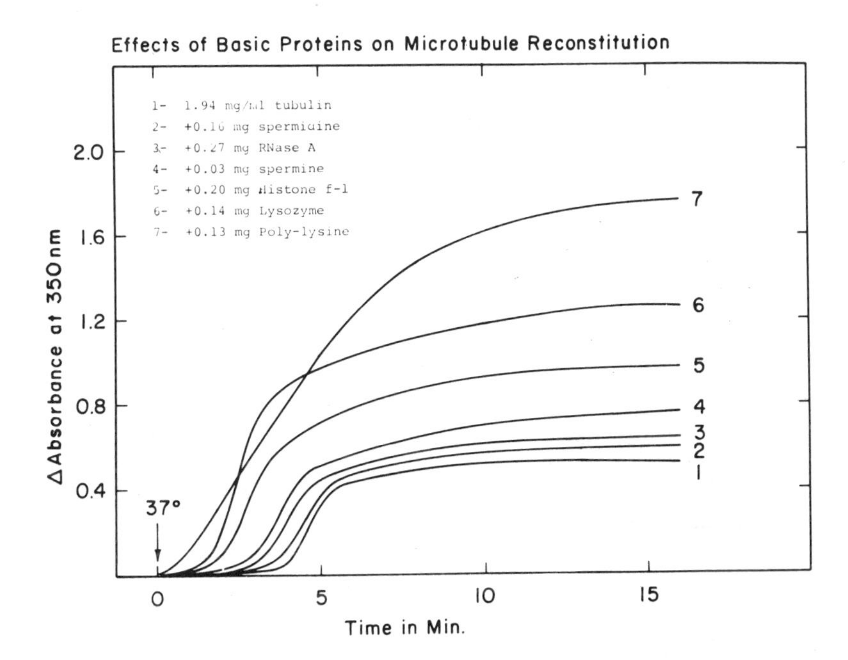

Fig. 9. Effect of basic proteins and polycations on the turbidity of purified tubulin when heated tp 37°C in PMG assembly buffer. The curves are identified on the figure. (reproduced from Ref. 64).

variety of polycations and basic proteins can induce an increase in turbidity when added to a constant amount of tubulin under assembly conditions. This turbidity was reversible by cooling to 10°, and electron microscopic examination showed the associated products to be microtubules, as seen in Figure 3C. The most effective polycation among those used was poly-L-lysine, addition of which lowered the critical concentration of purified tubulin to that of cycle tubulin at very low levels of PLL, as shown in Table 4. On the other hand, small amines, such as L-lysine and 1,4-butane-diamine had no effect, suggesting that cooperativity between basic groups is required to enhance the microtubule reconstitution reaction.

These results demonstrate unequivocally that a number of non-tubulin macromolecules can boost the microtubule assembly reaction, whether these macromolecules are MAPs or randomly added polycations. *In vitro*, none of the MAPs are essential for microtubule growth, although their involvement in nucleation (where traces might be sufficient) cannot be excluded due to the limits of analytical procedures, such as SDS gel electrophoresis. Whether they have any significance for the *in vivo* properties of microtubules is a question which cannot be answered by any of the observations based on reconstitution or other solution experiments described in the literature. On the other hand, several reports have presented circumstantial evidence favoring the proposition that the MAPs are microtubule-related. First, electron micrographs of microtubules reconstituted in the presence of these proteins display decorations similar to those observed on natural microtubules (16,18, 71). Second, immunofluorescence experiments have been interpreted as indicating that some of the MAPs are located in cells in networks similar to those formed by microtubules (72,73). Yet, none of these observations can be regarded as more than preliminary and suggestive and their possible significance must await further investigation. At present, it would seem most cautious to regard as open to investigation the question of the significance of MAPs for the microtubule system.

Summary

Our thermodynamic investigations of the self-assembly of purified tubulin into microtubules have established the following points. The minimal requirements are: pure tubulin, GTP or possibly some GTP analogs, probably magnesium ions and the absence of calcium ions. Thermodynamically, assembly can be described in terms of a cooperative polymerization process. The growth step involves the release of water molecules. The process can be enhanced by non-specific thermodynamic boosters, both those acting through indirect solvent effects, such as glycerol, and those complexing with the assembled structure, such as polycations. A similar effect can be exercised by MAPs, the relevance of which to the microtubule system is unclear at present.

SELF-ASSOCIATION TO SMALLER OLIGOMERS

Types of Oligomers Formed

In addition to self-associating to microtubules, tubulin is known to form a variety of other oligomeric structures, which have been described as single rings, double rings, sheets, ribbons, spirals,etc. The nature of the assembled structure seems to be a close function of solvent composition and of ligands present. Some of the oligomers described in the literature, mostly from electron microscopic observations, depend on interactions between tubulin and other proteins which co-crystallize with tubulin during the cycle preparation procedure, namely the MAPs. Two types of self-association have been described for purified tubulin: one is induced by the presence of magnesium (or calcium) ions (6,10,20) the other is mediated by the anti-mitotic drug, vinblastine (6,74). Our discussion will be limited to the last two systems, since they are the only ones which, up to the present, have been subjected to a rigorous examination. The method used in our studies was sedimentation velocity with the proper application of the Gilbert theory of polymerization (75,75,77,78,79) and the Cann-Goad theory of ligand-induced self-association (80,81).

Theory of the Sedimentation Velocity of Self-Associating Systems

Simple Polymerization of Proteins. The theory describing the behavior of moving boundaries in the case of rapidly reequilibrated polymerizing systems was first described in 1955 by Gilbert (75) who explained how such patterns could be interpreted properly. Using an analogy between transport in the ultracentrifuge cell and in a chromatographic column, Gilbert showed that the area distribution and the sedimentation profile under the schlieren diagram contained the information necessary to obtain both the stoichiometry and equilibrium constant of a self-association reaction. The experimental verification of this theory was provided five years later in our studies on the self-association of β-lactoglobulin dimers to octamers at low temperature (82).

Taking the reaction

$$nM \rightleftharpoons P \qquad (23)$$

where M is monomer, P is polymer and n is the degree of polymerization, Gilbert solved the continuity equation under a moving boundary, or front, with a neglect of diffusion, and showed that both monomer and polymer are present through the entire boundary. The protein concentration at any point, x, across the boundary, C_x, could be described by the sum of monomer and polymer in equilibrium

$$C_x = \left[k_a \left(\frac{1}{nk_a} \quad \frac{\delta}{1-\delta} \right)^{\frac{n}{n-1}} + \left(\frac{1}{nk_a} \quad \frac{\delta}{1-\delta} \right)^{\frac{1}{n-1}} \right] \qquad (24)$$

$$\delta = \frac{s(C) - s_M}{s_P - s_M}$$

where K_a is the association constant (in concentration units of mass per volume, usually gm/ml), s(C) is the sedimentaion coefficient at any point within the boundary at which the total concentration is equal to C, and s_M and s_P are the sedimentation coefficients of the monomer and polymer, respectively. The two terms on the right-hand side of equation 24 are the concentrations of polymer and of monomer, respectively, at the given point; δ is a normalized sedimentation coefficient relative to stationary monomer.

Equation 24 has several consequences. First, the shape of the boundary is a function of the degree of polymerization. When n = 2, the pattern is unimodal. For $n \geq 3$, the pattern becomes bimodal but the bimodality appears only at a given finite concentration determined by K_a. At all concentrations below this level, the protein sediments in a single peak with a sedimentation coefficient close to that of the monomer. Once bimodality sets in, all further increases in protein concentration contribute only to the areas under the rapid peak. Therefore, constancy of area under the slowly sedimenting peak, once bimodality has set in, is a good diagnostic criterion for identifying a Gilbert-type self-associating system. This is exemplified in Figure 10a, which shows simulation calculations by Cox (83) of the dependence of boundary shape on protein concentration for a hypothetical monomer-hexamer equilibrium. Patterns 3,4 and 5 display progressive increases in total area with no change under the slow peak. Second, the normalized sedimentation coefficient of the minimum position between the two peaks, $\delta_{min,}$ is determined only by the stoichiometry of the reaction, since, as shown by Gilbert (75),

$$\delta_{min} = \frac{n-2}{3(n-1)} \qquad (25)$$

A third consequence of equation 24 is found in the concentration dependence of the weight-average sedimentation coefficient, $\bar{s}$,

$$\bar{s} = \sum_i \left(s_i C_i\right) \Big/ \sum C_i \qquad (26)$$

where s_i is the sedimentation coefficient of species i and C_i is its concentration. In a Gilbert system, when n > 2, the weight average sedimentation coefficient of the protein, corrected for the hydrodynamic concentration dependence, increases in sigmodial fashion with protein concentration, as shown in the inset of Figure 10a, approach-

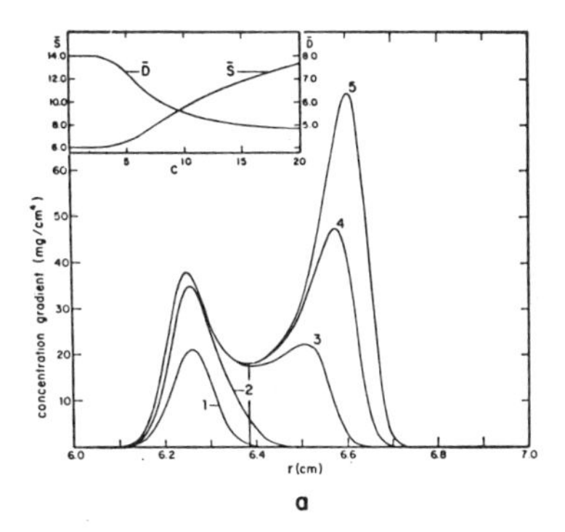

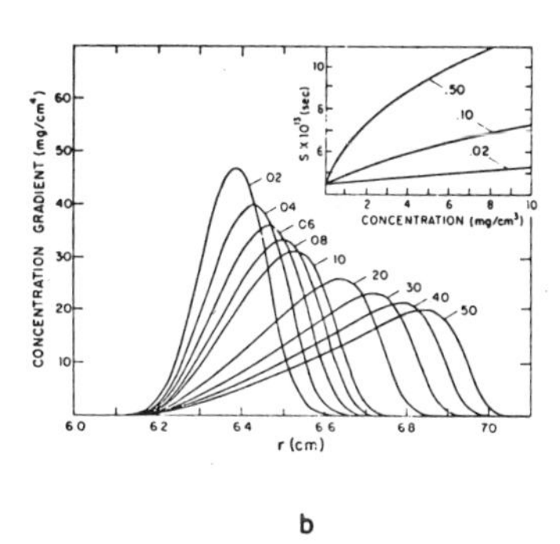

Fig. 10. Simulated sedimentation velocity patterns for self-associating systems in rapid equilibrium. a) Monomer-hexamer equilibrium; the successive patterns correspond to increasing total protein concentrations. b) Indefinite self-association; the numbers on the patterns indicate values of K_a. The insets give the weight-average sedimentation coefficients for the two reactions. (Reproduced from Ref. 83 and 85).

ing asymptotically the s value of the polymer. This is a consequence of the cooperative nature of the polymerization reaction described in equation 23. In general, such a concentration dependence is diagnostic of cooperative systems, and it can give rise trivially to a "critical concentration" above which the rapid peak appears, or indeed above which the association produce manifests itself in significant amounts, a situation which is equally true for highly cooperative systems such as described above for the self-assembly of tubulin into microtubules. The fact, however, that the association product is not detected either in a sediementation pattern as in a Gilbert system or by turbidity as in Oosawa type of polymerization does not mean that it is not formed at the given concentration. In fact, as is evident from equation 24, the association product must be present at all protein concentrations above zero, its amount becoming readily detectable and eventually predominant only above a given concentration determined by the association constant.

From the above considerations, it is evident that in a simple end-product polymerization as described by equation 23, knowledge of

s_M, s_P, the concentration dependence of the weight-average sedimentation coefficient, and the position of δ_{min} makes it possible to determine both the degree of polymerization and the association constant.

The Gilbert treatment has been extended to systems which self-associate according to an isodesmic mechanism (80,84,85) namely in which polymerization progresses indefinetly with no termination step, the association constant for the addition of subunits being identical for all steps of polymer growth according to the mechanism

$$
\begin{array}{lllll}
A & + & A & \rightleftharpoons A_2 & ; \quad K_2 \\
A_2 & + & A & \rightleftharpoons A_3 & ; \quad K_3 \\
\vdots & & & \vdots & \\
A_{n-1} & + & A & \rightleftharpoons A_n & ; \quad K_n \\
A_n & + & A & \rightleftharpoons A_{n+1} & ; \quad K_{n+1} \\
\vdots & & & \vdots &
\end{array}
\qquad (27)
$$

$$K_2 = K_3 = \ldots. = K_n = K_{n+1} = \ldots = K_a$$

Here K_i is in units of liters per mole. For such a system, the sedimentation pattern never becomes bimodal, but assumes a skewed shape, the degree of forward skewing being a function of the chain growth equilibrium constant, K_a. In this case, the concentration dependence of the weight-average sedimentation coefficient has a hyperbolic rather than sigmoidal shape. This is exemplified in Figure 10b, which shows simulated velocity sedimentation patterns calculated by Holloway and Cox (85) for a system which undergoes self-association according to the isodesmic scheme of equation 27.

In end-product polymerizations, equation 21 is only an idealized description of the overall reaction, since intermediates must form between the monomer and the terminal polymer. If the terminal polymeric structure is symmetric, it is reasonable to assume that the non-covalent bonds between the subunits are equivalent. The polymerization may then be assumed to proceed through a series of identical chain growth steps, according to an isodesmic pathway until the formation of an assembly with a degree of polymerization n-1. In order for the reaction to go to completion, the addition of the last subunit which closes the structure must occur with a greater association constant than that of the chain growth process. It is this last factor which imparts to the system its cooperative nature. This mechanism may be described by the scheme of equation 27, truncated at the n'^{th} step, with $K_2 = K_3 = --- = K_{n-1} < K_n$. This mechanism is, in fact, a combination of the reactions described by equations 23 and 27. Its sedimentation behavior may be expected, therefore, to exhibit the characteristics of both an isodesmic and an end-product polymerizations. At low concentrations, one should expect the generation of unimodal sedimentation patterns skewed forward, with a weight-average sedimentation coefficient which increases with protein concentration, i.e. a hyperbolic behavior, as shown in the inset of Figure 10b. At a given concentration, a second, more rapid peak, should appear, the system taking on the characteristics of a Gilbert pattern: all new material should contribute to an increase only of the area under the rapid peak, the area under the slow peak should remain constant and the concentration dependence of the weight-average sedimentation coefficient as a function of protein concentration should first exhibit a hyperbolic increase, followed by an inflection upward at the concentration at which bimodality sets in, and another hyperbolic region tending asymptotically toward the s value of the terminal polymer.

Exact analysis of such a polymerizing system is best carried out by numerical data fitting, using methods developed by Gilbert and Gilbert (77,78) and by Cox (83,85), with the original Gilbert theory (75) described by equation 24 being used as a guide in the first selection of approximate parameters which describe the reaction.

For the reaction of equation 27 terminated at step n, the association constants between any i-mer and the monomer, k_i, in units of ml/mg can be written out as expressed in equation 28 (6), where C_i is concentration of species i in mg/ml and $-RT\ln\gamma$ is the favorable additional free energy of formation of the terminal polymer. The weight-average sedimentation coefficient, $\bar{s}$, at any total protein concentration, C, is expressed in equation 29, where s_i° is the sedimentation coefficient of the i-mer extrapolated to zero protein concentration, C_1 is the concentration of monomer and g_i is the slope of the sedimentation *vs* concentration plot for the i-mer. A proper choice of the values of s_i° and g_i for the various aggregated species along the pathway of the self-association permits a numerical fitting of the data in terms of the three desired parameters, k_2, k_n and n (6). The final test of validity of the parameters deduced by such procedures is afforded by theoretical simulation of sedimentation patterns, namely patterns are calculated using the deduced parameters and compared with the experimental velocity profiles. This can be done using methods developed by Gilbert and Gilbert (78) and by Cox and co-workers (83,85,86).

$$k_2 = \frac{C_2}{C_1^2}$$
$$k_i = i\ (k_2/2)^{i-1} \qquad (28)$$
$$k_n = n\ (k_2/2)^{n-1}\gamma$$

$$\bar{s} = \sum_i s_i^o\ (1 - g_i C) k_i C_1^{\ i} \bigg/ \sum_i k_i C_1^{\ i} \qquad (29)$$

Ligand Mediated Self-Association of Proteins. Another mode of self-association is the more complicated system of ligand-mediated polymerization, which may be expressed by the general reaction scheme

$$nA + inX \rightleftharpoons (AX_i)_n \qquad (30)$$

where X is a ligand and i is the number of ligand molecules bound per molecule of protein monomer entering the reaction. This system has

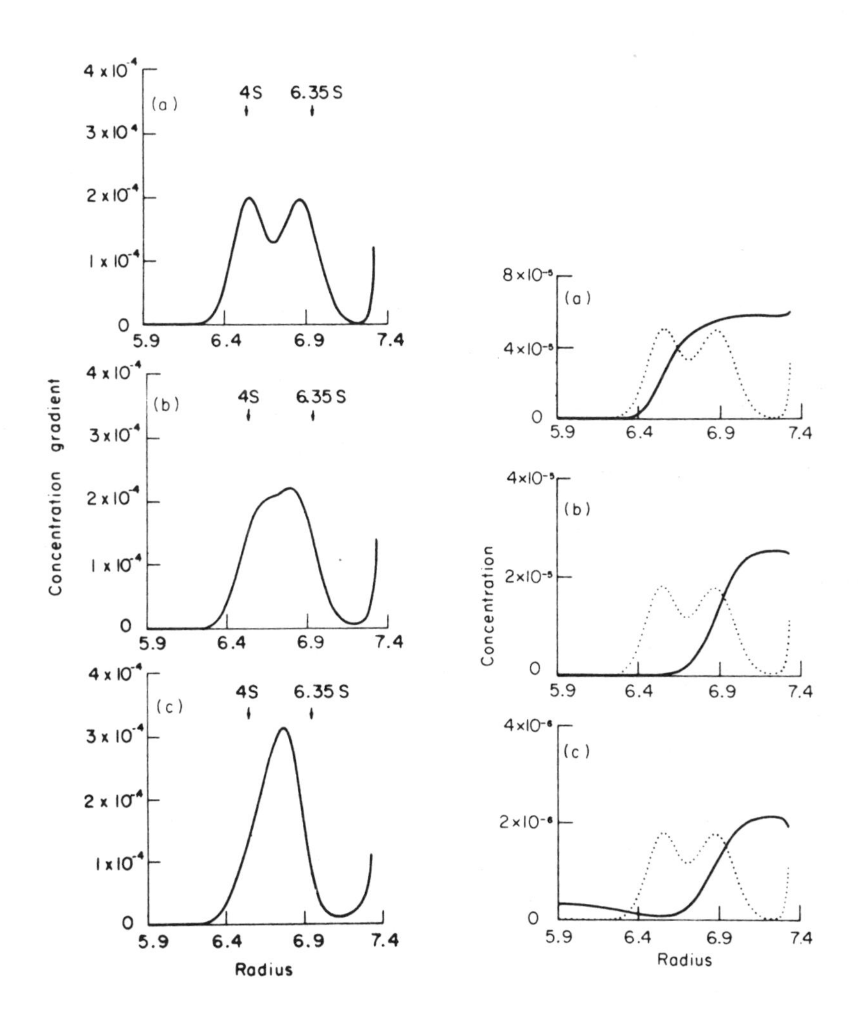

Fig. 11. Calculated sedimentation patterns for the ligand-mediated dimerization reaction $2M + X$ M_2X. Dependence of boundary shape at 5-% dimerization upon the initial concentration of unbound small ligand. Left: free ligand concentration: a) 2×10^{-6}M; b) 5×10^{-6}M; c) 3.5×10^{-5}M; total protein concentration: 1.4×10^{-4}M expressed as monomer. Right: Concentration distribution of species in top left pattern: a) monomer; b) dimer; c) free ligand. (Reproduced from ref. 80).

been analyzed systematically by Cann and Goad (80,81). No analytical expressions of the type of equation 24 may be written out for such systems, but the analysis must be carried out by simulation calculations using properly chosen values of n, i, the binding constant of ligand to protein and the polymerization association constant. Depending on the values of these parameters and the relative concentrations of protein and ligand, sedimentation patterns may become resolved into bimodal ones even for the cases of dimerization and indefinite polymerization. The Gilbert criteria are then no longer valid, and the area under the slow peak in an end-product polymerization need not be independent of protein concentration. Typical calculated patterns are shown in Figure 11 (80). From this figure it is evident that the complex sedimentation behavior depends on the formation of stable concentration gradients of free ligand across the sedimenting boundary in the centrifuge cell. This requires that the interaction between protein and ligand be strong, and the ligand concentration be of similar magnitude to that of the protein. Such behavior has been demonstrated for several protein systems, notable for the calcium-induced dimerization of hemocyanin described by Morimoto and Kegeles (87). When the protein-ligand interaction is weak, or when the ligand is present in large excess over the protein, ligand concentration gradients across the sedimenting reaction boundary cannot form and the system reduces to one which behaves in classical Gilbert manner. The application of these treatments to real systems will be exemplified in the next two sections on two modes of tubulin self-association.

Magnesium-Induced Self-Assembly of Tubulin to 42S Rings

Stoichiometry. The self-association of tubulin in the presence of magnesium ions has been observed first by Weisenberg and co-workers (8,10) in sedimentation velocity studies on porcine and calf brain tubulins. From their examination of the behavior of calf brain tubulin, Weisenberg and Timasheff (10) concluded that, in the presence of optimal concentrations of magnesium ions, the dimeric tubulin species appeared to be in rapid equilibrium with a rapidly sedimenting species and that the reaction was best thought of in terms of a monomer in a rapid equilibrium with a polymer of the same chemical composition. A detailed examination of this process was carried out by Frigon et al. (6,20,88) using sedimentation velocity as the method of analysis.

Photos and traced enlargements of typical sedimentation velocity profiles of tubulin as a function of Mg^{++} ion and protein concentrations are shown in Figures 12 and 13. Below 0.01 M $MgCl_2$, at a low

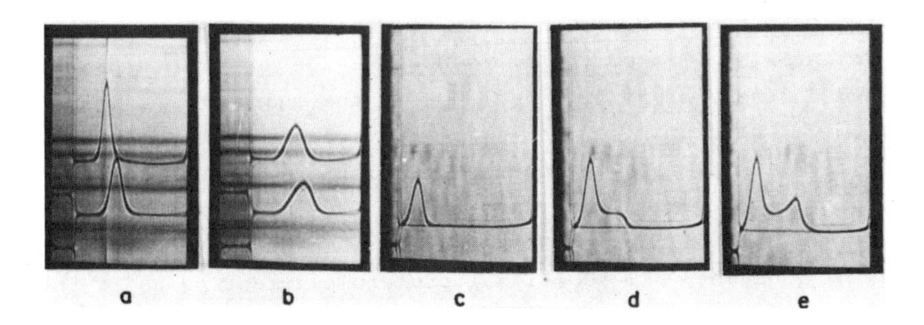

Fig. 12. Sedimentation patterns of tubulin in pH 7.0, 10^{-2}M phosphate, 10^{-4}M GTP buffer at 20°C. (a) and (b) protein concentration 8 mg/ml; (a) upper, no magnesium; lower, 0.0027M $MgCl_2$; (b) upper, 0.0055M $MgCl_2$ lower, 0.0082M $MgCl_2$; (c-e) 0.01M $MgCl_2$; protein concentration, 4,7, 10.4 and 15.5 mg/ml. (Reproduced from ref. 6).

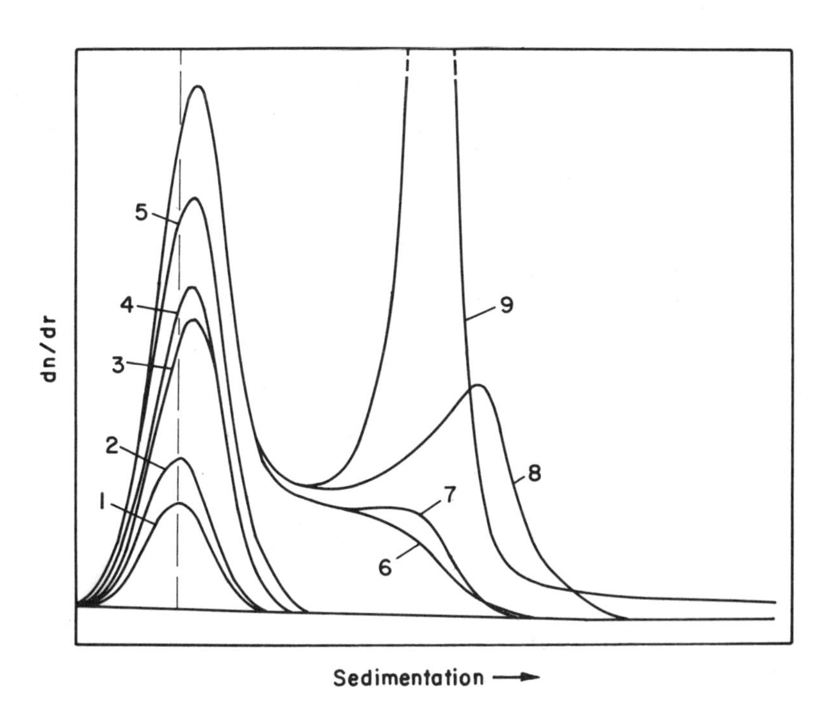

Fig. 13. Tracings of sedimentation velocity patterns of tubulin in pH 8.0, 10^{-2}M phosphate, 10^{-4}M GTP buffer containing 8 mM $MgCl_2$, at 20°C. Protein concentrations (in mg/ml) (1) 1.6; (2) 2.4; (3) 4.9; (4) 7.2; (5) 9.3; (6) 12.2; (7) 16.3; (8) 18.3; (9) 21.6; The vertical dashed line bisects the peak at the lowest concentration. Note that the velocity of the slow peak increases with concentration. (Reproduced from ref. 6).

protein concentration, an increase in Mg^{++} concentration results in an increase in the velocity of the single peak and in its spreading, as shown in Figure 12a-c. As the concentrations of Mg^{++} and protein are raised further, a pronounced change in the character of the sedimentation velocity profile, shown in Figure 12d and e and Figure 13 takes place. The sedimentation patterns, which are single peaked at low protein concentration, become bimodal above a given protein concentration. Once bimodality sets in, the area under the slower peak remains constant, all additional protein sedimenting under the rapid peak with a resulting increase in its area. Although this behavior is typical of a classical Gilbert system, a plot of the sedimentation coefficients of both the slow and rapid peaks as a function of protein concentration showed the situation to be more complicated. The velocity of the slow peak was found to increase with increasing protein concentration from a value of 5.8S, levelling off at a constant value of close to 9 S as the boundaries became bimodal. The rate of increase in the velocity increased with increasing Mg^{++} concentration, and the protein concentration at which the boundaries become bimodal decreased with increasing Mg^{++} concentration. The velocity of the rapid peak, on the other hand, first increased with protein concentration, then, after reaching a maximum, it slowly decreased with a further increase in protein concentration. This behavior is consistent with a self-associating system which proceeds to an end-state polymer via an isodesmic chain growth process, the free energy of formation of the final polymer being highly favorable. These patterns were analyzed in terms of the extended Gilbert theory (6,20). The initial values of s_M and s_P were taken as 5.8S and 41S, respectively. The last value was obtained by extrapolation to zero concentration of the sedimentation coefficients of the rapid peak at high protein concentrations, where $s_{20,w}$ decreases with protein concentration. This procedure is based on the fact, established by Josephs and Harrington (89), that, although neither the slow nor the fast peak in such a reaction boundary represents the sedimentation of any particular species, in cases where the degree of association is high and the association is strong, the sedimentation of the rapid peak is a close approximation of the behavior of the terminal polymeric species. Using these values of s_M and s_P, and assuming that the sedimentation of all intermediate species is related to that of the monomer, $s_M = s_1$, by (*)

$$s_i^\circ = s_1^\circ (i)^{2/3} \qquad (31)$$

the dependence of the weight-average sedimentation coefficient on protein concentration was subjected to the above-described numerical analysis. The results are shown in Table 5. The best value for the degree of polymerization, n, was found to be 26 $\pm$ 2, with $K_n >> K_2$ (6). The results of the data fitting are shown in Figure 14 for two magnesium concentrations. As is evident from Table 5, the addition of the last subunit is characterized by an association constant, K_n, which is much greater than that of the growth process, K_2. In fact $K_n \simeq (K_2)^3$.

TABLE 5

SELF-ASSOCIATION OF TUBULIN TO THE 42S POLYMER AT 20°C.

$C_{Mg^{++}}$ (M)	K_2 (M^{-1})	ΔG_2° (kcal/mole)	K_n (M^{-1})	ΔG_n° (kcal/mole)	ΔG_γ° (kcal/mole)
0.008	1.2×10^4	-5.5	1.6×10^{11}	-15.0	-9.5
0.16	2.2×10^4	-5.8	8.8×10^{12}	-17.3	-11.5

Why is the Addition of the Last Subunit Strongly Favored? The reason for the much greater association constant for the addition of the last subunit becomes clear if we examine a model for the addition of the last subunit in which the final structure is closed and cyclic, as depicted schematically in Figure 15 (90). In this mechanism, the isodesmic pathway is valid for the addition of all monomeric subunits, including the step $A_{25} + A \rightleftharpoons A_{26}$, as long as ring closure does not occur. Let us consider the addition of the last subunit in the $(n-1) + 1 \rightleftharpoons n$ step. We see that it involves the closing of two intersubunit bonds. Examining the formation of each bond separately, it becomes clear that they are different. The formation of the first bond may be regarded simply as a further addition to the chain, $A_{25} + A \rightleftharpoons A_{26}$, with an equilibrium constant, $K_a = K_2$. The second

---

* This relationship assumes spherical symmetry for all species; yet, in the absence of information on the geometry of the successive aggregates, it seemed that assumptions on their shapes would only unduly complicate the analysis without giving any additional information.

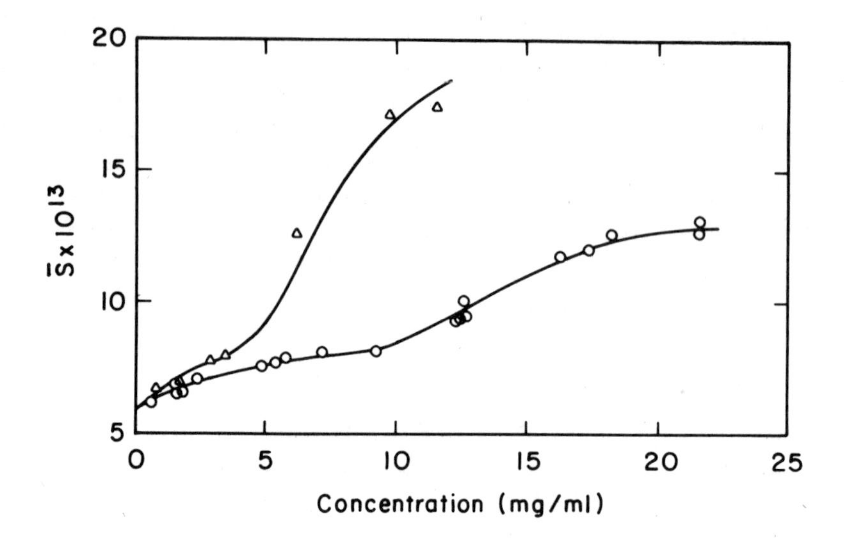

Fig. 14. Theoretical fitting of the concentration dependence of the weight-average sedimentation coefficient, $\bar{s}$, of tubulin in pH 7.0, 10^{-2}M phosphate, 10^{-4}M GTP buffer at 20°C. Cricles: 8mM $MgCl_2$; triangles: 16mM $MgCl_2$; solid lines: theoretical calculated curves. (Reproduced from ref. 90).

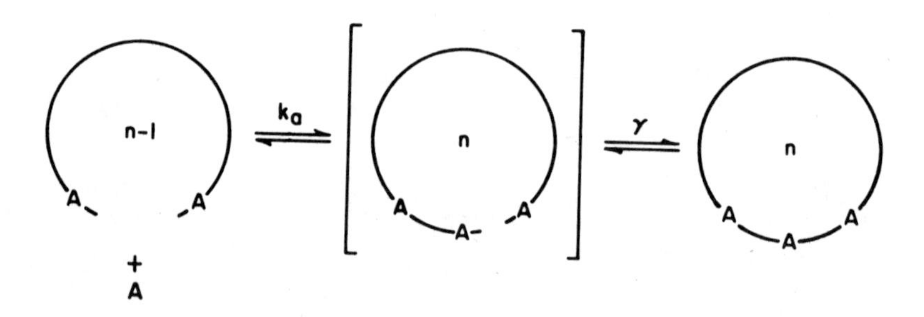

Fig. 15. Schematic representation of the addition of the last subunit, involving the formation of a closed cyclic structure when the last bond is formed

bond, however, differs from the others in that is is formed in a unimolecular reaction. Its free energy of formation may be expressed by ΔG°_γ, where

$$\Delta G^\circ_n = -RT\ln K_n = -RT\ln K_a \quad -RT\ln\gamma \qquad (32)$$

If the structure is symmetric, the free energy of formation of the last bond, ΔG°_γ, must be the sum of the intrinsic free energy of bond formation ΔG°_b, and of residual entropic factors involved in ring closure, ΔG°_R. The free energy of forming the last bond, ΔG°_b, can be related to the free energy of chain growth, ΔG°_a, by taking into account the statistical factor associated with the loss of freedom of ways in which the contact may be made, and the fact that the closing of the last bond does not entail the loss of entropy associated with the decrease in the number of particles in solution each time that an additional monomer molecule is added to the chain, namely it does not entail the loss of cratic entropy. If we assume that polymerization is not directional, then during chain extension, each step can occur at either end of the chain and either site on the added monomer, giving a statistical factor of 4. (Directional polymerization would reduce this factor to 2 or even one.) The cratic entropy term is equal within a close approximation to RTln $(x_n/x_{n-1}x_1)$, where x_i is the mole fraction of species i. Then, at unit molality of reacting components,

$$\Delta G^\circ_\gamma = \Delta G^\circ_a - RT\ln \frac{x_n}{x_{n-1}x_1} + RT\ln 4 + \Delta G^\circ_R \qquad (33)$$

As a result, the free energy of the addition of the last subunit with ring closure may be two to three times that of the chain propagation free energy. In the case of tubulin in 0.016 M $MgCl_2$, ΔG°_a =-5.8 kcal/mole and ΔG°_n = -17.3 kcal/mole.

Effect of State Variables. The chain growth reaction was also examined in terms of solution variables (20), namely temperature, pressure and ligands, using methods described above and properly taking into account that the association constant, K_a, was measured by a technique which depends on the mass distribution of species. It was found that the dimerization step is characterized by positive enthalpy and entropy changes, and that one additional magnesium ion is bound per tubulin dimer incorporated into the growing linear polymer. While implicating magnesium ions in the mechanism of tubulin self-association, this nevertheless does not establish their direct participation in the interprotein bond formation. It is equally probable that magnesium acts through an alteration of charge distribution on the surface of the protein molecules.

The effect of pressure on the self-association of tubulin was examined by ultracentrifugal techniques described by Josephs and Harrington (89,91). If self-association is accompanied by a change in molar volume, and if the end product is large, the formation of polymer becomes a sensitive function of pressure. In an ultracentrifuge cell, the equilibrium constant, K(P), of polymer formation at any pressure P_x, corresponding to position x in the cell, is related to the equilibrium constant under reference conditions (usually atmospheric pressure), K(O), by

$$\ln K(P) = \ln K(O) - \frac{1}{RT} \int_{x_o}^{x} \Delta V \left(\frac{\partial P}{\partial x}\right)_T dx \qquad (34)$$

where ΔV is the molar volume change of the reaction for the formation of 1 mole of polymer, P is the pressure, and x_o is the position of the meniscus in a centrifuge cell. If the pressure at any point in the cell is known, the change in association constant can be described by

$$\Delta G^\circ(P) - \Delta G^\circ(O) = RT[\ln K(O) - \ln K(P)] = M_p \Delta \bar{v} \, (P - P_o) \qquad (35)$$

where $\Delta G^\circ(P)$ and $\Delta G^\circ(O)$ are the free energies of the association reaction at pressure P and at atmospheric pressure, M_p is the molecular weight of the polymer, $\Delta \bar{v}$ is the change in partial specific volume on polymerization, R is the gas constant and T is the thermodynamic temperature. In the case of a self-associating system, a positive molar volume change results in polymer dissociation which increases progressively from the meniscus to the bottom of the centrifuge cell. This leads to negative gradients of polymer toward the bottom of the cell and convective disturbances within the solution are expected in the plateau region (92,93). In order to test whether the self-association of tubulin is affected by pressure, experiments were carried out in which the hydrostatic pressure over the protein solution was varied by layering varying amounts of oil over it. The results, showed that the effect is weak; $\Delta \bar{v}$ was found to have a value of ~ 2.5×10^{-4} ml/g. This corresponds to an increase of the molar volume by 750 ml when a 2.86×10^6 molecular weight polymer is formed.

At this point, it might seem of interest to examine the consequences that such a change in $\bar{v}$ would have on polymerization. An examination of equation 35 reveals that the free energy of the reaction has a linear dependence on the product $M_p\bar{v}$. The effect on the equilibrium constant is logarithmic. The pressure effect is a function not just of the change in $\bar{v}$, but also of the degree of polymerization of the aggregate. The effect on tubulin association to the 42S polymer was found to be weak. If the molecular weight is raised 100-fold into the realm of microtubules, however, the effect becomes considerably greater. This has been calculated for values of $\Delta\bar{v}$ of 2.5×10^{-4}, 5×10^{-4}, and 1×10^{-3} ml/g, all of which are within the range of values reported for a variety of protein self-association reactions (93). The results are presented in Figure 16 in terms of the pressure dependence of the critical concentration below which no microtubules exist in significant amounts.

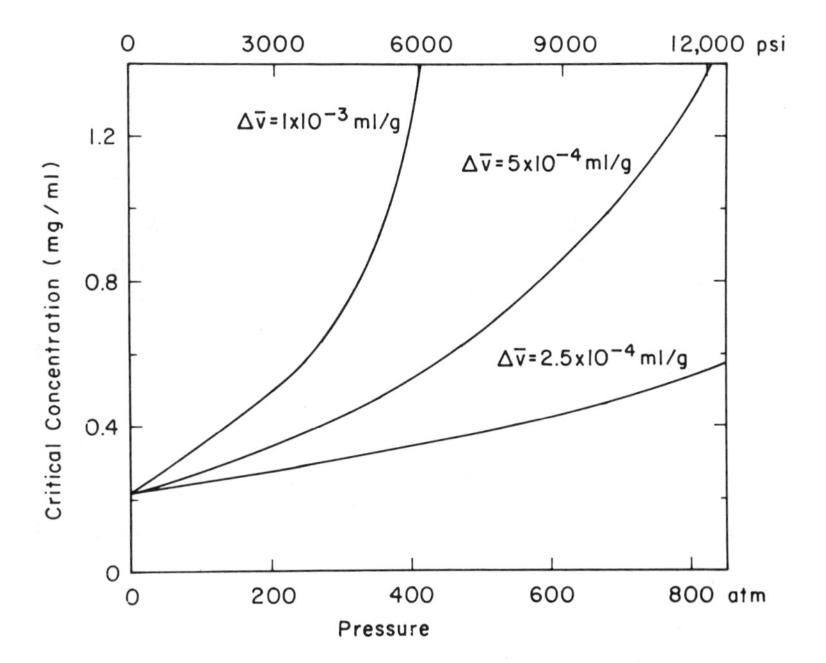

Fig. 16. Calculated effect of pressure on the stability of microtubules, assuming the validity of the Oosawa model. (Reproduced from ref. 90).

Setting arbitrarily the critical concentration of atmospheric pressure to 0.22 mg/ml, it is seen that an increase in pressure leads to an increase in critical concentration, the effect increasing rapidly with an increase in $\bar{v}$. The _in vivo_ and _in vitro_ depolymerizations of microtubules reported for pressures of 4,000 to 10,000 lb/in^2 (94,95) can then be accounted for simply interms of an increase in the partial specific volume of the protein when microtubules are formed of a magnitude which is normally encountered in protein system. Indeed, if this change in $\bar{v}$ on polymerization reflects in great part a release of water molecules when interprotein bonds are formed, it can be expected that this change would be considerably greater for microtubule formation than for the formation of 42 S rings, since in the former the intermolecular contacts must involve a considerably greater fraction of the protein surface.

Structure of the End-Product. The conclusion derived from thermodynamic arguments that the aggregate is cyclic in nature was further tested by an electron microscopic examination of the system. When tubulin samples prepared under the conditions of the ultracentrifuge experiments were examined with the electron microsocpe, results were obtained as shown in Figure 17 (6,21). The associated structure has the appearance of two closed concentric beaded rings with outside and inside diameters of 47 ± 3 and 27 ± 3 nm, respectively. The rings

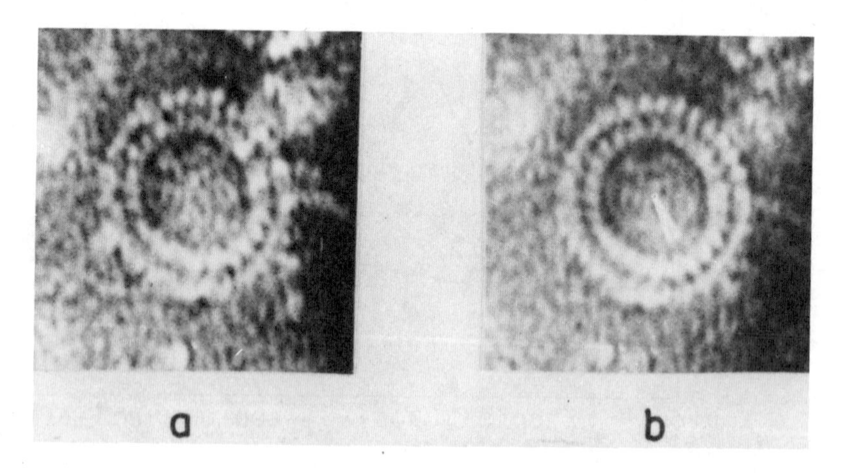

Fig. 17. Electron micrograph of the association product of tubulin in pH 7.0, 0.1M phosphate, 10^{-4}M GTP, 16mM $MgCl_2$ buffer, negatively stained with uranyl acetate. A is a direct view; B has been rotated about its center of symmetry by 13.8°, followed by double exposure. The primary magnification was X102,000. (Reproduced from ref. 6).

appear to be composed of subunits and the average center-to-center distance between adjacent units is 4.1 $\pm$ 0.3 nm. Testing the periodicity of this structure by the Markham procedure (96) resulted in an enhancement of detail after rotation through 13.8°, or 1/26 of the circumferential arc; namely, there appears to be a periodicity of 26 $\pm$ 4 identical pictorial associating subunits around the circle, or essentially double the number found around a cross section of a microtubule. This value is in good agreement with the most likely stoichiometry of polymerization deduced from the sedimentation studies. This analysis, however, does not establish the orientation of the 5.8S tubulin dimers in the ring, nor indeed their identity. Furthermore, the possibility has not been excluded that the ring is a double layer of the photographed structure, nor that it is deformed from its exact shape in solution by the fixing procedure.

The relation between the protein complex seen in the electron microscope and the tubulin polymer formed in solution was tested (6) by calculating its expected sedimentation coefficient by using the method of Kirkwood (97). According to the Kirkwood theory (97), if the complex is composed of an arbitrary array of connected identical subunits which may be fluctuating or fixed relative to one another, it is possible to calcualte its sedimentation coefficient by

$$s = \frac{M(1 - \bar{v}\rho)}{nN}\left(\frac{1}{f} + \frac{1}{6\pi\eta n}\sum_{\ell=1}^{n}\sum_{\substack{s=1\\ \ell\neq s}}\langle R_{\ell s}\rangle^{-1}\right) \qquad (36)$$

where f is the frictional coefficient of the subunit, n is the number of subunits in the array, and (R_{1s}) is the time-averaged distance between the 1 and s subunits. Simplifying the ring structure to that of a plane polygon with subunits of the complex lying at its vertices, the sedimentation coefficient for the structure shown in Figure 17 was calculated to be 43 $\pm$ 3S (6). Since the observed sedimentation coefficient is 42 $\pm$ 2S, it appears that the closed double-ring protein complex observed by electron microscopy describes well the structure of the polymer detected by sedimentation velocity when calf brain tubulin is induced to self-associate by the addition of magnesium ions to the system,and may indeed be a true picture of the structure in solution.

Are Double Rings Related to Microtubules? At this point, a few remarks seem desirable on the relation between the 42S double ring aggregates and microtubule self-assembly. In their examination of the cold depolymerization products of microtubules reconstituted from cycle preparations of tubulin, several authors have described double rings which are strikingly similar to those of Figure 17 (12,14). These authors have assigned to these structures the role of intermediates in the formation of microtubules (14,22,59). These conclusions, however, have been based solely on electron microscopic evidence, there being no direct connection between ring formation and that of microtubules. In our studies, it has been shown that microtubules can be formed simply by minor adjustments in the environment which favors ring formation. In fact, under our reconstitution conditions, rings can be seen in electron micrographs simultaneously with microtubules (21). Yet, none of this circumstantial evidence need indicate that rings are intermediates in microtubule formation. In fact, the 42S structure can be made in the presence of calcium ions (10), just as well as of magnesium ions, at similar levels of the order of 10^{-2}M; yet, calcium ions are known to inhibit microtubule formation already at a much lower level. Furthermore, this association reaction can proceed with ease in the presence of colchicine which is known to inhibit microtubule formation. While the inhibitory activity of calcium and colchicine may take place at a different stage of microtubule formation, for example, in the formation of lateral bonds, while double rings involve only longitudinal bonds, these observations suggest extreme caution in assigning any role to the double rings in the pathway of microtuble assembly. Indeed, the two processes may be totally independent modes of tubulin self-association which may occur simultaneously and it appears that, at present, the entire question of the nature of intermediates in microtubule reconstitution may best be regarded as completely open.

The Vinblastine-Induced Self-Association of Tubulin.

Another mode of protein self-assembly, namely ligand-mediated self-association is well exemplified by the vinblastine-induced polymerization of tubulin. *In vivo*, tubulin is known to form aggregates other than microtubules (98,99) when vinblastine is introduced into the system. This alteration of the mode of tubulin self-association by vinblastine imparts to it its clinical activity in the treatment of some cancers (7). Namely, the disruption of microtubules in the mitotic apparatus with the formation of physiologically incorrect assembled structures interferes with cell proliferation (100), making this alkaloid into an effective cancer chemotherapeutic agent. In order to understand the nature of the vinblastine effect on the molecular level, an ultracentrifugal examination was, undertaken of this process (74). In a preliminary study, Weisenberg and Timasheff (10) had concluded that the concentration dependence of the sedimentation properties of tubulin in the presence of high concentrations of vinblastine conforms best to a pattern consistent with an isodesmic indefinite type of self-association, described by equation 27. A more detailed examination by Lee et al. (74) revealed a complex pattern of self-association which is a strong function of the tubulin-vinblastine ratio in the system.

Typical results are presented in Figure 18, in which sedimentation patterns are shown for a set of experiments in which the tubulin-to-vinblastine ratio was varied at a low total concentration of vinblastine at 20°C in a pH 7.0 buffer consisting of 10^{-2}M phosphate and 10^{-4}M GTP. As the tubulin-to-vinblastine ratio increases, the sedimentation pattern changes from a skewed single peak at a ratio of 1.2 moles of tubulin per mole of vinblastine (pattern a) to a bimodal

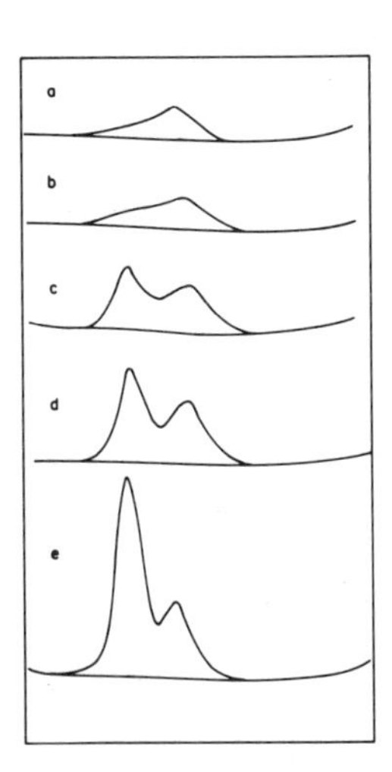

Fig. 18. Tracings of sedimentation patterns of tubulin in the presence of 2 x 10^{-5}M vinblastine in pH 7.0, 0.01M phosphate 10^{-4}M GTP buffer, at 20°C. The tubulin concentrations and tubulin to total vinblastine ratios (mole per mole) are: (a) 2.7, 1.2; (b) 3.2, 1.4; (c) 4.9, 2.2; (d) 6.8, 3.1; (e) 9.9, 4.5; (Reproduced from ref. 74).

profile as this ratio increases,and the relative area under the slow peak increases at higher tubulin-to-vinblastine ratios (patterns c,d and e). Such behavior is that expected from the theoretical predictions of Cann and Goad (80,81) for the ligand-mediated association of macromolecules. According to this theory, resolution of the sedimentation pattern into a bimodal one depends on the formation and maintenance of a free ligand concentration gradient along the centrifuge cell. This requires a strong ligand binding constant. A binding study carried out by Lee et al. (74) has shown that, under these conditions, vinblastine binds to tubulin dimers at two sites with identical binding equilibrium constants of 2.0×10^4 liters/mole, i.e. the binding is sufficiently strong to meet the criteria of the Cann and Goad theory. Raising the vinblastine-to-tubulin ratio to a value at which no ligand concentration gradients could form led to sedimentation patterns consisting of a single hypersharp boundary with a skewed trailing edge. This result is the one expected from the Cann and Goad theory which predicts that at ligand concentrations at which the concentration of unbound ligand is not significantly perturbed during sedimentation, the system shoud reduce to a pseudo-Gilbert one and may be analyzed as a simple polymerization.

The conclusion that the boundary resolution into two peaks is the result of the formation of stable vinblastine gradients across the ultracentrifuge cell was tested by determining the vinblastine distribution across the cell by means of the photoelectric scanner. The results are shown in Figure 19. In pattern A, where the protein-to-ligand ratio is 4.4 mole/mole, there is a drastic decrease in ligand concentration across the fast moving peak of the bimodal boundary, with little or no vinblastine under the slowly sedimenting peak. On the other hand, when vinblastine was present in excess (at a tubulin-to-vinblastine ratio of 0.22 mole/mole), the ligand concentration gradient decreased considerably. Simultaneously, bimodality almost disappeared from the ultracentrifuge pattern. These results are very similar to theoretical patterns calculated by Cann and Goad (80) (See Figure 11) and support the conclusion that vinblastine induces the self-association of tubulin by binding strongly to the protein.

The knowledge that this self-associating system behaves in the theoretically predicted manner made it possible to undertake a systematic study of this process at such conditions at which the system reduces to an isodesmic polymerizaiton (101). Analysis of the results by methods similar to those used in the magnesium ion induced self-association of tubulin to the 42S polymer, ~~described above,~~ established that, in the vinblastine-induced self-association, each step in the chain growth is accompanied by the binding of one ligand molecule per tubulin dimer (101). The chain growth association constant at 20°C is 1.5×10^5 liters/mole and the reaction is characterized by positive enthalpy and

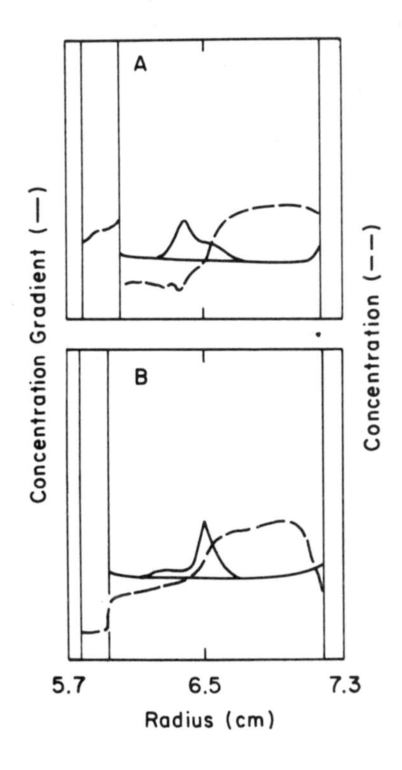

Fig. 19. Tracings of schlieren sedimentation patterns of tubulin and concentration profiles of vinblastine across the ultracentrifuge cell, in pH 7.0, 0.01M phosphate, 10^{-4}M GTP buffer, at 20°C. The tubulin concentration was 4.89 mg/ml. The vinblastine concentration and tubulin to vinblastine ratio (mole per mole) are: (A) 1×10^{-5}M, 4.4; (B) 2×10^{-4}M, 0.22. (Reproduced from ref. 74).

entropy changes, suggesting that the mode of association involves hydrophobic or electrostatic interactions, or both.

CONCLUSIONS

The three different self-associations of tubulin described in this chapter can be taken as good examples of the various modes of self-assembly that a protein can undergo and of the methods of analysis which are available for studying their equilibrium thermodynamics. The diversity in size and structure of the assembled products make it necessary to use different physical techniques for their elucidation. Yet, once the initial results, be they obtained by light scattering or by hydrodynamic techniques, have been unraveled by the theories pertinent to each method used and reduced to statements of stoichiometries and association constants, all the reactions become amenable to treatment in terms of fundamental thermodynamic relations. The universal applicability of reversible thermodynamics, of course, precluded the determination of detailed information on reaction pathways and mechanisms. From studies such as described in this chapter, it is not possible to arrive at unique models for a reaction, nor to establish the geometric nature of intermediates, nor indeed their sizes, in a self-assembly reaction. Yet thermodynamic analysis can lead to general structural conclusions. For example, in the magnesium induced self-assembly of tubulin, it was shown by a purely thermodynamic argument that the end product must be cyclic in nature. It is indeed satisfying that a later electron microscopic examination revealed in fact the formation of closed ring structures.

In thermodynamic analyses, such as described in this chapter, extreme care must be exercised in selecting the correct standard state for each reaction examined, and for each technique used. For example, results obtained by techniques which are based on mass distribution must be resolved into all the species of identical mass, but different conformation or degree of liganding. The Wyman linked function relations require that the system be reduced to a standard state of unliganded reactants. Similarly, great caution is required in the analysis of sedimentation velocity experiments. Otherwise much confusion may result if sedimentation patterns obtained with an associating system which involves ligands are analyzed in terms of the Gilbert theory even qualitatively, however tempting the idea might seem, without first ascertaining that the concentrations of ligands linked to the association are such that no gradients can be formed in the ultracentrifuge cell. This situation is rarely realized when the ligand in question is macromolecular in nature and in such systems analysis should be carried out by other approaches, such as the Gilbert-Jenkins theory (102) or by direct simulation calculations. On the other hand, when properly applied, the hydrodynamic and thermodynamic techniques described in this chapter can lead to the simplest and most reasonable model of the reaction under investigation, in terms of stoichiometry and equilibrium constants. Yet, even when such a model is deduced, it must be remembered that it is neither unique, nor necessarily the one operative in the particular self-association. Such information must be sought from other types of experiments, for example from an examination of the kinetics of the reaction. Proper variation of the solvent variables and judicious manipulation of solvent composition, with the proper application of multicomponent thermodynamic theory, may lead to an understanding of the general types of forces involved. Details of interactions in the sites of contact and identification of groups involved and of their mutual orientation, however, must be sought from other types of experiments. These may involve the use of genetic variants (103), specific chemical modifications (104) and the introduction of markers or reporter groups with an examination of the perturbation of their spectroscopic properties.

Spectroscopic techniques can also be used directly if proper chromophoric groups are present in the domain of the intermolecular contacts formed. The careful application of these various approaches can lead to details of the geometry of the sites of contacts, small changes in which on the micro scale may become translated into drastically different morphologies on the macro-scale.

REFERENCES

1. Grimstone, A.V. and Klug, A. (1966) J. Cell. Sci. 1, 351- .
2. Cohen, C., DeRosier, D., Harrison, S.C., Stephens, R.E. and Thomas, J. (1975) Ann. N.Y. Acad. Sci., 253, 53-59.
3. Erickson, H.P. (1975) Ann. N.Y. Acad. Sci., 253, 60-77.
4. Snyder, J.A., and McIntosh, J.R., (1976) Ann. Revs. Biochem. 45, 699-720.
5. Lee, J.C., Frigon, R.P. and Timasheff, S.N. (1973) J. Biol. Chem. 248, 7253-7262.
6. Frigon, R.P. and Timasheff, S.N. (1975) Biochemistry 14, 4559-4566.
7. Wilson, L. and Bryan, J. (1974) Adv. Cell. Molec. Biol. 3, 22-72.
8. Weisenberg, R.C., Borisy, G.G. and Taylor, E.W. (1968) Biochemistry 7, 4466-4479.
9. Weisenberg, R.C. (1972) Science 177, 1104-1105.
10. Weisenberg, R.C. and Timasheff, S.N. (1970) Biochemistry 9, 4110-4116.
11. Shelanski, M.L., Gaskin, F. and Cantor, C. (1973) Proc. Natl. Acad. Sci., U.S.A. 70, 765-768
12. Kirschner, M.W., Williams, R.C., Weingarten, M. and Gerhart, J.C., (1974) Proc. Natl. Acad. Sci., U.S.A. 71, 1158-1163.
13. Weingarten, M.D., Lockwood, A.H., Hwo, S.Y. and Kirschner, M.W., (1975) Proc. Natl. Acad. Sci., U.S.A. 72 1858-1862.
14. Erickson, H.P. (1974) J. Supramol. Struct. 2, 393-411.
15. Borisy, G.G., Olmstead, J.B., Marcum, J.M. and Allen, C. (1974) Fed. Proc., Fed. Am. Soc. Exp. Biol. 33, 167-174.
16. Murphy, D.B. and Borisy, G.G. (1975) Proc. Natl. Acad. Sci., U.S.A. 72, 2696-2700.
17. Murphy, D.B., Vallee, R.B. and Borisy, G.G. (1977) Biochemistry 16, 2598-2605.
18. Sloboda, R.D., Dentler, W.L. and Rosenbaum, J.L. (1976) Biochemistry 15, 4497-4505.
19. Cleveland, D.W., How, S.-Y., and Kirschner, M.W. (1977) J. Mol. Biol. 116, 207-225.
20. Frigon, R.P., and Timasheff, S.N. (1975b) Biochemistry 14, 4567-4573.
21. Frigon, R.P., Valenzuela, M.S. and Timasheff, S.N. (1974) Arch. Biochem. & Biophys. 165, 442-443.

22. Weingarten, M.D., Suter, M.M., Littman, D.R. and Kirschner, M.W., (1974) Biochemistry, 13 5529-5537.

23. Keates, R.A.B., and Hall, R.H. (1975) Nature (London) 257, 418-420.

24. Kuriyama, R., (1975) J. Biochem. (Tokyo) 77, 23-31.

25. Borisy, G.G. and Olmsted, J.B. (1972) Science 177, 1196-1197.

26. Lee, J.C., Hirsh, J., and Timasheff, S.N. (1973) Arch. Biochem. and Biophys., 168, 726-729.

27. Lee, J.C. and Timasheff, S.N. (1975) Biochemistry, 14, 5183-5187.

28. Gaskin, F., Cantor, C.R., and Shelanski, M.L. (1974) J. Mol. Biol. 89, 737-756.

29. Berne, B.J. (1974) J. Mol. Biol. 89, 756-758.

30. Fournet, G. (1951) Acta Crystallogr. 4, 293-301.

31. Kirkwood, J.G., and Mazur, J. (1952) C.R. Acad. Sci. Reun. Annu. Soc. Chim. Phys., 143-146.

32. Olmsted, J.B. and Borisy, G.G. (1973) Biochemistry 12, 4282-4289.

33. Himes, R.H., Burton, P.R., Kersey, R.N., and Pierson, G.B., (1976) Proc. Natl. Acad. Sci., U.S.A. 73, 4397-4399.

34. Herzog, W. and Weber, K. (1977) Proc. Natl. Acad. Sci. U.S.A. 74, 1860-1864.

35. Oosawa, F. and Kasai, M. (1971) in "Biological Macromolecules" (S.N. Timasheff and G.D. Fasman eds.), Dekker, N.Y., Vol. 5, pp. 261-322.

36. Oosawa, F., and Higashi, S. (1967), Prog. Theor. Biol., 1, 28-164.

37. Oosawa, F., and Kasai, M. (1971) Biol. Macromol., 5, 261-322.

38. Oosawa, F., and Kasai, M. (1962), J. Mol. Biol., 4 10-21.

39. Wyman, J. (1964) Adv. Protein Chem. 19, 224-285.

40. Tanford, C. (1969), J. Mol. Biol. 39 539-544.

41. Casassa, E.F., and Eisenberg, H. (1964) Adv. Protein Chem., 19, 287-395.

42. Inoue, H., and Timasheff, S.N. (1972) Biopolymers 11, 737-743.

43. Lee, J.C. and Timasheff, S.N. (1977) Biochemistry 16, 1754-1764.

44. Timasheff, S.N., Gorbunoff, M.J., and Hinz, H.-J., (1978) Fed. Proc. 27, 1970.

45. Timasheff, S.N. (1973b), Protides Biol. Fluids, Proc. Colloq. 20, 511-519.

46. Zimmerman, A.M. and Marsland, D. (1964), Exp. Cell. Res., 35, 293-302.

47. Salmon, E.D. (1975a) J. Cell. Biol. 66, 114-127.

48. Stephens, R.E. (1973) J. Cell. Biol. 57, 133-147.

49. Timasheff, S.N., and Townend, R. (1968), Protides Biol. Fluids Proc. Colloq. 16, 33-40.

50. David-Pfeuty, T., Erickson, H.P. and Pantaolni, D. (1977) Proc. Natl. Acad. Sci., U.S.A. 74, 5372- .

51. David-Pfeuty, T., Laporte, J. and Pantaloni, D., (1978), Nature 272, 282-284.

52. Timasheff, S.N. (1970) in "Biological Polyelectrolytes, (A. Veis ed.) New York, N.Y. Marcel Dekker, chapter 1.

53. Henkens, R.W., Watt, G.D., and Sturtevant, J.M. (1969) Biochemistry 8, 1874-1878.

54. Anderegg,G. (1964) Helv. Chim. Acta. 47, 1801-1814.

55. Na, C. and Timasheff, S.N. (1978) Fed. Proc. 37, 1791.

56. Gekko, K. and Timasheff, S.N. (1979) to be published.

57. Pittz, E.P. and Timasheff, S.N. (1978) Biochemistry 17 615-623.

58. Salmon, E.D. (1975) Science 189, 884-886.

59. Kirschner, M.W., Honig, L. and Williams, R.C. (1975) J. Mol. Biol. 99, 263-276.

60. Johnson, K.A. and Borisy, G.G. (1975) in "Molecules and Cell Movement" (Inoue, S. and Stephens, R.E., eds.), Raven Press, N.Y. pp. 119-141.

61. Borisy, G.G., Marcum, J.M., Olmsted, J.B., Murphy, D.B. and Johnson, K.A. (1975) Ann. N.Y. Acad, Sci., 253, 107-132.

62. Bryan, J., Nagle, B.W. and Doenges, K.H. (1975) Proc. Natl. Acad. Sci., U.S.A. 72, 3570-3574.

63. Erickson, H.P. and Voter, W.A. (1976) Proc. Natl. Acad. Sci. U.S.A. 73, 2813-2817.

64. Lee, J.C., Tweedy, N. and Timasheff, S.N., (1978) Biochemistry, 2783-2790.

65. Haga, T., and Kurokawa, M. (1975) Biochim. Biophys. Acta., 392, 335-345.

66. Berkowitz, S.A., Katagiri, J., Binder, H.K. and Williams, Jr., R.C. (1977) Biochemistry, 16, 5610-5617.

67. Levi, A., Cimino, M., Mercanti, D., Chen, J.S. and Calissano, P., (1975) Biochim. Biophys. Acta. 399, 50-60.

68. Erickson, H.P. (1976) Cell Motility (Goldman, R.,Pollard, T. and Rosenbaum, J. Eds.) Part C, pp 1069-1080, Cold Spring Harbor Laboratory, Cold Spring Harbor, N.Y.

69. Behnke, O. (1975) Nature (London) 257, 709-710.

70. Jacobs, M., Bennett, P.M. and Dickens,M.J. (1975) Nature (London) 257 707-709.

71. Amos, L.A. (1977) J. Cell. Biol. 72 642-654.

72. Sherline, P. and Schiavone, K. (1977) Science 198 1038-1040.

73. Connolly, J.A., Kalnins, V.I., Cleveland, D.W. and Kirschner, M.W. (1977) Proc. Natl. Acad. Sci. U.S.A. 74, 2437-2440.

74. Lee, J.C., Harrison, D. and Timasheff, S.N. (1975) J. Biol. Chem. 250 9276-9282.

75. Gilbert, G.A. (1955), Discuss. Faraday Soc. No. 20, 68-71.

76. Gilbert, G.A. (1959), Proc. R. Soc. London, Ser. A, 250 377-388.

77. Gilbert, G.A. (1963), Proc. R. Soc. London, Ser. A, 276, 354-366.

78. Gilbert, L.M. and Gilbert G.A. (1973), Methods Enzymol. 27, 273-296.

79. Gilbert, L.M. and Gilbert, G.A., (1978) Methods in Enzymology, Volume XLVIII, Part F. (C.H.W. Hirs & S.N. Timasheff, eds.) Academic Press, N.Y. p. 155

80. Cann, J.R. (1970), Interacting Macromolecules: Theory and Practice of Their Electrophoresis, Ultracentrifugation and Chromatography, New York, N.Y. Academic Press.

81. Cann, J.R. and Goad, W.B. (1972) Arch. Biochem. Biophys. 153. 603-609.

82. Townend, R., Winterbottom, R.J. and Timasheff, S.N. (1960), J. Am. Chem. Soc., 82 3161-3168.

83. Cox, D.J., Arch. Bicohem. Biophys. (1969) 129, 106-123.

84. Nichol, L.W., Bethune, J.L. Kegeles, G. and Hess, E.L., (1964), in The Proteins, 2nd edition, (Neurath, H.ed.) New York, Academic vol. 2, p. 305-403.

85. Holloway, R.R., and Cox, D.J., (1974), Arch. Biochem. Biophys. 160, 595-602.

86. Cox, D.J., (1978) Methods in Enzymology, Volume XLVIII, Part F, (C.H. W. Hirs and S.N. Timasheff, Eds.) Academic Press, N.Y. p. 212.

87. Morimoto K. and Kegeles, G. (1971) Arch. Biochem. Biophys., 142, 247-257.

88. Frigon, R.P. Valenzuela, M.S. and Timasheff, S.N. (1974) Arch. Biochem. Biophys. 165, 442-443.

89. Josephs, R. and Harrington, W.F. (1968) Biochemistry 7, 2834-2847.

90. Timasheff, S.N., Frigon, R.P. and Lee, J.C., (1976), Fed. Proc. 35, 1886-1891.

91. Josephs, R., and Harrington, W.F. (1967) Proc. Natl. Acad. Sci. U.S.A. 58, 1587-1594.

92. Kegeles, G. (1970), Arch. Biochem. Biophys. 141, 68-72.

93. Harrington, W.F. and Kegeles, G. (1973) Methods Enzymol. 27 306-345.

94. Salmon, E.D., (1975), Biophys.J., 15, 77a.

95. Tilney, G., Hiramoto, Y. and Masland, D. (1966), J. Cell. Biol. 29, 77-95

96. Markham, R.M., Frey, S. and Hills, G.J. (1963), Virology 20, 88-102.

97. Kirkwood, J.G. (1954), J. Polym. Sci. 12, 1-14.

98. Bensh, K.G. and Malawista, S.E. (1969), J. Cell. Biol. 40, 95-107.

99. Bryan, J. (1972), J. Mol. Biol. 66, 157-168.

100. Wilson, L., Bamburg, J.R., Mizel, S.B., Grisham, L.M. and Creswell, K.M. (1974) Fed. Proc. 33, 158-166.

101. Na, C. and Timasheff, S.N., (1978) Biophysical Journal, 21, 22a.

102. Gilbert, G.A. and Jenkins, R.C., (1959), Proc. Roy. Soc. A253 420-

103. Kumosinski, T.F. and Timasheff, S.N., (1966) J. Am. Chem. Soc. 88, 5635-5642.

104. Gorbunoff, M.J., Fosmire, G. and Timasheff, S.N. (1978), Biochemistry 17 , 4055-4064.

ACKNOWLEDGEMENTS

This work was supported in part by National Institutes of Health Grants GM 14603, and CA16707, and National Science Foundation Grant PCM 72 02572 A03.

Catsimpoolas, ed. Physical Aspects of Protein Interactions

DETECTION OF HETEROGENEITY IN SELF-ASSOCIATING SYSTEMS

DAVID A. YPHANTIS, JOHN J. CORREIA, MICHAEL L. JOHNSON* AND GAY-MAY WU**
Biochemistry and Biophysics Section, Biological Sciences Group and Institute of Materials Science, University of Connecticut, Storrs, Connecticut 06268

ABSTRACT

Heterogeneity in self-associating systems is discussed and extant methods of detecting such heterogeneity are reviewed. A proposal is made to detect heterogeneity by the variation in the apparent association constants, $\hat{K}$, observed in sedimentation equilibrium of different loading concentrations, $c°$. The sensitivity of this procedure is illustrated. It is shown that a correct model of the self-association examined is not needed, provided the ultracentrifuge data for all the loading concentrations is truncated so as to span the same observation concentration, $c(r)$, range for all the $c°$. Expressions are derived for the quantitative estimation of small extents of heterogeneity both from the diagnostic lack of overlap of curves of $M_k(r)$, the apparent molecular weights, vs $c(r)$ for the different $c°$ and from the variation of $\hat{K}_{app}$ for the different $c°$. These expressions are strictly valid only for systems where there is no hetero-association between the main self-associating component and the contaminating component. For other systems the expressions presented provide lower bounds for the extent of heterogeneity.

INTRODUCTION

The estimation of standard free energies of association (and derived parameters) from observed apparent association constants in self-associating systems can be influenced strongly by the presence of small amounts of contamination by molecules similar to the component of interest[1]. Generally this fact would be considered to be a nuisance, especially when interest is focussed on the properties of the main component. However, this effect can provide a sensitive indication of heterogeneity in certain specific cases, allowing detection of small quantities of molecules closely related to the main component.

Heterogeneity in self-associating systems may have one or more of the following sources:

a) the presence of adventitious contaminating solutes with a molecular weight that is generally, different than that of the monomer of the component of interest.

*Present address: Clinical Endocrinology Branch, National Institute of Arthritis, Metabolism and Digestive Diseases, National Institutes of Health, Bethesda, MD 20014

**Present address: Department of Biochemistry, Brandeis University, Waltham, MA 02154

b) the presence of irreversible aggregates of the main component;
c) the presence of "incompetent monomer" incapable of self-association;
d) the presence of other chemical forms of the self-associating solute with, perhaps, slightly different sequence, differing location of amides, variations in carbohydrate substitutents, etc., as in isozymes, with differing association constants but with effectively the same monomer molecular weights; and
e) the presence of metastable conformers of the self-associating solute that are not in rapid equilibrium with each other over the time course of the experiment and that exhibit differing association behavior.

The first two types may be considered as "trivial heterogeneity" and generally can be removed from the sample by suitable preparative procedures. For example, irreversible aggregates in biochemical preparations frequently may be eliminated by simple gel exclusion chromatography immediately before the self-association experiment.

The association behavior of heterogeneous systems of the last three types, where the contaminating component, B, has the same monomer molecular weight as the component of primary interest, A, generally can be represented by equilibria of the form:

$$nA \rightleftharpoons A_n \qquad K_{A,n} = \frac{m_{A,n}}{m_{A,i}^{n}} \tag{1}$$

$$aB \rightleftharpoons B_q \qquad K_{B,q} = \frac{m_{B,q}}{m_{B,q}^{q}} \tag{2}$$

$$rA + pB \rightleftharpoons A_rB_p \qquad K_{rp} = \frac{m_{A_rB_p}}{m_{A,1}^{r}\, m_{B,1}^{p}} \tag{3}$$

with various n, p, q and r, where $m_{i,j}$ is the molarity of the molecular species defined as the j-mer of the i-th component.

The problem treated here is the detection of small amounts of component B in the presence of large amounts of component A. To make this problem tractable we ignore the possible presence of mixed equilibria and set $K_{rp} = 0$ in eq. (3). [Since distinction between different components is based completely on their association constants, it is possible, with both K_{rp} and $K_{B,m}$ exactly equal to the corresponding $K_{A,n}$, to make the heterogeneity completely invisible to the observer. In general the presence of significant contributions from mixed equilibria will decrease the sensitivity of detection of heterogeneity.] For simplicity we treat in detail only systems where a single simple equilibrium, eq. (1), is effective; the presence of successive equilibria, with different values of n, can lead to interesting effects, but these will be discussed elsewhere. It can be shown that contributions from the self-association of small enough amounts of component B can be completely ignored. Thus we consider in detail here only systems for which the assumptions

$$K_{B,m} = K_B = 0 \tag{4a}$$

$$K_{rp} = 0 \tag{4b}$$

are valid and for which there exists only a single value of n for the self-association of component A.

Unless otherwise specified, the equilibrium constants will be the molar association constants defined in equations (1) and (2) for the ideal self-association of the two components. The weight concentrations corresponding to the $m_{i,j}$ are the $c_{i,j}$ and are given in g/L. The heterogeneity is specified by the parameter θ ($=c_{B,T}/c_{A,T}$), the ratio of the total amount of component B to the total amount of component A in the solution being examined.

It is instructive to examine the error introduced into estimates of the standard free energy of association from the presence of a nonassociating component of the same size as the associating monomer. Figure 1 presents the differences between the values of $\Delta F°/RT$ ($= -\ln \hat{K}$) calculated from $\hat{K}$, the apparent association constant that would be observed for such a heterogeneous

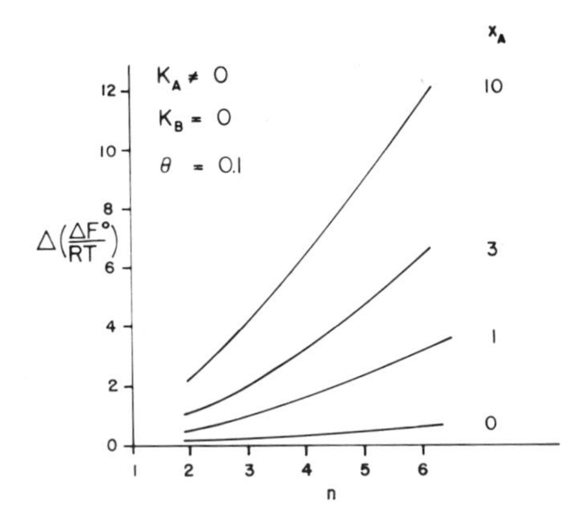

Fig. 1. Differences between true and apparent values of the standard free energy of association calculated from the corresponding association constants for systems with 10% of a non-associating component B, as a function of n, the degree of association, and for several extents of association, as measured by x_a, the molar ratio of n-mer to monomer for the self-associating component A. The values presented as $\Delta(\Delta F°/RT)$ are equal to $-\ln(\hat{K}/K_A)$, where $\hat{K}$ is the apparent association constant estimated for the total system and K_A is the true association constant of component A.

system, and the values of $\Delta F°/RT$ (= $-\ln K_A$), calculated for K_A, the actual association constant of the main component. These differences are presented for associations ranging from monomer-dimer to monomer-hexamer, for varying extents of association as specified by x_A (= $m_{A,n}/m_{A,1}$), the ratio of polymer to monomer molarity for component A, and for heterogeneity specified by $\theta = 0.1$. It is obvious that a) rather large errors in $\Delta F°$ can be introduced by this modest heterogeneity, b) these errors in $\Delta F°$ increase with increasing degree of polymerization, n, and c) these errors become much larger with higher extents of association.

Clearly it is essential to have procedures available to detect and, if possible, quantitate such heterogeneity so as to be able to avoid potential pitfalls associated with this effect. We start by noting that the observed variation of apparent association constant with extent of reaction (Figure 1) may provide one such procedure to detect heterogeneity. Accordingly, we next derive expressions for the effects of heterogeneity on the apparent association constant.

EFFECTS OF HETEROGENEITY ON APPARENT ASSOCIATION CONSTANTS

The apparent association constant, $\hat{K}$, is defined as

$$\hat{K} = \frac{m_{T,n}}{m_{T,1}^{n}} = \frac{m_{A,n} + m_{B,n}}{(m_{A,1} + m_{B,1})^{n}} = \frac{K_A m_{A,1}^{n}}{(m_{A,1} + m_{B,1})^{n}} = \frac{m_{A,1} x_A}{(m_{A,1} + m_{B,1})^{n}} \qquad (5)$$

where $m_{T,1}$ is the total molarity of all j-mers and where we have substituted relations (1) and (2) for the equilibria, made assumptions (4), and introduced the symbol x_i for the molar ratio of the n-mer to monomer for component i. We note that m_i^o, the total molarity of the monomers of component i, is given by

$$m_A^o = m_{A,1} + nm_{A,n} = m_{A,1}(1 + nx_A) \qquad (6a)$$

$$m_B^o = \theta m_A^o = m_{B,1} \qquad (6b)$$

where the last relation of eq.(6b) is valid only for $K_B = 0$. Substituting the values of the $m_{i,1}$ from eqs.(6) into eq.(5) and eq.(1) we obtain

$$\hat{K} = \frac{x_A m_{A,1}}{[m_A^o(1 + \theta) - nm_{A,1}x_A]^n} \qquad (7a)$$

$$K_A = \frac{x_A m_{A,1}}{[m_A^o - nm_{A,1}x_Z]^n} \qquad (7b)$$

Dividing eq.(7b) by eq.(7a), we obtain

$$\frac{K_a}{\hat{K}} = \frac{m_A^o - nm_{A,1}x_A + \theta m_A^o}{m^o A - nm_{A,1}x_A}^n = [1 + \frac{\theta m_A^o}{m_A^o - nm_{A,1}x_A}]^n \qquad (8)$$

Noting from eq.(6a) and the definition of x_A that

$$m_{A,1} = m_A^o/(1 + nx_A) \qquad (9)$$

we then obtain the relation

$$\ln(\hat{K}/K_a) = -n \ln (1 + \theta<1 + nx_A>) \qquad (10a)$$

as generally applicable under assumptions (4). For θ sufficiently small we expand eq.(10) to obtain

$$\ln (\hat{K}/K_A) \simeq -n\theta (1 + nx_A) \qquad (10b)$$

Equations (10) predict the increasing differences between $\ln \hat{K}$ and $\ln K_A$ with increasing n and increasing extent of association illustrated in Figure 1.

The variations in apparent association constant with extent of association could, in principle, be used to infer association heterogeneity in a preparation. One simple method would be to compare the shapes of the curves for an apparent molecular weight average, say the apparent weight average molecular weight, of the sample as a function of concentration. However, such inferences would only be valid if there were no other source for variation of the apparent association constants (or, equivalently, of departure of the curve of the apparent molecular weight vs concentration from the curve for a homogeneous simple system) than heterogeneity. Unfortunately, one rarely has such assurances. Most systems exhibit the effects of some nonideality, which, at least qualitatively, can generate somewhat similar variations in $\hat{K}$ (or in shapes of curves) as generated by heterogeneity. It is fairly common to find sequential self-association reactions, with increasing values of n. The variation of the apparent association constants observed for such complex systems readily may mask the contrary effects of modest amounts of heterogeneity. Even in the absence of such common complications, the sensitivity of detection of heterogeneity from variations in shape of the molecular weight curves (or from variations in $\ln \hat{K}$ with association) may be fairly low. For example, the curves for $M_w(c)$ vs c for a monomer dimer system with $\theta = 0.1$ may be fit, for $K_Ac < 1$, rather well (with maximum deviations in $M_w(c)$ less than about 1%), assuming a single constant value of the association constant ($K \simeq 0.768$). Since experimental error is generally larger than these deviations, the heterogeneity would be virtually invisible to this approach. A more general procedure is needed, preferably one that does not depend on the assumption of the complete correct model for the association of the main component of the system.

DETECTION OF HETEROGENEITY BY EQUILIBRIUM SEDIMENTATION

Such a procedure was suggested by Squire and Li[2] for sedimentation equilibrium experiments. It is based on the phase rule; for a single-phase system with two chemical components, the solvent and the self-associating solute, there are three degrees of freedom. At constant temperature and pressure the behavior of the system is completely determined by the solute concentration and all the properties of the system must be a function of concentration only. In practice most systems fulfill the requirement of two components at constant pressure by approximation: Although the solvent component itself usually consists of more than one component - for biochemical systems typically this would be water and one or more buffer components - the solvent may be considered as a single component in the thermodynamic sense if its composition is effectively invariant over the conditions of interest; similarily if the volume change on associations is small, as it usually is, then the requirement of constant pressure may be relaxed. (If the volume change is known not to be negligible, then the effects of pressure may be compensated for by calculation).

Since all properties of the system depend solely on the solute concentration, the various point average molecular weight moments, such as $M_w(r)$ and $M_z(r)$, the weight and z-average molecular weights at radius r, must be functions only of c(r), the concentration at the point r. Accordingly for solutes that consist of a single thermodynamic component the curves of $M_k(r)$ vs. c(r) for different loading (initial) concentrations, $c°$, must be exactly superposable. Conversely, divergence of these curves implies that the solute consists of more than one component and thus must be heterogeneous. By similar reasoning one may compare the curves of $M_w(r)$ vs. c(r) for the same loading concentration but observed for sedimentation equilibrium at differing angular velocities, again with lack of coincidence indicating the presence of more than one thermodynamic component. Generally the procedure of examining different loading concentrations in the same rotor and at the same time would be preferred since it tends to minimize extraneous variations in experimental conditions, such as differences in speed, in temperature, and in history between different experiments.

A calculated example of this procedure to detect heterogeneity in associating systems is furnished by Figure 2. The several curves of apparent weight average molecular weight shown have been calculated for the indicated loading concentrations, assuming a heterogeneous associating system with two solute components: main component A, which dimerizes reversibly with $K_A = 1$, and component B, present at the level of 1/10 that of component A, and which is incapable of any association. The moderate heterogeneity of this system is obvious.

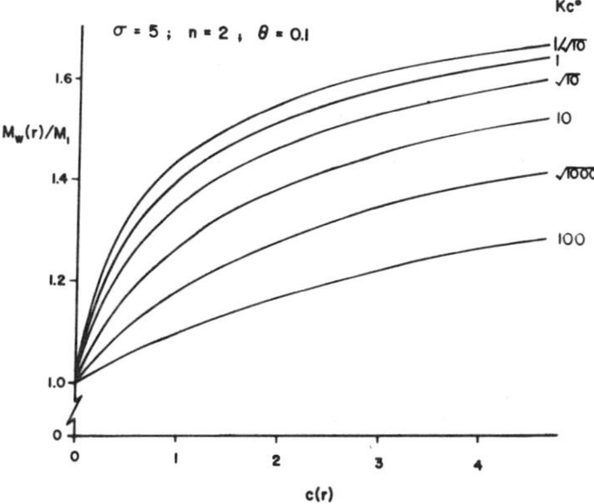

Fig. 2. Apparent weight-average molecular weights, at each point, $M_w(r)$, presented as a function of the concentration at that point, c(r), calculated for several loading concentrations, c°, for a monomer-dimer system with $K_a = 1$, $\sigma_1 = 5\ cm^{-2}$, $\Delta\xi = 1.965\ cm^2$ and with $\theta = 0.1$, corresponding to the presence of about 9% of "incompetent monomer". The lack of overlap of these curves is evidence that the solute consists of more than one thermodynamic component, provided that there is no pressure dependence of the association and provided that the solvent can be represented as a single thermodynamic component.

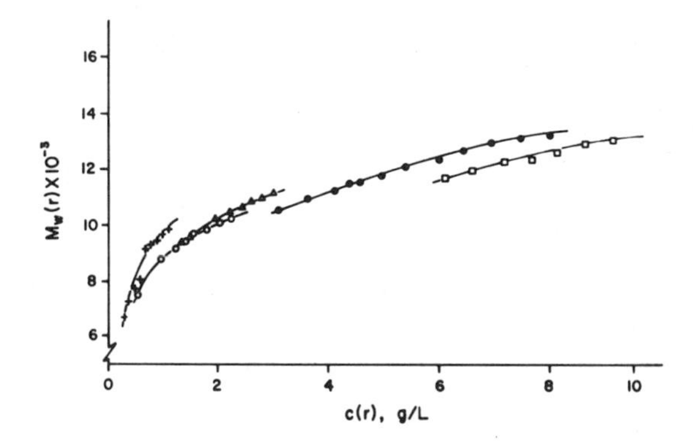

Fig. 3. Apparent weight average molecular weights, $M_w(r)$, as a function of c(r) observed by Jeffrey and Coates[3] for insulin at pH 2.6, 0.1 ionic strength at 25°C. The divergence of the curves for different loading concentrations indicates heterogeneity in association behavior.

An illustration of the application of this procedure to data from the literature is given in Figure 3. Here we present data from Jeffrey and Coates[3] for the self-association of insulin at pH 2.6, replotted to distinguish data corresponding to the different loading concentrations and the different equilibrium speeds. The lack of coincidence of the several curves indicates significant association heterogeneity.

The detection of association heterogeneity in β-lactoglobulin A by this procedure and the subsequent partial resolution of the protein into nearly homogeneous fractions with differing self-association constants has been described by Roark and Yphantis[4]. Numerous other applications of this procedure can be found in the literature.

A useful variant of the procedure of comparing curves of an apparent molecular weight moment, $M_k(r)$, vs the observation concentrations, $c(r)$, for different loading concentrations is the replacement of the observation concentrations by their corresponding concentration gradients, $\frac{dc}{d\xi}(r)$. These concentration gradients are given, for two component systems, by

$$\frac{dc}{d\xi}(r) = \frac{1}{r}\frac{dc}{dr} = \frac{c(r)\, M_w(r)\, (1-\bar{v}\rho)\, \omega^2}{RT \left\langle 1 + c \frac{\partial \ln y}{\partial \ln c} \right\rangle_{P,T}} = c(r)\, \sigma_w(r) \qquad (11)$$

where $\xi = r^2/2$, $\bar{v}$ is the partial specific volume of the solute, ρ is the solvent density, ω the angular velocity, R the gas constant, T the absolute temperature, y the activity of the solute on the c-scale, P the pressure and $\sigma_w(r)$ is the effective reduced weight average molecular weight of the solute. It is obvious from eq.(11) that for homogeneous systems $\frac{dc}{d\xi}(r)$ is a function of $c(r)$ alone, provided only that variations in $\bar{v}\rho$ may be neglected, since both $M_w(r)$ and y are functions of concentration alone. This variant is useful when the concentrations $c(r)$ are not explicitly and absolutely known, for example when the assumption of mass conservation throughout the ultracentrifuge cell may not be reliable because of solute adsorption, precipitation, etc. An example of the application of this variant may be found in studies of the self-association of tyrocidine-B[5].

The several curves of Figure 2 correspond to differing values of the apparent dimerization constants, K_2, ranging from 0.664 at the lowest loading concentration to 0.060 at the highest c° presented. This variation of apparent association constant with loading concentration appears to be useful as a sensitive indicator of heterogeneity in self-associating systems. The degree of sensitivity depends on two factors: the uncertainty in the estimated values of the association constant, $\hat{K}$, obtained from the sedimentation equilibrium experiments and the magnitude of the change in the apparent equilibrium constants with variations in loading concentrations.

The intrinsic variation in $\hat{K}$ with loading concentration depends on the properties of the associating system under study and on the product $\sigma_1\Delta\xi = (\xi_b-\xi_a)\sigma_1$, where ξ_a and ξ_b are the values of $r^2/2$ at the meniscus and at the base of the solution column, and where σ_1 is the effective reduced molecular weight $(=\omega^2 s/D=\omega^2 M(1-\bar{v}\rho)/RT)$-where s and D are the sedimentation and diffusion coefficient of the ideal monomer. In addition to these factors, the uncertainties in the $\hat{K}$ critically depend on the details of the experimental procedures used to estimate the $\hat{K}$. The parameters and conditions used in our estimates of these uncertainties naturally mirror our experimental techniques:

The ultracentrifuge cells[6] used in our experiments provide six channels for the observation of three solution-solvent pairs. This facilitates the simultaneous examination of different loading concentrations in the same rotor at the same time, speed and temperature. These cells may be cleaned without disassembly and can provide reproducible patterns even for speeds above 50,000 rpm. The high pressure mercury-arc lamp usually used as interferometer light source has been replaced by a pulsed argon-ion laser[7,8,9] fired in phase with the ultracentrifuge rotor. A minicomputer system controls synchronization of the laser with the rotor so as to illuminate the appropriate cell or counter-weight at the designated time intervals with the selected number of flashes, and advances the motorized film transport. The resultant Rayleigh interferograms have been measured by one of three automatic image measurement systems: The first system, patterned after that of DeRosier, Munk and Cox[10], scans the images mechanically in a raster pattern and records relative light intensities as a function of position; fringe shifts are then calculated by Fourier transforms from these recorded intensities. Unfortunately our mechanical raster scanning system is time consuming, requiring almost 90 minutes for the scan of an interferogram at 300 radial points within one solution channel. Accordingly, this system has been replaced by a television based system[11] that utilizes the T.V. raster scan to provide the "y-scan" across the interference fringes. This speeded up the data acquisition by nearly two orders of magnitude. Unfortunately this T.V. system exhibited only limited resolution at moderately high gradients and has recently been replaced, in turn, by our currently operational system. This third system utilized a 1024-element self-scanned photodiode array to perform the fringe scan[12]. It has been set up using a modified "real time" Walsh transform to obtain the phase of the fringes. This system measures, calculates and records data from two radial points per second with a reproducibility of better than 10^{-3} fringe on appropriate photographic exposures.

The overall reproducibility of our measurement system, for the net fringe displacements of the solutions corrected for blank distortions, ranges from 1 to 3 μm fringe displacement ($0.3 - 1 \times 10^{-2}$ fringe). For our discussions we assume the value of 3 μm, since such reproducibility can be obtained by careful manual measurements on interferograms obtained with the usual AH-6 mercury arc light source[6,10,13].

These measured fringe displacements are then used to characterize the system of interest through one or more of our data analysis programs: BIOSPIN, ORTHO and NONLIN. All these programs accept data in the form of fringe displacements as a function of comparator x-coordinates, for both the run itself and its corresponding blank, obtained by running water vs water in the same cell, without disassembly, before and/or after the run itself. The blank is subtracted, after appropriate interpolation, and the blank corrected data further processed. BIOSPIN[2] calculates several point average apparent molecular weight averages, such as the number, weight, z and (z+1)-averages. ORTHO[14] requires the user to supply a value of σ_1 for the monomer, and a set of m integers, n_i, to specify the various n-mers thought to be present in the system being studied. This program then fits the observed fringe displacements, y_i, to the form

$$y_j = A_o + \sum_{i=i}^{m} A_i \exp\left(n_i\sigma_1[\xi_j - \xi_1]\right) \tag{12}$$

where A_o is the program's estimate of the fringe offset at the first data point and the A_i are the concentrations in terms of fringe displacement at the first data point (where $\xi = \xi_1$) that are estimated for the various n_i-mers. The various K_i are then obtained from these A_i by applying the appropriate formulation for the equilibrium constant desired. As the name of the program implies, the actual fitting is by means of orthogonal functions generated from the set of the exp $(n_1\sigma_1[\xi_j - \xi_1])$ over all the data points ξ_j by a Gram-Schmidt orthonormalization[15]. It should be noted that nowhere is the questionable assumption of mass conservation made. The program NONLIN[16,17] fits observed fringe displacements, y_i, to the more general form

$$y_j = A_o + \sum_{i=i}^{m} A_i \exp\left(\lambda_i[\sigma_1\langle\xi_j - \xi_1\rangle - \sum_{k=1}^{4} \frac{k+1}{k} B_k c^k(\xi_j)]\right) \tag{13}$$

where the λ_i, the analogs of the n_i, need not be integers. The value of σ_1, the effective reduced molecular weight of the monomer, is not required here, but can be estimated, along with the λ_i, by the program. In addition the program can be used to estimate values of the several virial coefficients, B_k. These estimations may be carried out for single channels of data or for several channels of data analyzed jointly. Applications of this program to nonideal and/or associating systems have been reported[13,18,19,20].

We now wish to estimate the uncertainty in the association constants. With all measurement systems the maximum practical gradient is limited to concentration gradients, $(dc/d\xi)$, below about 15 mm fringe displacement/cm^2, in 12 mm thick centerpieces, by the effects of Wiener skewing[21]. This limitation to a definite maximum gradient that can be measured reliably in an equilibrium experiment must be taken into account in any examination of the reproducibility of determining the values of the K.

The introduction of this restriction proceeds most straightforwardly using the formulation associated with ORTHO to propagate errors and to keep track of analyses in terms of a linear system of equations. For such analyses one usually knows (or can determine) the value of σ_1. One then has the minimum number of parameters to determine; one more than the number of molecular species analyzed for. For a linear system it can be shown that the root mean square (rms) uncertainty in the $\ln \hat{K}$, which will be denoted by $\langle \ln K \rangle$, is proportional to $\langle y \rangle$, the rms differences between the observed fringe displacements and their calculated values--or standard deviation of the y_i from the fitting. The program ORTHO then readily provides the value of $\langle \ln K \rangle / \langle y \rangle$, the ratio of the effective standard deviation in $\ln \hat{K}$ to that of y, for various specific experimental arrangements, including the number and disposition of data points, the number and type of molecular species for which analysis is to be carried out, and the value of $\sigma_1 \Delta\xi$--from the relation

$$\frac{\langle \ln K_n \rangle}{\langle y \rangle} = \left(\frac{\beta_n}{K_n m_1^n}\right)^2 + \left(\frac{\beta_1}{m_1}\right)^2 \qquad (14)$$

where the $\beta_n = \partial \langle A_i \rangle / \partial \langle y \rangle$ are coefficients available from the program and the value of m_1 is the monomer molarity at the first data point, ξ_1. Values of m_1 and K_n to minimize the value of $\langle \ln K \rangle / \langle y \rangle$ subject to the imposition of a maximum gradient $(dc/d\xi)_{max}$ at the base of the solution column may be chosen using undetermined LaGrangian multipliers. The resulting values of m_1 and K_n then specify conditions for the best possible estimate of K_n that can be obtained for the particular experimental arrangement considered and determine the magnitude of $\langle \ln K \rangle$ for any $\langle y \rangle$. Figure 4 presents such values of the uncertainty in $\ln \hat{K}_2$ for a typical experimental arrangement: 101 equispaced data points in a 3 mm high solution column and an rms fitting error, $\langle y \rangle$, of 3 μm. The errors in $\ln \hat{K}_2$ are presented on a logarithmic scale, as a function of σ_1 over the experimentally useful range from 0.1 to 10 cm^{-2}, for the estimation of a monomer-dimer equilibrium alone and, using formulations similar to eq.(14), for the simultaneous estimation of dimerization and trimerization equilibria, etc., up to the simultaneous determination of equilibria among monomer, dimer, trimer, tetramer and hexamer. It can be seen that the values of $\ln \hat{K}_2$ can be obtained

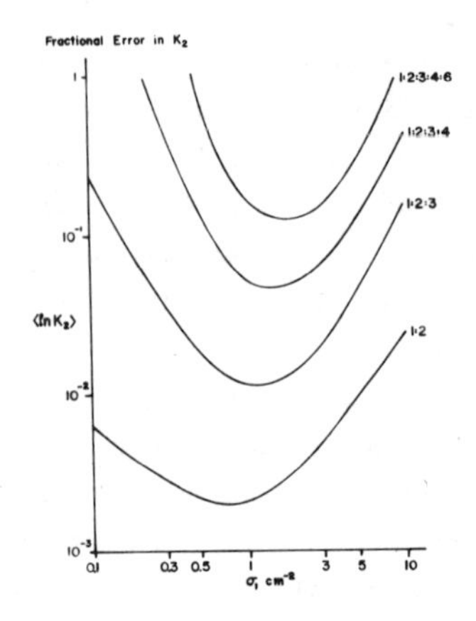

Fig. 4. Estimated minimal fractional uncertainty in apparent dimerization constants originating from an rms fitting deviation of 3 μm in fringe displacement, presented as a function of σ_1, the effective reduced molecular weight of the monomer. The calculations assumed that the fringe displacements used as data were obtained at 101 (=N) equispaced radial points ranging from 6.4 to 6.7 cm and that these data were subject to the restraint that the maximum concentration gradient, $dc/d\xi=(dc/dr)/r$ = 15 mm fringe displacement/cm^2. The several curves correspond to the estimated uncertainties when the system was analyzed as containing the indicated species, e.g. the top curve corresponds to a system analyzed for the simultaneous presence of monomer, dimer, trimer, tetramer and hexamer. These estimated uncertainties are minimal values, corresponding to optimal values of c° and of the several values of the association constants, K_i. They represent the random errors associated with the best possible estimates of K_2 under the specified conditions for σ_1, N, $\Delta\xi$, and $(dc/d\xi)_{max}$.

with an uncertainty of ~3 x 10^{-3} for σ_1 ~1, increasing gradually with either increasing or decreasing σ_1, and strongly increasing, by a factor of about 4, for each additional component being simultaneously determined. The behavior for a monomer-tetramer equilibrium is quite similar. The main differences are in the locations of the minima which are shifted to lower values of σ_1, with relatively minor quantitative differences between the actual values of $\langle \ln K_4 \rangle$ and of $\langle \ln K_2 \rangle$.

If the concentration gradient at the base of the solution column is made smaller than the limiting value of 15 mm fringe displacement/cm^2 that was used here, then the estimates of K_2 become less well-defined, with larger values of $\langle \ln K_2 \rangle$. Similarly if the values of K_2 differ from the optimal values for which Figure 4 was drawn, then the values $\langle \ln K_2 \rangle$ also increase. However, the rate of increase of $\langle \ln K_2 \rangle$ with changes in K_2 is relatively slow, as shown in Figure 5 which presents values of $\langle \ln K_2 \rangle$ for typical values of σ_1 = 1, 3, and 5 cm^{-2}, in the most useful range of σ_1, as a function of $\log_{10} K_2$. It can be seen that the minima are relatively broad and that there is a moderate latitude in the values of K_2 since $\langle \ln K_2 \rangle$ is no more than twice its minimum value for about a 100-fold range in K_2. A similar graph for the equivalent monomer-tetramer system shows

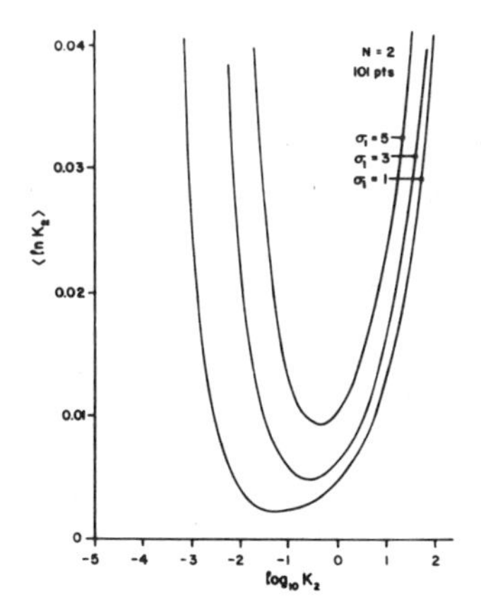

Fig. 5. Estimated minimal fractional uncertainties in the apparent dimerization constants for systems analyzed as consisting only of monomer and dimer presented as a function of ln K_2. Procedure and restrictions are the same as in Figure 4, except for values of K_2.

slightly greater latitude in K_4, with about a 200-fold range K_4 for a doubling of the minimum uncertainty in K_4.

The uncertainties in the estimated association constants show almost the expected proportionality to the reciprocal of the square root of N, the number of points: over the range from 31 to 301 equi-spaced data points the values of $\langle \ln K_2 \rangle$ are proportional to $N^{-0.48}$.

These simple estimations appear to be usefully applicable to estimates of the uncertainties of the association constants from NONLIN. Table I presents comparisons of the average estimated values of K_2 obtained from NONLIN, keeping the value of σ_1 fixed at the appropriate value, for various σ_1. The data to be

TABLE I

Comparison of observed variations in ln $\hat{K}_2$ for different noise perturbations with the predicted variations. All calculations are for monomer-dimer systems with K_2 = 1.

σ_1 cm^{-2}	$\frac{dc}{d\xi}$ max	$\hat{K}/K_A$	$\langle \ln K_2 \rangle_{obs}$	$\langle \ln K_2 \rangle_{pred}$
0.5	1.38	0.996	0.041	0.031
1	4.7	1.005	0.016	0.010
1.5	15	1.008_8	0.003_6	0.004_1
3	10	0.999_1	0.007_2	0.008_4
3	2.5	0.996	0.037	0.030
5	8.3	0.991	0.015	0.017
5	15	1.005	0.012	0.011
10	12	1.002	0.049	0.036

fitted was calculated for the indicated ideal dimerization. These "perfect" data were then modulated by adding to the data a set of Gaussian random numbers, with mean of zero and standard deviation of 3 μm, so as to simulate typical experimental error. This "noisy" simulated data was then fitted by NONLIN, and the procedure repeated five to ten times for each case shown. The mean of the values of the estimated values of ln K_2 is compared to the true value of ln K_2 (=ln K_A) in the third column of the table and the standard deviation of these values from their mean is presented in the fourth column. The corresponding estimations of $\langle \ln K_2 \rangle$ were then carried out for the number of data points and for the maximum gradient present and measurable in the simulated solution column (any points with $\frac{dc}{d\xi}$ >15 mm fringe displacement/cm^2 were discarded and not used in the fitting procedures.) These "predicted" estimates in column 5 compare reasonably with the observed uncertainties in ln K_2. Similar comparisons for other monomer-oligomer systems were likewise successful in predicting the observed uncertainties. Accordingly we use these simple estimations of the uncertainties in $\langle \ln K_A \rangle$ for examining sensitivity to association heterogeneity.

The expressions already developed for the effect of heterogeneity on apparent association constants, eqs.(10), are only valid for the ultracentrifuge under conditions where there is no sensible redistribution of solute, i.e. for small values of $n\sigma_1\Delta\xi$, as for equilibrium in short columns at low speeds or for experiments at very low speeds. Other expressions are required for more general conditions. These may be obtained as follows: starting with the formal statement of mass conservation for each component within a sectoral solution column, and invoking the definition of $\sigma_n(r)$, the apparent number average molecular weight of that component we write

$$c_k{}^\circ\Delta\xi = \int_a^b c_j{}^\circ d\xi = \int_a^b c_j(\xi)d\xi = \left.\frac{c_j(\xi)}{\sigma_{n,i}(\xi)}\right|_a^b \tag{15a}$$

substituting for the c_j and $\sigma_{n,i}$ we obtain

$$c_j{}^\circ\Delta\xi = M_1\left[\frac{m_{j,i} + nM_{j,n}}{\sigma_1\left(\dfrac{m_{j,1} + nM_{j,n}}{m_{j,1} + m_{j,n}}\right)} - \frac{\lambda m_{j,1} + n\lambda^n m_{j,n}}{\sigma_1\left(\dfrac{\lambda m_{j,1} + n\lambda^n m_{j,n}}{\lambda m_{j,1} + \lambda^n m_{j,n}}\right)}\right]$$

$$= M_1/\sigma_1\ \{m_{j,1}\ (1 + x_{j,n}) - \lambda m_{j,1}(1 + x_{j,n}\lambda^{n-1})\} \tag{15b}$$

where M_1 is the molecular weight of the monomer, $m_{j,i}$ is the molarity of the i-mer of the j-th component at the base of the solution channel, λ is exp $(-\sigma_1\Delta\xi)$, the ratio of the monomer concentration at the meniscus to the monomer concentration at the base of the solution column, and $x_{j,n}$ is $m_{j,n}/m_{j,i}$, the ratio of the n-mer to the monomer molarity at the base for the j-th component. Substituting the parameter β for $(1-\lambda^n)/(1-\lambda)$ this becomes

$$c_j{}^{\circ}\Delta\xi = \frac{m_{j,1} M_1 (1-\bar{v})}{\sigma_1} (1 + \beta x_{j,n}) \tag{15c}$$

With the substitution $\gamma_j = \sigma_1 \Delta\xi\, c_j{}^{\circ}/[M_1(1-\lambda)]$ we csn express the monomer molarity of component j as

$$m_{j,1} = \gamma_j/(1 + \beta x_{j,n}) \tag{16}$$

and this expression, in turn, can be used to obtain the value of the apparent association constant for a heterogeneous system with components A and B as

$$\hat{K} = \frac{m_{T,n}}{m_{T,1}} = \frac{m_{A,n} + m_{B,n}}{(m_{A,1} + m_{B,1})^n} = \frac{K_A m_{A,1}^n + K_B m_{B,1}^n}{(m_{A,1} + m_{B,1})^n} = \frac{K_A + K_B(m_{B,1}/m_{A,1})^n}{(1 + m_{B,1}/m_{A,1})^n} \tag{17}$$

Dividing by K_A and introducing eq.(16) for the $m_{j,1}$ we obtain

$$\hat{K}/K_A = \frac{1 + (K_B/K_A)(\gamma_B/\gamma_A)^n[(1 + \beta x_{A,n})/(1 + \beta x_{B,n})]^n}{[1 + (\gamma_B/\gamma_A)(1 + \beta x_{A,n})/(1 + \beta x_{B,n})]^n} \tag{18a}$$

When $\theta = \gamma_B/\gamma_A = c_B^{\circ}/c_A^{\circ}$ is sufficiently small, this expression can be approximated by

$$\hat{K}/K_A = 1/(1 + \theta\langle 1 + \beta x_{A,n}\rangle)^n \tag{18b}$$

or

$$\ln(\hat{K}/K_A) = -n \ln(1 + \theta\langle 1 + \beta x_{A,n}\rangle) \simeq -n\,\theta\,(1 + \beta x_{A,n}) \tag{19a}$$

with the relation on the right valid in the limit as $\theta \rightarrow 0$. When $n\sigma_1\Delta\xi \ll 1$, i.e. for sufficiently low speed and/or short enough solution columns, β approaches n and $x_{A,n}$ approaches $x_{A,n}^{\circ}$ the molar ratio of n-mer to nonomer in the original solution and

$$\ln(\hat{K}/K_A) = -n \ln(1 + \theta\langle 1 + n x_{A,n}^{\circ}\rangle) \simeq -n\,\theta\,(1 + n x_{A,n}^{\circ}), \tag{19b}$$

expressions identical to eqs.(10). On the other hand, for high speed equilibrium conditions with $\sigma_1 \Delta\xi >> 1$, the value of β approaches unity and eq.(19a) becomes

$$\ln(\hat{K}/K_A) = -n \ln(1 + \theta < 1 + x_{A,n} >) \simeq -n\,\theta\,(1 + x_{A,n}) \tag{19c}$$

Expressions (19) provide simple direct estimates of $\ln \hat{K}/K_A$ for heterogeneous systems with sufficiently small amounts of heterogeneity. More generally eq.(17) may always be used if the values of $m_{A,1}$ and $m_{B,1}$ are known. The required values of $m_{A,1}$ and $m_{B,1}$ may be estimated by successive approximations for any given system. Figure 6a presents values of $\ln(\hat{K}/K_A)$ for monomer-dimer systems with $K_B=0$ as a function of θ, the relative amount of the contaminating non-associating component B. In this particular figure the value of σ_1 was taken as 3 cm^{-2}, a value corresponding to intermediate speeds, as recommended by Teller _et al_[1] for studies of self-associating systems, and the value of $\Delta\xi$ = 2 cm^2, approximately the value corresponding to a 3 mm high solution column near the center of the ultracentrifuge cell. The several curves correspond to different values of $K_A c°$ and therefore various loading concentrations. The ranges indicated on the lower left hand corner of the figure correspond to ranges of four times the expected standard deviation (± 2 standard deviations) in $\ln \hat{K}$. The different ranges correspond to different values of K_2 (over a 10^4 range) for the associating component, as indicated by the label $\log_{10} K_2$. From this figure it appears possible, at least in favorable cases, to detect considerably less than 1% heterogeneity using experiments with a 3- or 10-fold range of initial concentrations.

a b

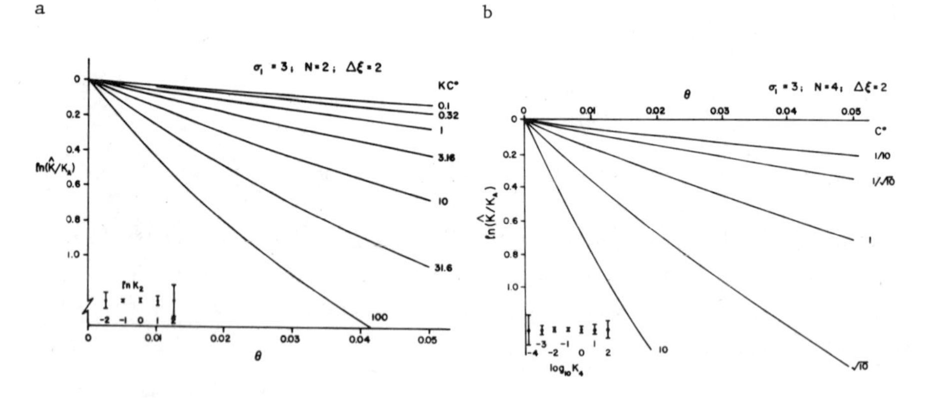

Figs. 6. Values of apparent dimerization (panel a) and apparent tetramenization (panel b) constant $\hat{K}$, presented in the form $\ln(\hat{K}/K_A)$, as a function of θ for a system with K_A = 1, K_B = 0, σ = 3 cm^{-2} and $\Delta\xi$ = 2 cm^2. The several curves shown correspond to different initial (loading) concentrations. The ranges indicated on the lower left hand corner are 4 times the predicted optimal error in $\ln \hat{K}$ (± 2 standard deviations) for data sets with 101 equispaced points and 3 μm rms noise.

Figure 6b presents the same type of information for the equivalent monomer-tetramer system. The sensitivity to heterogeneity appears to be significantly greater than for the monomer-dimer case, over a considerably larger range in K_4 (a range of about 10^6).

The sensitivity to heterogeneity has been tested using calculated concentration distributions for several heterogeneous systems and the program NONLIN to analyze these distributions after they had been modulated by the addition of sets of Gaussian random numbers of known variance so as to simulate typical random errors. Five to ten sets of such noisy simulated data were analyzed for each distribution. The uncertainties of the derived parameters were taken as the standard deviations of the derived parameters from the mean of their values. Two such examples are given in Figures 7 for monomer-dimer equilibria, systems for which it is most difficult to detect association heterogeneity. Values of the K are presented as a function of θ for three loading concentrations and for high speed conditions, $\sigma_1 = 5\ \text{cm}^{-2}$, in Fig (7a) and for moderate speeds, $\sigma_1 = 1.5\ \text{cm}^{-2}$, in Fig. (7b). The radii of the circles correspond to the standard deviation of the values of $\hat{K}$. From Figure (7a) we can infer that a 2% contamination by "incompetent monomer" would barely be detectable under these high speed conditions. Comparison with Figure (7b) shows that considerably more sensitivity in detecting heterogeneity would be expected at moderate speeds, largely because of the increased precision in obtaining values of $\hat{K}$.

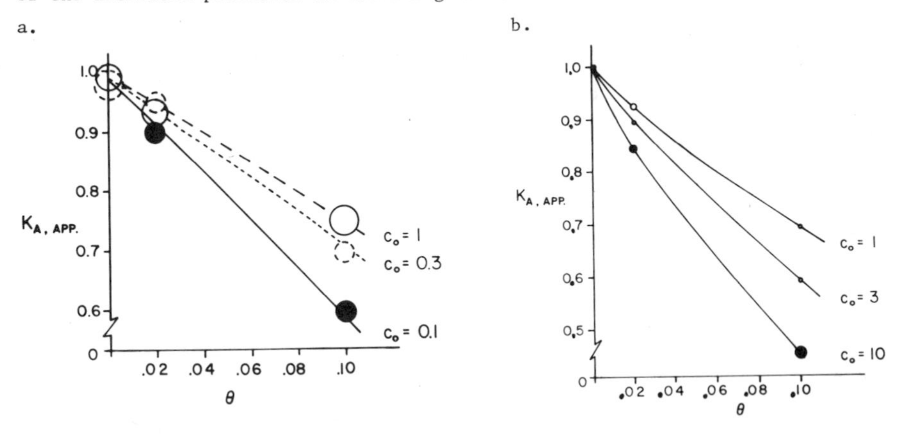

Figs. 7. Illustration of the detection of heterogeneity with simulated data. Concentration distributions were calculated at 100 equispaced points from r = 6.4 to 6.7 cm for $K_2 = 1$ and the indicated values of c°. All data points corresponding to concentration gradients, $dc/d\xi$, greater than 15 mm fringe displacement/cm^2 were discarded and the remaining data modulated by random numbers with a variance of 3 μm to simulate random experimental error. Five to ten of these "noisy" simulated data sets were individually fitted by NONLIN to the appropriate model. The mean and standard deviation from the mean of the estimated values of K_2 are presented as the points and their indicated ranges as a function of θ. Panel (a) corresponds to $\sigma_1 = 5\ \text{cm}^{-2}$ and panel (b) to $\sigma_1 = 1.5\ \text{cm}^{-2}$.

The procedure of comparing association constants for equilibrium experiments at different initial loading concentrations as outlined so far depends on the model used to obtain an apparent association constant from the experimental observations. For example, if a homogeneous monomer-dimer system were analyzed as a monomer-trimer system, then the apparent association constant that would be obtained with such an incorrect model would depend strongly on the region of the data analyzed, with the apparent trimerization constant decreasing with higher observation concentrations. Thus even a homogeneous system would give rise to different apparent constants with different loading concentrations unless the concentration ranges of the data used to estimate the apparent ameration constants for the different c° were maintained exactly the same for all estimations. Such truncation of the data sets to be used to a common concentration range insures that the errors from the choice of an incorrect model affect all the estimations of K for the different loading concentrations equally. (Strictly speaking, the distribution of data points for all the experiments at the different loading concentrations should be identical, but this requirement may be relaxed if the point density is sufficiently high.)

This procedure of using a common concentration span for determining the K is strongly urged since one rarely can be absolutely sure that the model of the associating system that is being used to analyze the data is completely correct. The effectiveness of this procedure for analyzing a system for heterogeneity even with an incorrect model, provided only that the data used span a common observation concentration range, is illustrated in Figures 8. The first of these

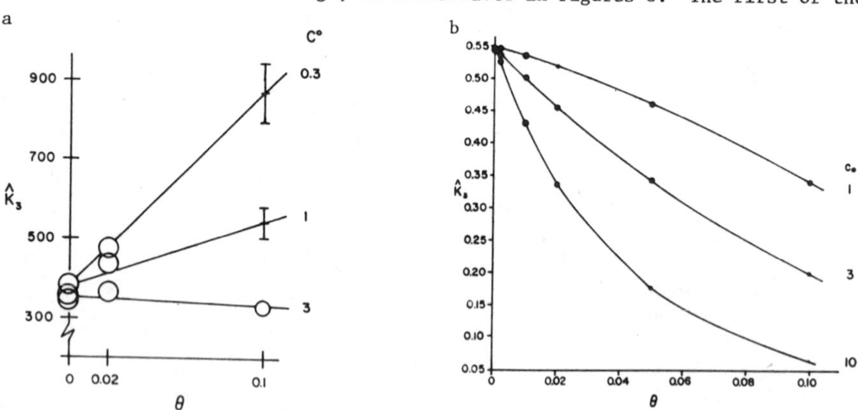

Figs. 8. Analyses of simulated self-associating systems for heterogeneity using incorrect models: a) monomer-dimer system with $K_2 = 1$ and $\sigma_1 = 2\ cm^{-2}$ analyzed as a monomer-trimer system; and b) monomer-dimer-tetramer system with $K_2 = K_4 = 1$ and $\sigma_1 = 2\ cm^{-2}$ analyzed as a monomer-trimer system. In both figures five to ten "noise" perturbed (rms = 3 μm fringe displacement) data sets with about 100 equispaced data points for which $dc/d\xi < 15$ mm fringe displacement/cm^2 were analyzed as described in the text.

(Fig. 8a) presents the results of an analysis of a heterogeneous monomer-dimer system with $\sigma_1 = 2\ cm^{-2}$ in terms of a trimerization reaction -- clearly an incorrect choice of model. The procedure used for "mis-analyzing" this system was to first truncate the data for all three initial concentrations so that this data spanned a common concentration range. Then this truncated data was fit by NONLIN, using the data from all three channels jointly, to determine the "best value" of σ_1 for a monomer-trimer fit of the complete trimmed data set. Each channel of data was then analyzed separately using NONLIN, as a monomer-trimer system with this specific communal "best value" of σ_1 and the resultant K_3 noted. This procedure was repeated for all the initial concentrations, in each case using five different sets of Gaussian random numbers to mimic the effect of random noise (3 μm) on the calculated distributions. Figure 8a presents the values of the apparent trimerization constants thus obtained, with their standard deviation from the mean indicated by the dimensions of the points, as a function of θ, the extent of heterogeneity. Clearly one can distinguish the effects of heterogeneity for $\theta \geq \sim 0.02$, although the sensitivity appeared distinctly less than the nearly comparable case illustrated in Figure 7b for a correctly analyzed monomer-dimer system.

Figure 8b presents a similar analysis for a monomer-dimer-tetramer system with $\sigma_1 = 2\ cm^{-2}$ again analyzed, as a monomer-trimer system. In this case one can clearly detect the presence of less than 2×10^{-3} of a non-associating monomer even though an incorrect model was used for the analysis. Part of the reason for the increased sensitivity in this case is the much better fitting of the monomer-dimer-tetramer data by a trimerization as compared to the monomer-dimer data of Figure 8a: in both cases the perturbation by random numbers to simulate experimental error corresponded to an rms of 3 μms; for the monomer-dimer data the rms of the fit was about 7 μm, while for the monomer-dimer-tetramer data the rms of the fit was about 3 μm, indicating a nearly perfect description of the data by the assumed trimerization.

An application of this procedure to experimental data is illustrated by some of our work with bovine insulin. The sample of insulin used had been chromatographically purified [22,23], first on SP-sephadex in 10M acetic acid and then on QAE-Sephadex at pH 9 in 50% dimethylformamide. This insulin preparation appeared to have less than one part per thousand present of any contaminant, including desamido-insulins, as determined by disc-gel electrophoresis and chromatography. Figure 9 presents the apparent z-average molecular weights, $\underline{M_z}(r)$, observed for two loading concentrations of this insulin at two angular velocities as a function of the normalized concentration gradient $\frac{dc}{d\xi}(r)/\omega^2$. The solvent was 0.1 M NaCl, 0.1 M TRIS and 0.001 M EDTA pH 7 at 25°C, conditions under which insulin shows extensive association[24]. This preparation clearly appears to be heterogeneous in association behavior, as indicated by the divergence

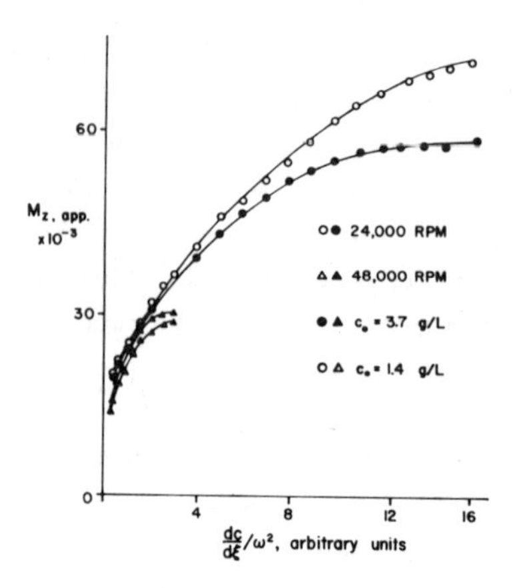

Fig. 9. Apparent z-average molecular weight presented as a function of the normalized concentration gradient for insulin at pH 7 in 0.1 M NaCl, 0.1 M TRIS and 0.001 M EDTA at 25°C.

of the several curves. However, at very low concentrations, where the equilibria appear to involve primarily monomer, dimer and tetramer[23], the apparent molecular weights are sensibly independent of loading concentration or angular velocity. Table II presents values of the apparent dimerization constants determined for insulin from two replicate interferograms under these conditions. These values were obtained by truncating the experimental observations to fringe displacements less than 1 mm in a 30 mm thick centerpiece corresponding to concentrations less than 6.9×10^{-5}M (in monomer units). This truncation spans most of the range where the apparent molecular weights appear to be independent of both concentration

TABLE II

Test of homogeneity of initial association stages of insulin. Experimental conditions: pH 7.0, 0.1 M TRIS, 0.1 M NaCl, 10^{-3} M EDTA, 25°C, 48,000 rpm in 30 mm thick centerpiece.

	Plate G305		Plate G308	
c° g L^{-1}	RMS of fit μm	$\hat{K}_2$* Lg^{-1}	RMS of fit μm	$\hat{K}_2$* Lg^{-1}
0.34	3.0	310	2.8	306
1.01	2.5	277	2.7	291
2.44	4.2	313	4.6	306

*Fit as a monomer-dimer-tetramer system for $\hat{K}_2$, $\xi(c_a)$, c,(a), keeping $\hat{K}_4 = 6.36 \times 10^3\ L^3g^{-3}$, over the concentration range corresponding to 0-1 mm fringe displacement (0-6.9 x 10^{-5}M in monomer equivalent molarity).

and speed and also avoids regions where there is a significant participation by oligomers greater than the tetramer. The model used in fitting the experimental observations was that of a monomer-dimer-tetramer equilibrium with the value of the tetramerization constant fixed at a value consistent with our analyses of this system[23]. The lack of significant variation in $\hat{K}_2$ with loading concentration that is shown in Table II suggests that this insulin sample is indeed homogeneous, within experimental error, in its association properties for this initial region where only the smallest oligomers participate. The heterogeneity at higher concentrations is thus ascribable to the participation of higher degree of polymerization.

It would be useful to obtain quantitative estimates of the heterogeneity actually present in such experiments. As shown in Figure 10, even as little as 1% heterogeneity can generate significant divergence between curves of the apparent molecular weight averages of different loading concentrations if $n = 4$. This divergence between such curves increases strongly for larger n. Clearly it would be useful to have procedures for making quantitative estimates of the extent of heterogeneity from curves of apparent molecular weight averages vs concentration, or from values of $\hat{K}$, obtained with different $c°$. Accordingly we present here, as a first step towards such analyses, some considerations applicable to single ideal systems that can be described by a single value of n.

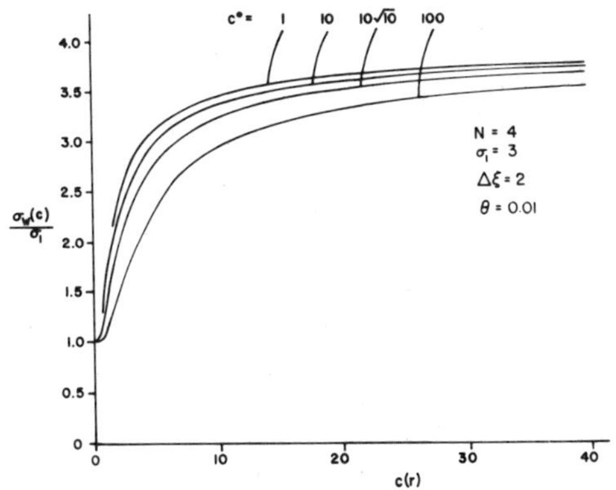

Fig. 10. Apparent reduced weight average molecular weight as a function of observation concentration for several loading concentrations, as indicated. These curves were calculated for a monomer-tetramer system. The effects of the small extent of heterogeneity ($\theta = 0.01$) are clearly discernible, especially at the higher loading concentrations.

First we examine the sensitivity to heterogeneity of the procedure of examining curves of weight average molecular weight, $M_w(r)$, vs $c(r)$ for different loading concentrations. Indications of heterogeneity are provided by the differences between different curves at the same $c(r)$ but for different values of $c°$. Accordingly we wish to evaluate the derivative of the weight average molecular weight with respect to initial concentration, c°, keeping association constant, heterogeneity and observation concentration constant. It is convenient to express the derivative in logarithmic form and to split the desired derivative into two parts according to the chain rule

$$\left(\frac{\partial \ln M_w(r)}{\partial \ln c^\circ}\right)_{K_A,\theta,c(r)} = \left(\frac{\partial \ln M_w(r)}{\partial \ln \hat{K}}\right)_{K_A,\theta,c(r)} \left(\frac{\partial \ln \hat{K}}{\partial \ln c^\circ}\right)_{K_A,\theta,c(r)}$$

$$= \left(\frac{\partial \ln M_w}{\partial \ln \hat{K}}\right)_c \left(\frac{\partial \ln \hat{K}}{\partial \ln c^\circ}\right)_{K_A,\theta} \quad (20)$$

The partial derivative of $\ln M_w(r)$ with respect to $\ln \hat{K}$, at constant K_A, θ, and $c(r)$, is identical to the derivative

$$\left(\frac{\partial \ln M_w}{\partial \ln \hat{K}}\right)_c = \left(\frac{\partial \ln M_w}{\partial \ln \hat{K}}\right)_{m_1} + \left(\frac{\partial \ln M_w}{\partial \ln m_1}\right)_{\hat{K}} \left(\frac{\partial \ln m_1}{\partial \ln \hat{K}}\right)_c \quad (21)$$

where for any ideal association equilibrium the total concentration, c, is given in terms of the monomer concentration, m_1, by

$$c = M_1(m_1 + n\hat{K}m_1{}^n) \quad (22)$$

and where the weight average molecular weight, M_w, is given by

$$M_w = M_1(m_1 + n^2\hat{K}m_1^n)/(m_1 + n\hat{K}m_1^n) \quad (23)$$

Here we consider m_n and m_1 to be the total molar concentration of n-mer and monomer present. The derivatives appearing in eq.(21) may be evaluated from relations (22) and (23) to obtain

$$\left(\frac{\partial \ln M_w}{\partial \ln \hat{K}}\right)_c = \frac{n^2\hat{K}m_1^{n-1}}{1 + n^2\hat{K}m_1^{n-1}} - \frac{n\hat{K}m_1^{n-1}}{1+ n\hat{K}m_1^{n-1}} - \left(\frac{n^2(n-1)\hat{K}m_1^{n-1}}{1 + n\hat{K}m_1^{n-1}} - \frac{n(n-1)\hat{K}m_1^{n-1}}{1 + n\hat{K}m_1^{n-1}}\right) \frac{n\hat{K}m_1^{n-1}}{1 + n^2\hat{K}m_1^{n-1}}$$

$$= \frac{n(n-1)\ \hat{x}(r)}{<1 + n^2\hat{x}(r)>^2} \quad (24)$$

where, as before, we have substituted $\hat{x}(r) = Km_{T,1}^{n-1} = m_{T,n}/m_{T,1}$ for the apparent molar ratio of n-mer to total monomer at the point r. The derivative represented by eq. (24) has the maximum value of $(n-1)/4n$ at $\hat{x}(r) = 1/n^2$. The other derivative needed in eq. (20) may be obtained by noting that the apparent association constant, $\hat{K}$, is a function of K_A, θ and $c°$, and that the total initial loading concentration $c°$ can be considered to be a function of K_A, θ and $m_{A,1}$, the molarity of the monomer at the base of the solution column. Thus we may write

$$\left(\frac{\partial \ln \hat{K}}{\partial \ln c°}\right)_{K_A,\theta} = \left(\frac{\partial \ln \hat{K}}{\partial \ln m_{A,1}}\right)_{K_A,\theta}\left(\frac{\partial \ln m_{A,1}}{\partial \ln c°}\right)_{K_A,\theta} \tag{25}$$

The derivatives on the right may be evaluated by noting that $x_{A,n}$, the molar ratio of n-mer to monomer for component A at the base of the solution column, is equal to $K_A m_A^{n-1}$. Combining this relation with eq. (15c) we obtain

$$\left(\frac{\partial \ln m_{A,1}}{\partial \ln c°}\right)_{K_A,\theta} = \frac{m_{A,1} + \beta K_A m_{A,1}^n}{(m_{A,1} + n\beta K_A m_{A,1}^n)} = \frac{\sigma_1}{\sigma_n(b)} \tag{26}$$

where we have substituted the number average reduced molecular weight at the base of the cell for its explicit defintiion. Similarly we obtain from eq. (19a) the other derivative needed for eq. (25) as

$$\left(\frac{\partial \ln \hat{K}}{\partial \ln m_{A,1}}\right)_{K_A,\theta} = -\frac{\theta\beta n\ (n-1)\ x(b)}{1 + \theta\langle 1 + \beta\ x(b)\rangle} \tag{27}$$

where we write $x(b)$ for $x_{A,n}$, the molar ratio of n-mer to monomer for component A at the base of the solution column, changing notation slightly so as to distinguish this ratio from the x of eq. (24) which refers to the molar ratio of n-mer to monomer corresponding to the observation concentration, $c(r)$. Combining eqs. (20) and (24-27) we obtain the desired result:

$$\left(\frac{\partial \ln M_w(r)}{\partial \ln c°}\right)_{K_A,\theta,c(r)} = = \frac{n^2(n-1)^2\theta\beta\hat{x}(r)\ x(b)\ \sigma_1}{\langle 1+n^2\hat{x}(r)\rangle^2(1+\theta\langle 1+\beta x(b)\rangle)\sigma_n(b)} \tag{28a}$$

Generally we are interested in the detection and quantitation of small extents of heterogeneity. Accordingly we consider the limit

$$\lim_{\theta\to 0}\left(\frac{\partial \ln M_w(r)}{\partial \ln c°}\right)_{c(r)} = -\frac{\theta n^2(n-1)^2\beta\hat{x}(r)\ x(b)\ \sigma_1}{\langle 1+n^2\hat{x}(r)\rangle^2\ \sigma_n(b)} \tag{28b}$$

This expression for small θ may be further simplified for high speed equilibrium experiments when $\sigma_n(b)/\sigma_1 \simeq n$ to

$$\left(\frac{\partial \ln M_w(r)}{\partial \ln c^\circ}\right)_{c(r)} \simeq -\frac{\theta n(n-1)^2 \hat{x}(r)x(b)}{<1+n^2\hat{x}(r)>^2} \tag{28c}$$

At the point where $\hat{x}(r) = 1/n^2$ the value of $M_n(r)$ is $M_1(n+1)/2$ and eq. (24) has a maximum; thus eqs. (28a,b,c) will also have maxima at this point. For this point of maximum sensitivity we then obtain the simple relation

$$\left(\frac{\partial \ln M_w(r)}{\partial \ln c^\circ}\right)_{c(r)}\Bigg|_{M_n = M_1(n+1)/2} = -(n-1)^2\theta x(b)/4n \tag{28d}$$

valid for high speed equilibrium ($\sigma_1\Delta\xi>>1$) with $x(b)>>1$, and for small extents of heterogeneity.

Combining eqs.(25-27) we obtain the variation of the apparent association constant for a simple heterogeneous system with loading concentration as

$$\left(\frac{\partial \ln \hat{K}}{\partial \ln c^\circ}\right)_{\theta,K_A} = -\frac{\theta\beta n(n-1) \quad x(b) \quad \sigma_1}{(1+\theta<1+\beta x(b)>) \quad \sigma_n(b)} \tag{29a}$$

As before, this complete expression may be simplified for specific cases of interest, e.g. for sufficiently small extents of heterogeneity as:

$$\left(\frac{\partial \ln K}{\partial \ln c^\circ}\right)_{\theta,K_n} = -\theta\ \beta n(n-1)\ x(b)\sigma_1/\sigma_n(b) \tag{29b}$$

and for high speed equilibrium we obtain the simple relation

$$\left(\frac{\partial \ln K}{\partial \ln c^\circ}\right)_{\theta,K_a} = -\ (n-1)\theta x(b) \tag{29c}$$

valid when $x(b)$, $\sigma_1\Delta\xi>>1$ and $\theta<<1$.

Expressions (28) and (29) all contain the term $x(b)$. A useful approximation for $x(b)$ in terms of the initial concentration, c°, may be obtained when $\theta<<1$ so that $x(b) \simeq \hat{x}(b)$, and when $x(b)$ is large, as is often the case of high speed equilibrium experiments. Under these conditions $c(a)<<c(b)$, $c_1(b)<<c_n(b)$, $\sigma_n(b) \simeq n\sigma_1$, and eq.(15), as written for the main component, becomes applicable to all of the solution:

$$c^\circ\Delta\xi \simeq \frac{c(b)}{\sigma_n(b)} \simeq \frac{c_1(b) + c_n(b)}{n\sigma_1} \simeq \frac{c_n(b)}{n\sigma_1} = \frac{\hat{K}_c c_1^n(b)}{n\sigma_1} \tag{30}$$

where $\hat{K}_c$ $(=\hat{K}/M_1^{n-1})$ is the concentration (gL^{-1}) scale based apparent association constant, and c, c°, $c_n(b)$ and $c_1(b)$ refer to all components jointly. Equation (30) is then solved for $c_1(b)$, which, in turn, is used to obtain the desired approximation as

$$\hat{x}(b) = \frac{m_{T,n}(b)}{m_{T,1}(b)} \simeq \frac{\hat{K}_c c_1^n(b)}{n c_1(b)} = \left(\frac{\hat{K}_c}{n}\right)^{\frac{1}{n}} (c^\circ \sigma_1 \Delta\xi)^{\frac{n-1}{n}} \tag{31}$$

which is valid provided that both x(b) and $\sigma_1\Delta\xi >> 1$. This approximation may be inserted into the appropriate eqs. (28) and (29) to obtain simple limiting relations in terms of the readily accessible variables n, σ_1, $\Delta\xi$, K_c and c°

$$\left(\frac{\partial \ln M_w(r)}{\partial \ln c^\circ}\right)_{c(r)} \Bigg|_{M_n = M_1(n+1)/2} = -\,\theta(<n-1>^2/4n)(\hat{K}_c c^{\circ\, n-1}/n)^{\frac{1}{n}}(\sigma_1\Delta\xi)^{\frac{n-1}{n}} \tag{28D}$$

and

$$\left(\frac{\partial \ln K}{\partial \ln c^\circ}\right) = -\theta(n-1)(\hat{K}_c c^{\circ\, n-1}/n)^{\frac{1}{n}}(\sigma_1\Delta\xi)^{\frac{n-1}{n}} \tag{29C}$$

that are useful for estimation of small values of θ from high speed sedimentation equilibrium experiments. Eqs. (28a) and (29a) are valid for large values of θ, say θ>0.1, however, these expressions may be difficult to use since values of $x(b) = x_{A,n}$ can differ significantly from the apparent $\hat{x}(b)$ estimated from the $\hat{K}$ when θ cannot be assumed to be small. In such cases we note that for $K_B \simeq 0$

$$\hat{x}(b) = \frac{x_{T,n}}{x_{T,1}} = \frac{x_{A,n} m_{A,1}}{m_{A,1} + m_{B,1}} = \frac{x_{A,n}}{1 + \theta(1 + \beta x_{A,n})} \tag{32}$$

Solving eq. (32) for $x_{A,n}$ and inserting the result in eqs. (28a) and (29a) we obtain

$$\left(\frac{\partial \ln M_w(r)}{\partial \ln c^\circ}\right)_{c(r)} = -\,\frac{n^2(n-1)^2\theta\beta\,\hat{x}(r)\,\hat{x}(b)}{<1+n^2 x(r)>^2} \tag{28A}$$

and

$$\left(\frac{\partial \ln K}{\partial \ln c^\circ}\right) = -\,\theta\beta\, n(n-1)\,\hat{x}(b)\,\frac{\sigma_1}{\sigma_n(b)} \tag{29A}$$

which is exactly the same expression as eq. (29b) except that x(b) has been replaced by its apparent value, x(b), the apparent molar ratio of n-mer to total monomer at the base of the solution column. The approximation of eq.(31) is still valid for high speed equilibrium and for x(b)>>1, and may be used here also to obtain the needed estimates of x(b).

The data of Table II may be used in conjunction with eqs.(29) to obtain quantitative estimates of the heterogeneity of the initial association of our insulin preparation. This data leads to a least squares estimate of the derivative $(\partial \ln \hat{K}/\partial \ln c°) = (+0.1 \pm 3.9) \times 10^{-2}$. In experiments at 48,000 rpm the value of $\sigma_1 \Delta\xi = 3.05$, making the use of eq.(29c) marginal. Accordingly we calculated points for a graph of $\hat{x}(b)$ vs $c°$ by assuming several values of c_1 and then combining eqs.(30) and (31). From this curve we estimated the value of $\hat{x}(b)$ for the geometrical mean $c°$ $(=0.94 \text{ g L}^{-1})$ and the geometrical mean $\hat{K} = 300 \text{ L g}^{-1}$ as $\hat{x}(b) = 20.5$; the corresponding value of $\sigma_1/\sigma_n(b) = 0.512$. Eq.(29b) then leads to the estimate $\theta = 0.000 \pm 0.002$. [Equation (29c), which is strictly applicable only for both high speed equilibrium and extensive association, yields the same estimate.]

Unfortunately the heterogeneity for insulin evident in figure 9 is more difficult to evaluate properly since the higher stages of associating depicted there are as yet poorly characterized and since theory for evaluating heterogeneity in systems with several sequential equilibria is not available. However, the maximum value of the derivative $(\partial \ln M_w/\partial \ln c°)_{dc/d\xi}$, analogous to the derivatives of eqs.(28) is 0.2-0.3. When n is large eq.(28d) - or its analog that may be derived for the z-average molecular weight - suggests that this value might arise from heterogeneity of a few percent or so, provided $x(b)$ were sufficiently large. Clearly more work is needed before any definite conclusions can be reached about the heterogeneity apparent with these larger oligomers.

We illustrate the estimation of heterogeneity with two further examples, taken from our work with *Limulus polyphemus* hemocyanin[20]. Figure 11a presents the apparent weight average molecular weights observed as a function of $c(r)$ for three loading concentrations at pH 5, in acetate buffer, 0.05 ionic strength with 0.45 M KCl added at 25.8°C and at 6800 rpm. Under these conditions, *Limulus* hemocyanin exhibits a slightly nonideal reversible equilibrium between the 24S dodecamer and a 36S 24-mer. These curves are systematically displaced towards lower values with increasing loading concentration indicating heterogeneity. Separate fits of the initial regions, for fringe displacements up to 1 mm, to a monomer-dimer equilibrium gave the following apparent association constants: $\hat{K}_2 = 0.213 \text{ L g}^{-1}$ for $c° = 0.2 \text{ g L}^{-1}$; 0.189 L g^{-1} for 0.4 g L^{-1}; and 0.119 L g^{-1} for 2 g L^{-1}. The typical rms error for these fits corresponded to about 3 μm fringe displacement. From these data we estimate the derivative $(\partial \ln \hat{K}/\partial \ln c°)$ as -0.3 (+0.1, -0.2). The value of σ_1 is 4.32 cm^{-2} and $\Delta\xi = 1.965 \text{ cm}^2$. We take the geometric mean (corresponding to the arithmetic average of the $\Delta F°$) of the values of $\hat{K}_2$ as 0.168 L g^{-1} and of the loading concentration as $\overline{c°} = 0.68_3$, giving an effective value of $\hat{K}c° = 0.115$. Equation (31) gives poor estimates

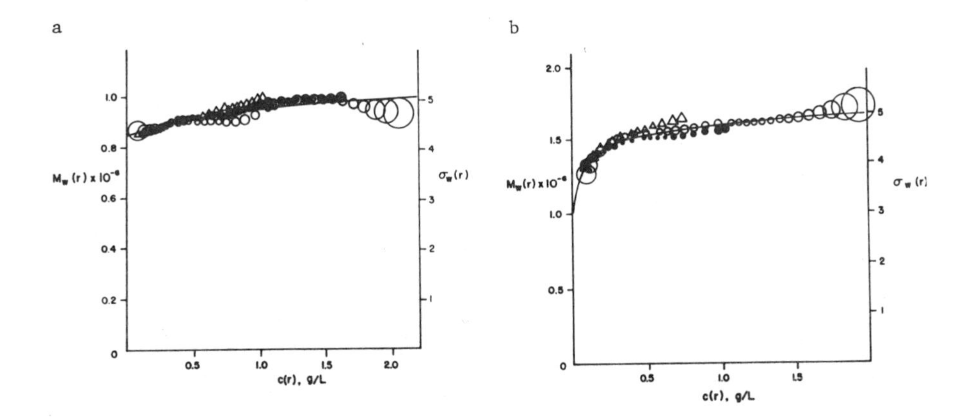

Figs. 11. Apparent weight average molecular weights, $M_w(r)$, (and values of $\sigma_w(r)$ as a function of c(r) observed for Limulus polyphemus hemocyanin by Johnson and Yphantis[20]. Panel a) presents the data obtained at pH 5, 0.5 M total ionic strength at 25.8°C and 6800 rpm. The data can be represented by a slightly nonideal dimerization of the 24S dodecamer to the 36S molecular species. Panel b) displays the observations at pH 6.1, 26°C and 5200 rpm. Here the hemocyanin is more strongly associated with both dimers (36S) and tetramers (60S) apparently participating in the equilibria. In both cases the triangles correspond to initial concentrations, c,° of 0.2 g L^{-1}, the filled circles to c° = 0.8 g L^{-1} and the open circles to c° = 2 g L^{-1}.

of $\hat{x}(b)$ since the extent of association at the base is relatively small. Accordingly we again estimated values of $\hat{x}(b)$ (=0.265) and of $\sigma_N(b)/\sigma_1$ (=1.21) by assuming several values of c_1 and thus generating curves of $\hat{x}(b)$ and of $\sigma_N(b)$ vs c°, Equation (29A) then yields the estimate of θ as 0.70 (from 0.47 to 1.16). Clearly this estimated value of θ is not small and thus eq.(29b) is not strictly applicable. The appropriate equation for large θ is eq.(29A) which provides the even larger estimate of θ = 2.3 (from 0.9 to no defined upper limit).

Figure 11b presents $M_w(r)$ as a function of c(r) for the same three loading concentrations as before (0.2, 0.8 and 2 g L^{-1}), but at pH 6.1, with 0.05 ionic strength phosphate buffer and 0.45 M KCl added. The temperature was 26°C and the ultracentrifuge speed 5200 rpm. Under these conditions hemocyanin is associated somewhat more strongly than at pH 5 with the apparent participation of both a dimer and a tetramer of the 24S dodecamer. The curves of Figure 11b diverge more distinctly than do those of Figures 11a, clearly demonstrating heterogeneity. Evaluation of the apparent dimerization constants from fringe displacements truncated to 0.7 mm for the three channels separately gave values of $\hat{K}_2$ of 92.6 L g^{-1} at c° = 0.2 g L^{-1}; 42.8 L g^{-1}; at 0.8 g L^{-1}, and 23.8 L g^{-1} at 2 g L^{-1}. The geometrical average value of $\hat{K}_2$ was 45.5 L g^{-1}, the corresponding c° was again 0.683 g L^{-1}, σ_1 = 2.5 cm^{-2} and $\Delta\xi$ = 1.965 cm^2. Here the association is sufficiently extensive and the value of $\sigma_1\Delta\xi$ (=4.93) sufficiently large that eq.(29c) appears to be quite valid. The value of $(\partial \ln \hat{K}/\partial \ln c°) = -0.59$ (-0.49 to -0.76) leads to the estimate $\theta = 0.068$ (0.056 to 0.087). At first glance it is surprising that the much more evident divergence of the curves in Figure 11b compared to Figure 11a, and the significantly larger spread of the values of $\hat{K}_2$ at pH 6 compared to pH 5, correspond to significantly smaller estimates of heterogeneity. Examination of eqs.(28) and (29) quickly convinces one of the crucial role played by the extent of association in the superficial appearances of heterogeneity, with the divergences both in the point average molecular weight curves and in the values of $\hat{K}_2$ being essentially proportional to x(b).

DISCUSSION

It should be pointed out that in all three experiments examined here the data were severely truncated to minimize the effects of higher association stages or nonideality. In general, considerably wider ranges of data are available for determining the K and thus the precision and sensitivity of estimating heterogeneity should be greater than in the cases presented. Furthermore, the cases analyzed have all been effectively monomer-dimer systems which are least sensitive to the effects of association heterogeneity.

Conditions conducive to maximum sensitivity include high centrifugation speed and long solution columns, to maximize the extent of association; choice of systems with large n; and the use of a large range of c° for increased precision in determining the derivatives of eqs.(28) and (29). The sensitivity to heterogeneity can be multiplied by utilizing separation procedures based on association behavior, such as gel exclusion chromatography under associating conditions with comparisons of early and retarded fractions. Conversely the effects of heterogeneity may be minimized by utilizing small values of $\sigma_1\Delta\xi$, thus effectively approaching determinations with no sensible differential redistributions of solutes, and the choice of conditions so as to keep Kc_o^{n-1} small. Again separation procedures can be used to isolate relatively homogeneous fractions on the basis of their association behavior.

In closing it must be emphasized that the relations derived depend completely on the assumption that the hetero-association of different components to form mixed oligomers can be neglected. Generally that is not so, and the estimates provided here must therefore be considered as minimal estimates of association heterogeneity. Similarly, heterogeneity can be underestimated in cases where there exists a considerable amount of a strongly self-associating minor component. Indeed it may be shown that if θ and K are both large enough the derivatives of eqs.(28) and (29) go through zero and become positive. Occasionally such cases have been seen experimentally.

SUMMARY

Some types of heterogeneity in self-associating systems can be detected with high sensitivity. Expressions to quantitate the effects of such heterogeneity show that the usual subjective impressions obtained from the lack of coincidence of the $M_K(r)$ vs c (c) for different c° can be highly misleading since sensitivities depend greatly on both the extent and degree, n, of association present.

ACKNOWLEDGEMENTS

This work has been supported by grants from the National Institutes of Health (AM-18001) and from the National Science Foundation (PCM 76-21847). We thank Raymond Kikas for his technical assistance.

REFERENCES

1. Teller, D. C., Horbett, T. A., Richards, E. G. and Schachman, H. K. (1969) Ann. N.Y. Acad. Sciences, 164, 66-101.
2. Squire, P. G. and Li, C. H. (1961) J. Am. Chem. Soc., 83, 3521-3528.
3. Jeffrey, P. D. and Coates, J. H. (1966) Biochemistry, 5, 489-498.
4. Roark, D. E. and Yphantis, D. A. (1969) Ann. N.Y. Acad. Sciences, 164, 245-278.
5. Williams, R. C., Jr., Yphantis, D. A. and Craig, L. C. (1972) Biochemistry, 11, 70-77.
6. Ansevin, A. T., Roark, D. E. and Yphantis, D. A. (1970) Analytical Biochem., 34, 237-261.
7. Paul, G. H. and Yphantis, D. A. (1972) Analytical Biochem., 48, 588-604.
8. Paul, C. H. and Yphantis, D. A. (1972) Analytical Biochem., 48, 605-612.
9. Laue, T. M., Domanik, R. A., Rhodes, D. and Yphantis, D. A. (1977) Biophys. J., 17, 101a (Abstract), and in preparation.
10. DeRosier, D. J., Munk, P. and Cox, D. J. (1972) Analytical Biochem., 50, 139-153.
11. Domanik, R. A., Yphantis, D. A. and Lue, T. M. (1977) Biophys. J., 17, 101a (Abstract) and in preparation.
12. Yphantis, D. A., Domanik, R. A. and Laue, T. M. (1978) In preparation.
13. Szuchet, S. and Yphantis, D. A. (1976) Arch. Biochem. Biophys., 173, 495-516.
14. Correia, J. J. and Yphantis, D. A. (1978) In preparation.
15. Davis, P. and Rabinowitz, P. (1954) J. Assoc. Computing Machinery, 1, 183-191.
16. Johnson, M. L. (1973) Ph.D. Thesis, University of Connecticut.
17. Johnson, M. L. and Yphantis, D. A. (1978) In preparation.
18. Szuchet, S. and Yphantis, D. A. (1973) Biochemistry, 12, 5115-5127.
19. Szuchet, S. (1976) Arch. Biochem. Biophys., 177, 437-460.
20. Johnson, M. L. and Yphantis, D. A. (1978) Biochemistry, 17, 1448-1455.
21. Yphantis, D. A. (1964) Biochemistry 3, 297-317.
22. Wu, G.-M. (1974) Ph.D. Thesis, State University of New York at Buffalo.
23. Wu, G.-M. and Yphantis, D. A. (1978) In preparation.
24. Pekar, A. H. and Frank, B. H. (1972) Biochemistry, 11, 4013-4016.

Index